MW00332321

The Myth of the Moral Brain

Basic Bioethics

Arthur Caplan, editor

A complete list of the books in the Basic Bioethics series appears at the back of this book.

The Myth of the Moral Brain

The Limits of Moral Enhancement

Harris Wiseman

The MIT Press
Cambridge, Massachusetts
London, England

This book was set in Stone Sans and Stone Serif by Toppan Best-set Premedia Limited. Printed and bound in the United States of America.

Library of Congress Cataloging-in-Publication Data

Names: Wiseman, Harris, author.
Title: The myth of the moral brain : the limits of moral enhancement / Harris Wiseman.
Description: Cambridge, MA : MIT Press, 2016. | Series: Basic bioethics | Includes bibliographical references and index.
Identifiers: LCCN 2015038270 | ISBN 9780262033923 (hardcover : alk. paper)
Subjects: LCSH: Ethics. | Brain. | Bioethics.
Classification: LCC BJ58 .W57 2016 | DDC 171/.7—dc23 LC record available at http://lccn.loc.gov/2015038270

10 9 8 7 6 5 4 3 2 1

Contents

Series Foreword

Glenn McGee and I developed the Basic Bioethics series and collaborated as series coeditors from 1998 to 2008. In Fall 2008 and Spring 2009, the series was reconstituted, with a new Editorial Board, under my sole editorship. I am pleased to present the forty-fifth book in the series.

The Basic Bioethics series makes innovative works in bioethics available to a broad audience and introduces seminal scholarly manuscripts, state-of-the-art reference works, and textbooks. Topics engaged include the philosophy of medicine, advancing genetics and biotechnology, end-of-life care, health and social policy, and the empirical study of biomedical life. Interdisciplinary work is encouraged.

Arthur Caplan
Basic Bioethics Series Editorial Board
Joseph J. Fins
Rosamond Rhodes
Nadia N. Sawicki
Jan Helge Solbakk

Acknowledgments

First and foremost, I wish to thank my family for supporting me through this time. There are certain persons without whose support and faith this project would have been impossible. The idea of scaffolding is very important to this book, and there can be no question that my family has been the scaffolding without which this book would not have come about. Similarly, gratitude must go to Fraser Watts. Without Fraser's ongoing faith and support in my work, I would not have been enrolled at Queens' College, Cambridge. Ultimately, it was Fraser's willingness to support my application for the funding that came to pay for this project that gave the Faraday Institute, by means of the Templeton Foundation, confidence enough to invest in my ideas as part of their "Uses and Abuses of Biology" project. Again, without Fraser's support, none of this would have been possible. Finally, I would like to thank Philip Laughlin at MIT Press for his honest, no-nonsense approach as acquisitions editor, Christopher Eyer as his indefatigable and ceaselessly helpful assistant in answering my constant barrage of formatting questions, and Katherine Almeida for her patience during the copyediting process.

1   The New Problem of Evil: Gloomy Preamble

The wickedness of mankind presented itself nakedly before him, and he became obsessed with gloomy thoughts.
—F. Voltaire (2006, 58)

The headlines are big; they are bold. The mainstream news reports Paul Zak's various proclamations that he has discovered the chemical in the brain that "makes people good" (Zak 2011a); Liat Clark (2014) writes in *Wired* that scientists have discovered how to cure "binge-drinking rats" of their alcohol dependency. On the subject of hormone-producing "love pills," Hookem-Smith (2012) writes: "Forget about marriage counseling, the survival of our future relationships could be as simple as popping a pill." Apparently we are entering a "golden age of neuroscience" (Goldberg 2014), where the findings pouring out from the "the moral laboratory" (ibid.), are giving us great insights into how the brain "does" morality. Such insights promise to give us ever-greater understanding of how to manipulate this "moral brain," how to control it. Here a magnetic pulse to different regions of the brain at just the right moment can create instantaneous changes to the kinds of value judgments people make, apparently, and to what they consider to be morally significant.

Yet even among the most scholarly, dubious analogies worthy of the tabloids are drawn—Molly Crockett paints a picture of serotonin as some manner of strange philtre preventing our rational and cool-headed "inner Jekylls" from being overrun by our primitive and savage "inner Hydes" (2008), and this story is played out at the highest levels of neuro-scientific and philosophical study. Yet the contrast of such garish headlines with those things actually presented to us as offering potential for moral enhancement are not exactly what one might expect—nothing too futuristic, and very little that is shiny and exciting at all, in fact, but instead some rather crude ideas for how all this might work: nasal sprays, pills, brain scans, gene therapy, and magnetic contraptions attached to the sides of persons' heads.

The aspiration to find some means for altering the character of man, and the supposedly urgent need to do so, has a long history. Far back before the advent of modern technologies and the world-destroying dangers we can see that they pose to us, the sense has been that a time bomb exists within man, woven into the fabric of his being. Indeed, the techno-apocalyptic fear can itself be understood as a modern retelling of this same age-old story, that the seeds of humanity's destruction are to be found nowhere else but in humanity itself, and that man, by his very nature, regardless of all his good qualities, has some essential corruption and inborn tendency toward wickedness and destruction. The proposition here is that even good men and women can be easily overwhelmed by pride, stupidity, sloth, and circumstance, the combination of which can make otherwise well-meaning persons, persons who treat their neighbors and various strangers with kindness and respect, into assenting participants in genocides and crimes too unspeakable to mention. So whether it be by passivity, willed ignorance, or through active wickedness, there is some common sense among many observers that there lies in the very fabric of the human being an unassailable gravity, a self-moving and living shadow, looming constantly over all of humanity's efforts to civilize itself.

So prevalent and ancient is this observation that terms for it can be discerned practically everywhere human writings are to be found. We are well familiar with those who speak of the stain of original sin, the echoes of the first disobedience of man before God resounding through the generations, written since then into his very nature. "All is vanity," cries Ecclesiastes, the voice of the pinnacle text of biblical Jewish wisdom literature: "man hath no pre-eminence above the beasts" (Ecclesiastes 3:19, ERV). We have a long tradition of English biologists who are in perfect agreement with him. Here man, understood as "the squalid ape," is seen then to be no different and no better than a beast. To this Mary Midgley (2002) offers a less than comforting reappraisal of that proposition, finding the application of the term "beast" as a condemnation to be rather unfair to the creaturely kingdom, most of whose citizens seem to behave considerably more cooperatively than human beings do—the only species to hold grudges over generations, to kill, torture, and rape for pleasure. We have Freud's (1991) talk of the dynamic unconscious—self-moving, amoral forces which give animation to the human creature. This picture is of the human being as a creature whose conscious existence exists only to negotiate and find compromises between social prohibitions and endless aggressive and sexual needs, and of civilization itself as a vehicle for doing precisely the same. And let us not forget the radical ecologists whose primary conviction is that the human race, by its very nature, is the most heinous plague the planet has ever had the misfortune to endure, a misfortune however that will inevitably, and happily, turn out to be a quite brief chapter of this planet's history, fated as we are for self-destruction, apparently soon to cease our gluttonous rape of the earth's abundant and rich but not inexhaustible generativity. The list of such pronouncements goes on seemingly without end.

Indeed, there is much suspicion regarding our natures and passions, as well as their power to overwhelm good sense and reason, to subdue all higher motives under their desires. Be it religion, then, philosophy, psychology, science, literature, politics, or economics—ideas about the dark forces in man's character, whatever else one thinks of man, have impressed themselves upon our theories and stories of what the human being is. Perennial observations regarding these unsightly tendencies which lie forever just beneath the surface, which require no training, no cultivation, but which spring naturally from the human's very being, are to be found everywhere.

*

But we, like Voltaire's *Candide* quoted above, also vacillate constantly between our despair in humanity and our optimism, our faith that, despite everything, things are yet worthwhile, that there is just enough good in the human being to sustain a hope that he might grow, or be nurtured, into something great and good. Indeed, the mere fact that humans have managed to continue to exist is itself cause to think that a purely pessimistic picture of human nature is false, that there must be some bedrock upon which the transformation of man, the humanization of the human race, can be facilitated. And as Midgley (2002) continues, no species can last long on this earth if it has no natural propensity for cooperation. Cooperation, then, must also be in our nature, too—a gift of our animal inheritance. Pursuing this insight regarding the cooperative, and even self-sacrificial, tendencies inherent in nature is much in vogue in recent evolutionary theory and the philosophical constructs of the nature of life they are built upon.

Hope remains, then, some sense that humanity can be "saved from itself." But what means for such salvation are there? Precisely how might man's moral nature be "perfected," or at least augmented—nudged gently but firmly in the right direction? There are long traditions of religious thought which suggest only God is capable of making this change, the apocalyptic prophet's sense that the stain on man's nature is just too overwhelming a power. The sense here is that humanity can, and indeed must, work on its moral character, but we must also know that this can never be enough. In the end, or so it is popularly understood, some great fire or some great flood, some overwhelming power must come, one which can cleanse away the mess that has been made, to restore a new Heaven and a new Earth, a new Kingdom, where peace and justice reign at last. Alien, fantastical, or outright ridiculous as this apocalyptic hope may sound to many, a significant quantity of hard-nosed secular thinkers in the moral enhancement domain do think in surprisingly similar ways and share a similar set of presuppositions that are found within this apocalyptic sense: that man's will is not adequate to the task of improving his moral character, that too much inborn darkness lurks within, that traditional means of moral teaching have failed, and must fail—above all, that *intervention is necessary*. If it is not God to rescue us, it must then be some God Machine.

Advances in science afford other opportunities, and biology in particular has been a most congenial vehicle for those wishing to advance ideologies of all kinds. For example, eugenics, with the belief that good character and moral degeneracy alike were functions of heredity, spawned the search for tending a better mankind through good breeding—"more children from the fit, less from the unfit" the resounding call (Larson 2010, 174). And to this day there are those in prominent university chairs who advocate there is nothing less than a moral obligation to carry on a new vision of "positive" eugenics, voluntarily done for the good and welfare of our future children.[1] Most recent of all we have the present fascination with the brain, the beginnings of a "golden age of neuroscience," and with it this notion that "we are our brains." For us it is this brain, its anatomy and its chemistry— "the moral brain," no less—that holds the key to altering man's nature, of affecting in us a change that might transform us into the sorts of benevolent, altruistic, peaceable creatures that would be so much to the benefit of all persons, and creation as a whole. So biology has swapped "nature" for "neurology," and posits the power of neurobiological explanations with the very same authority and confidence that they previously invested in genetic causes, and as if they had always claimed it to have been the truth. "Neuroscience," we are told, "is about discovering our limits, then hacking to get around them" (Crockett 2013, in Clark 2013). Here we are hoping to find "hacks that help make us better at doing the right thing" (ibid.). And this is exactly what moral bioenhancement is all about.

So the aspiration toward moral enhancement, in one form or another, is nothing new at all. We might well picture it, this supposedly futuristic goal, in the rather backward terms that have already been presented us by history. For this "urgent imperative to improve man's moral character"

(Persson and Savulescu 2008, 162) has been attempted over and over in recent memory and can be seen in terms of the various crude interventions from the past: the lobotomization of aggressive children; the sterilization of the criminal and mentally disturbed; the purification of the blood and removal of "the red gene"; segregation and holocaust laid upon the mentally disabled, gypsy, and Jew (all of which have been perpetrated in the past forty to eighty years). The various attempts to cure those with perceived "moral sickness," in whatever form it may take, by whatever rubric, to segregate or to separate out those of poor character (we use more clinical terms for this now, such as "anti-social personality disorder" and the like), those of low birth, from any "higher" or more "desirable" types that might possibly be produced for the welfare and good of our future children: we still find calls for nothing less than a *moral obligation* to continue this project (Savulescu 2001, 2005, 2010).

Are such evils the price of manufacturing a morally worthwhile race of persons? Must some moral eggs be broken if the moral omelet is to be made? Or is the very idea of such moral enhancement itself an evil so tremendous that it ought to be avoided at all costs? Would it be anything other than massive overextension of the apparatus of state control into the fundamental biology and neurology of the individual, the ultimate power to shape the very nature of the human creature put under the control of the state, all in the name of "saving us from ourselves," even if there is but "the slimmest chance" that our destructive natures might be thereby assuaged?

The truth is likely to be a great deal more complex than any of these positions allow. Might new advances in science offer more velvet means for augmenting our moral powers? Exclusively horrific as moral bioenhancement's prehistory is, our very present context is exceedingly congenial by way of contrast. Much less is written of eugenics and the like, explicitly at least, and actually, moral enhancement tends to talk with highly avuncular tones and displays the sort of warm and smiling face one would associate with the family doctor. Potential arises, it is suggested, in terms of nasal sprays to nudge one toward greater empathy, to facilitate trust and trustworthiness; supposed pharmacological interventions to improve self-control, or for resolving a bad temper (DeGrazia 2013); enhancers for our motivation (Kjaersgaard 2015); manifold techniques aimed at treating addictions of all kinds, treatments for our gluttony and all our various excesses (Hughes 2012b). On the face of things, given the history of the idea of producing better men, or a new Kingdom, whatever it may look like—contemporary moral enhancement discourse is, to a considerable extent, a relatively mundane, sedate, and pleasant affair.

*

Of course, as is so often the case, everything depends on how we define our terms. What do we even mean by "moral enhancement"? And what exactly is being enhanced? Is it our abilities for moral reasoning that are to be improved? Our impulses to do good and bad things? And how? By drugs? By various technologies and machineries? By covert psychological manipulation? And who is to be mandating this moral enhancement? Is it the state? Is it a matter for medical professionals to determine? Or, is the whole thing to be a matter of personal choice? And what are the ethical differences between choosing such enhancements for oneself, and the various compulsory ways of foisting such interventions upon ourselves, or our children, for that matter? Endless questions plague the discourse. What we need to note above all is that there is no consensus over how moral enhancement is even to be defined, and commentators often get confused when having their conversations because the foundations for common dialogue have not been made clear beforehand. What one person calls moral enhancement, another person denies and calls mere behavioral control. What another person demands by way of moral enhancement, the generation of "moral identity," another will suggest is too narrow a definition to give a representative picture of the potentiality of the field, shutting out possibilities that are at once exciting, vexing, and worthy of discussion. Some commentators define moral enhancement so loosely that practically anything can be included; some define it so strictly that nothing at all can count as moral enhancement. A middle way is not always best, and we shall see that both narrow and wide approaches to looking at the subject matter have merit—just so long as one keeps clear what one has in mind.

And so, that is the task that behooves us now: gaining an initial clarity on the domain, setting the foundations straight so that conversation can be had at all—so that, when asked, "What is moral enhancement?" one can answer, if not concretely, then at least with some indication of the intricacies of the territory, the place upon which one is staking one's claim. Perhaps the only concrete bedrock that most commentators agree upon is that biomedical moral enhancement[2] should be understood in the following terms:

Some technological or pharmacological means of affecting the biological aspects of moral functioning, to boost what is desirable, or remove what is problematic.

And that is as far as I will go in defining the key term of the book. Leaving things in such a vague state may seem shocking, but this actually points us to two of the core points to be made throughout, that (a) there is no one thing that goes by the name of moral enhancement, and (b) we misrepresent the range of meaningful possibilities for the domain by trying to combine it under a single unified conceptual paradigm. There are just so many ways of going about understanding the term that attempting to capture it in a sentence or two not only does an injustice to moral enhancement, it radically mischaracterizes what is going on here. Namely, that moral enhancement can be construed in a rather large number of ways, since it embraces a tremendous diversity of potential meanings, and it is very important for the domain that we make room for a diverse range of approaches when looking at the subject matter. Let us begin with a series of distinctions that will help orient the conversation to come, to make clear these important differences so key to getting our conversation off on the right foot. And, as we will see, the various permutations of these distinctions proffer considerably different proposals, which raise very different ethical issues.

Background Concepts for Understanding Moral Enhancement

In Which Moral Philosophy Are We Invested?

Which moral standpoint are we coming from? Before we even talk about what values we think are good and bad, we need to ask after which criteria we are using to assess whether a moral improvement has been made. Persons not only have different values, but they have different ways of measuring acts as moral or not. These ethical questions are interminable. For example, some judge things by consequences, some by intent, some in terms of whether a person is acting from a strong disposition to be good (and even within these broad categories there is a complex range of gradations and types); some persons are individualists, some are universalists, some are communitarians. Some claim to have one view intellectually, but then find themselves having moral intuitions in practice that are of a completely different nature. So, it makes a great difference in each case what one's starting point is. For a consequentialist, the primary concern with moral enhancement may be precious little more than a weighing of overall harms prevented against the various costs involved. For a person interested in motives, a moral enhancement might be measured by the extent to which it contributes to a person *wanting* to be a morally better person, or discovering better reasons for being moral. For a virtue-based proponent, a moral enhancement would have to contribute to the formation of "moral habits," generating an amount of moral growth—a durable transformation in the personality or character of the agent in question. Unfortunately, reviewing each and every proposition to be made here from the perspective of each theory would create a tome far too ponderous to be of interest. On occasion I will subject certain propositions to analysis from these various positions, but I would ask the reader to keep in mind throughout how this variety of ethical standpoints might modulate the interpretations and conclusions here drawn.

Voluntary versus Compulsory Moral Enhancement

Without doubt, the most fundamental distinction drawn in the domain is that between "compulsory" and "voluntary" moral enhancement. This is understandable. It makes a world of difference, for example, if one is making a case for a particular enhancement which is to be forced upon us by the state, something legislated for and enforced with the full power of the state's police apparatus and judiciary; or whether we are simply talking about the sort of product one might buy over the counter at a dispensing pharmacy, as one purchases antihistamines, or some cream for a rash. We shall see that this split at once impinges upon a range of issues of permissibility, yet on the other hand, is not nearly as clear-cut as it first appears. There are numerous ways in which what is voluntary and compulsory can overlap, and that subtle but powerfully manipulative social forces can shift what appears to be voluntary into a shady middle ground—and this is not always a bad thing. Moreover, we shall see, contrary to the dystopian fear that compulsory state-sponsored moral enhancement instantly evokes, there may very well be instances where a compulsory moral enhancement might actually be more responsible of the state than leaving such interventions up to individual choice. In fact, present treatment of convicts with drug or alcohol problems already comes particularly close to this sort of ideal.

Hard versus Soft Moral Enhancement

While the above distinction has the dominating importance in the philosophical discourse, in the present book we will contend that there is a yet more important distinction to be drawn. There is a fundamental distinction relating to whether we think moral enhancement is constituted in terms of an explicit project on the part of the individual to improve themselves in a specifically moral fashion ("hard")—that is, whether a person is undergoing moral enhancement with the specific aim of becoming a more moral person, and whether a state is compelling an intervention explicitly on the grounds "to make you more moral"; or whether we are talking about capacities merely related to moral functioning, regardless of whether any intent exists on the part of the person undergoing the intervention to morally improve ("soft"). This is a crucial distinction. If one thinks that moral enhancement is constituted only when persons are purposively engaged in a project of explicitly attempting to improve their moral capacities, then this narrows the discussion to a considerable degree. The problem here is that most persons do not actually sit down and explicitly attempt to become "more moral persons." Instead, moral functioning is complexly diffused within daily living, and manifests in fluid and subtle ways perhaps not best captured by the traditional image of an individual sitting down like Rodin's statue and reflecting upon the nature of "the Good."

Opening up discussion to include what shall from herein be called the "morally related"—that is, interventions which do impact upon a person's moral functioning, even if it be indirectly, and which might additionally result in an overall reduction of harm in the world, but are not necessarily done for explicitly moral reasons (for example, an intervention to reduce aggression in a pathologically aggressive patient might be done for public safety reasons, but by implication would involve there being less harm in the world, and may have a corollary impact on the patient's moral functioning in certain contexts)—facilitates the inclusion of a great deal more in our discussion. For some, such a broad definition will include too much, but it is necessary to be broad if we are to get a grasp of all the important embracing conceptual background to moral enhancement. In fact, we will be utilizing both "hard" and "soft" definitions, since the former facilitates a stricter, more rigorous way of looking at the debate, whereas the latter opens matters up to the important diaspora of related phenomena from which, it will be contended, the debate cannot be readily severed—namely, paternalism, social control, behavioral control, neuroeconomic attempts to manipulate behavior through policy, and the larger ideological and political backdrops against which a moral enhancement project will have to be constructed. None of these things can be properly broached if one takes too strict a view of what moral enhancement consists in, yet all of these phenomena form an essential background to moral enhancement debate. In fact, the present book can be understood as the story of these two dominant modes of moral enhancement playing out in a range of different, morally significant forms.

The Realistic versus the Fantastical

Another distinction which is particularly significant for our present purposes is the distinction between "realistic" and "fantastical" forms of moral enhancement. Many of the fantastical ideas proposed, such as highly calibrated neurodevices created to read thoughts and override motor functioning when evil intentions are formed, serve as interesting thought experiments and are worthwhile for that. In the end they distract attention away from the real substance of the matter, the more realistic interventions that are proposed. Perhaps the biggest theme throughout the book will be the attempt to bring practical realities to bear upon the subject (and we shall see that practical reality is something that many of the philosophical enthusiasts in the domain have refused outright to engage with). Talking about moral enhancement as if it is entirely speculative and futuristic, when the subject has in fact a relatively long prehistory made up of crude attempts to alter man's character (i.e., is past and present, not just future), is not only outright false, but it also trivializes that which is considerably important within the domain and worthy of sustained reflection.

Positive versus Remedial Moral Enhancement

There is also a very important distinction to be drawn between the kinds of usage we are putting moral enhancement to, whether the object is "remedial" or "positive" (Agar 2013). That is, it matters a great deal whether we are attempting to deal with a kind of immoral or destructive behavior that is pathological in nature, for example, psychopathy, antisocial personality disorder ("remedial"); or whether we are talking about taking normal, healthy persons and trying to build upon their already existing capacities, for example, their capacities for empathy, altruistic impulses, and so forth ("positive"). Of course, there is a huge controversy over the treatment/enhancement distinction in medical contexts, but we shall be defending a pragmatic form of this distinction and showing its considerable importance when talking about moral enhancement. For depending on the context of the debate, there are very different issues of permissibility and practice involved in, say, attempting to heal a damaged person whose malady carries over to their moral functioning, as opposed to the attempt to "raise the bar," ethically speaking, for the citizenry at large.

Maximal and Minimal Accounts of the Good

The "maximal"/"minimal" distinction refers to the question of how pervasive one thinks a program of moral enhancement ought to be, and speaks to how varied a pluralism one is willing to accept in moral enhancement. This is a highly subjective distinction, but an important one nonetheless. A "maximal" vision of moral enhancement would be very confident in what it takes to be the nature of "the Good," very confident about its ability to decide which values are worthwhile, and the scope of the values which it feels can be rightly dictated. This would likely include a set of values that many would not agree with, them being idiosyncratic to the respective culture or personal philosophical vision of those making their claims. One could imagine, say, a member of the Taliban, or an extreme fundamentalist Christian, or a radical communist, as having very strong opinions regarding what counts as good and what goods should be enhanced. This would be a highly "maximal" view of moral enhancement. In contrast, much more modest and pluralistic accounts are possible, accounts less willing to dictate particular values and goods over a wide range. As we shall see, there are many who believe at least some common ground can be found between "reasonable moral visions," and that this common ground, even if fragile and narrow, proffers a kind of bedrock upon which moral enhancement can gain some purchase. Goods like altruism, or not murdering members of one's kin group for no good reason, might well fall into this category of goods that pretty much everyone can agree upon. This would be a "minimal" account of moral enhancement.

Strong and Weak Accounts

Finally, it makes a difference whether one is thinking about moral functioning in terms of isolated moral or immoral acts (e.g., helping an elderly person cross the road), or whether we are thinking in terms of a disposition toward certain kinds of behavior (e.g., talking of an "honest" person who, over the time one has known him, has demonstrated genuineness and truth-telling as something more like a character trait). It makes a difference, then, whether one is talking about a kind of moral enhancement that is aimed at making mid- to long-term progress toward developing a disposition toward some good, the creation of some "moral identity" or "moral character," on the one hand, and those interventions which might be more focused upon dealing with individual moral acts, on the other. Some of the interventions aimed at our more petty excesses might be construed in this way, not as something taken with the aim of "becoming a better man" or the like, but only to temporarily help deal with a particular problem or weakness one might have. This distinction is separate from the hard/soft distinction mentioned above, which refers to whether such enhancement would be done in an explicitly moral way or not, whereas this "strong"/"weak" distinction refers more to the *durability* and extension of the intended effects, to create an entire disposition or to achieve some short-term gains and benefits.

*

All these distinctions, at the very least, must be kept in mind whenever we are having our discussions about the subject, for even but a brief inspection of these distinctions will make clear that the range of possible permutations of kinds of project that might be construed as moral enhancement is incredibly broad, and each kind of project brings with it different issues and are desirable or dangerous to different degrees and for different reasons. A *compulsory* moral enhancement for treating *remedial* cases (e.g., drug addiction) might be more attractive than a *compulsory* moral enhancement to enhance certain *positive* moral traits (e.g., generosity). A *voluntary* project of moral enhancement aimed only at a *minimal* range of values that most can agree are good (e.g., altruism) is a very different proposal to one in which we are being *compelled* to enhance an entire *maximal* spectrum of values that we might not be in agreement with (e.g., piety, nationalism). A project focused on some prospect which is radically *fantastical*—for example, a well-pleased James Hughes (2013, 31) suggests it is "quite likely" that in the future we will have neural interfaces which can read and augment our moral neural impulses, thoughts, and behaviors in real time, that "eventually we will have the capacity to change genes that affect the brain permanently, and install neurodevices that constantly monitor and direct our thoughts and behavior" (ibid., 31)—would be very different from a much more *realistic* prospect; for example, a highly limited attempt to deal with aggressiveness through pharmacological means as part of a mental health intervention. We could go on endlessly like this, analyzing possible permutations. All this, of course, is treating of the discourse in a rather abstract conceptual way, as if from a distance, and this is precisely what we are attempting to avoid. Yet some generalities must be considered, and it is important to talk about the tremendous number of various permutations of moral enhancement project that even this partial list of distinctions make possible so that we do not seek to constantly reduce the whole of moral enhancement to a singular conceptual paradigm.

In other words, it is not always appropriate to talk of moral enhancement "per se," as if it is "one thing" about which purely general comments will suffice. We shall see that numerous such commentaries do exist—for example, John Harris' (2011) concern that moral enhancement poses a threat to our "moral freedom," along with all the various critical responses to that concern—can be seen to involve many of the fallacies that taking massively generalized overviews of a subject can produce. Likewise, the various arguments over whether moral enhancement (always "per se," as if one big thing) should be made voluntary or compulsory without asking first which moral enhancement intervention we have in mind, in which context, with respect to which kind of moral issue, and using which means of intervention (as if merely positing that moral enhancement "per se" should be voluntary or compulsory has some magical power in and of itself to tell us whether such a project is morally permissible or not). Without specificity, virtually everything of importance is lost from the debate—above all, the specific practical realities to be found on the ground level, which are not at all incidental but the very realities around which the abstractions of the debate must be made to shape themselves (not the other way around). This "practical-realities first" approach is at the core of the book. *All of the abstractions that make up moral enhancement discourse must be held to account by the practical realities on the ground, and be shaped and relativized throughout by them.*

What must be understood is that no two moral enhancement proposals are exactly the same with respect to how they seek to influence us. So, the sort of purely generalized overview type approach which refuses to respect the incredible diversity of potentiality within the domain, which refuses to "get its hands dirty" with the particularities of any specific given instance, is precisely what we will be putting to pasture in this book. While it is necessary to make some general claims, all of this must arise, first and foremost, out of a detailed look at things on a case-by-case basis, treating the issues that are raised by each particular prospect on their own terms.[3] There are just too many possible permutations, too many prospects which vary with respect to how desirable and defensible they might be, to think that grand overview analyses can suffice here.

By extension, value judgments as to whether moral enhancement "in general" is a good or a bad thing simply cannot stand. Many of the prospects to be considered overlap considerably with what we already experience and accept as normal in medical or mental health practice (we shall see this in the section on treating alcoholism). Some of these prospects are too innocuous and safe to justify apocalyptic worries, some are too desirable in the right context to justify turning our noses up at them, some are monstrous. Some are simply ludicrous. It is therefore impossible to be reasonably against moral enhancement as a totality (there is no such totality), and the reverse judgment is equally true. No doubt we all come to the table with prejudiced ideas about enhancement generally. I would ask readers, in light of the broad range of possibilities that moral enhancement encompasses, to temporarily suspend any global judgments they might have, and instead begin to look at things on an individual basis. This is the only way we can do justice to the potentialities of the field.

Biomedical Moral Enhancement: The Shape of the Book

Before we launch into the main body of the text, it is important to have some sense of the structure of the book and where the argument is going. The discourses, scientific and philosophical, that moral enhancement draws upon are prodigious in size and volume. It is important to make clear that the aim of the present book is not to offer a comprehensive history of everything that has been said, but rather to find paradigmatic exemplars whose work, first of all, represents important trends in the respective domains; and, secondly, which point toward larger issues of moral concern, which when woven together point then toward the larger thesis that will be contended. This last point bears repeating: the subject matter we will be dealing with is huge, and exclusions are necessary (and even defending the exclusions in some detailed way would require too much space); thus, our focus will extend only to factors which impact on the particular thesis to be argued. It is to be kept squarely in mind that the various contributors drawn upon are brought in solely because their views point toward more general concerns relevant to the overall picture to be painted. If certain commentators have been omitted, this is not an indictment on their significance or the quality of their work but necessitated by the overall shape and nature of the argument to be presented.

As for the structure of the text, though the book is split into four very distinct parts which deal separately with the insights garnered from quite different disciplines—philosophy, biology, theology, and clinical psychology—using very different languages and orders of explanation, there is a linear progression from beginning to end, and the same philosophical eye will be brought to bear on each separate part. We will begin with the philosophical rationales and aspirations for moral enhancement and the practical realities they must contend with in order to sift through what is worthwhile in the philosophical discourse and where the various limits arise (part I). The idea constantly lurking in the background here is *"ought implies can."* When practical realities are put at the forefront of the matter, the suggestion to be kept firmly in mind is this: if enhancement enthusiasts cannot present even the semblance of a realistic proposal for what they are advancing as morally obligatory, *then they have not really presented anything at all.*

Forgetting to put practical realities first happens a great deal in moral enhancement discussion. This is a problem. It is all well and good saying, for example, as Persson and Savulescu (2008, 162) have, that there is an "urgent imperative to improve man's moral character." Or similarly, as

Rakić has put it, that "we must create beings with moral standing superior to our own" (Rakić 2015, 58). All this depends on the existence of something which can actually perform such a task, something which might realistically be brought about in the world that we actually live in, given the various political (and related) constraints of this present world. The thesis of the present book is that moral functioning is travestied when approached primarily through biological lenses. Given that, a close look at practical realities is going to put a rather severe dampener on the proceedings, limiting considerably whatever alleged "moral obligations" to morally enhance that may or may not exist.

After these philosophical chapters, we will move on to the empirical work conducted on "the moral brain," specifically the various studies conducted with respect to oxytocin, serotonin, and dopamine's effects on behavior and judgment, as well as looking at the various problems with this science to see whether the philosophical aspirations are justified and why a biologically privileged approach to moral functioning is so problematic (part II). We will then move on to the subject of traditional moral education and its more social-environmental underpinnings, a rich picture of which is to be found in the theologies of the various faith traditions (though our focus will be predominantly upon Christian faith, albeit viewed through secular lenses, and a lengthy justification of all this will be elaborated in chapter 6). We will also be confronting the fact, much neglected in moral enhancement discourse and the empirical enquiry into the moral brain, that we live in a world shaped by religious thought, belief, and practice, that these factors cannot be bracketed out of the discourse, and that any nonvacuous account of moral enhancement, or research into "the moral brain," must take account of the idiosyncratic nature of the faith traditions of which billions are members if they are to be realistically applied in the present world (part III). This will lead toward the concluding practical chapters, which are an attempt to rebalance moral enhancement discourse so that a rich, integral, bio-psycho-social approach to such moral enhancement might be presented. The book will end with a detailed picture of what is realistic and desirable in moral enhancement, and a rebuke of what remains.

As such, while this book refers to the contents of many disciplines, let us be clear what it actually is—it is a piece of applied moral philosophy, an effort to bring a critical-analytic eye to the various claims made in various interrelating disciplines, to assess their quality and validity, to see how well the various claims involved support or contradict each other, and, ultimately, to assess the worth of the various ideas of moral enhancement that are supported thereby. The core point is that moral functioning involves phenomena that have many, many interacting influences, and if moral enhancement is to be presented in a nonsuperficial manner, then some account of these complexly woven influences must be provided. One helpful way to do this is to consider the matter in a closely interdisciplinary manner, thus providing a mirror to the nature of the subject matter itself.

As for the argument itself, what we have here is a narrative which begins and ends with the insistence that biomedical moral enhancement, and the biological dimensions of moral functioning more generally, constantly need to be understood as having a dynamic and integral but, most importantly, *auxiliary* relationship with the various psycho-social and environmental backgrounds in which any given intervention is to be embedded. That is, by nature of their sheer numerousness, the tremendous plurality of potential social-environmental influences on moral functioning will be presented as significantly more important in shaping moral functioning than the various biological or neurobiological influences they are interwoven with.

One of the very first things to be brought into dispute here is the idea that biological manipulation of the moral capacities of the human creature holds some world-salvatory potential. The fundamental thesis, to be worked out across the various parts of the book, using the very different languages of the respective disciplines involved, is that presenting moral functioning in exclusively or even predominantly biological terms is to provide an impoverished account of the reality of moral functioning and its various influences, an immensely complex reality which is best characterized as a mélange of influences, a hodgepodge, a complex and messy mix of overlapping factors not readily separated from each other and not appropriately dealt with in primarily biomedical terms.

There is no question at all that when it comes to moral functioning, most, if not all, of its dimensions are biologically mediated. We are after all biological beings, and this fact must be respected. But so too must all of the many, many other influences on moral functioning which interweave with these biological influences, all of which modulate the other in variously incalculable ways. Certainly the idea of direct "biological causality" with respect to moral functioning will be subjected to vigorous critique.

So, the thesis of this book involves a constant and very difficult balancing act between

(a) respecting that moral functioning is biologically mediated, and thus, in principle, opportunities might be proffered for influencing moral functioning in some fashion via biological mediation; and

(b) disputing as vigorously as possible this idea that these biological influences are in some way the only elements that matter, nor even the primary elements, nor even equal partners with the supervening social-environmental and personal influences which are highly numerous and varied in nature.

When thinking about this balancing act, it is a good idea to keep sight of the broader philosophical puzzle all this plays into, namely the existential questions of *facticity and transcendence*. Put more simply, we are born into a world and a body which constrain us, but which also leave room for us to determine ourselves within those constraints. Most kinds of reductionism attempt to get the audience to forget their transcendence, their capacity to choose among options that are available, to shape themselves and their environment, to stand back from the facts which surround and direct them, to choose otherwise.

Whether we are talking about genetic reductionism, neuroreductionism, or even cultural determinism (the idea that we are merely the products of our environment), we are continually forgetting, at tremendous cost, this capacity for transcendence. As such, the polarity of facticity and transcendence should always be lingering at the back of the reader's mind. It is the crux of the argument that will play itself out throughout the whole book, and it represents a dialectic that must be continually maintained. We will see throughout that biological and neurological facts about our being influence and constrain us in various ways. This must be taken very seriously. Moreover, when we talk, as we will, continually, about our capacity to shape ourselves, and how this defies the various forms of reductionism that we will be confronting, it must be made very clear that we are not referring to an absolute freedom either. We are influenced, nudged toward certain kinds of judgment and behavior in various ways, and our society does constrain us in many ways. This is all part of our facticity, and fighting it or shaping these influences is not always easy.

It can be comforting to believe that we have no capacity for transcendence, that all our decisions and motivations can be biologically reduced, or that we are merely mechanisms, or the products of our environment. But some of what is absolutely valuable in the moral life comes directly out of our capacity for transcendence. More precisely, it is the exercise of our capacities for transcendence in the face of the various forces that make up our facticity, and this in relation to our various determining influences, that can make for the most praiseworthy moral goods of all. Whenever we meet any kind of reduction of moral functioning to the level of determining forces, anything that seeks to ignore transcendence, our capacity to "rise above," we have missed something fundamental. For it is in the relationship between our constraints and our choices that the real intricacies of moral living arise.

It is this *relationship* that must be kept in mind. In considering the biological components of moral functioning, we have to respect that they are but one element of our facticity, one influencing factor among many, many others, all of which have to be explored in relation to our capacity to shape ourselves within those limits and in relation to our capacity to rise above. This is one very good reason for being suspicious, from the outset, of biologically dominant explanations of our moral functioning. For even if it were the case that the human being is a mere slave to the various deterministic forces which work upon him, biology would still only be but one of those forces, and by no means necessarily the most decisive element in that interweaving of forces.

This can be illustrated by examining the common popular fiction generally known as "the nature/nurture debate." This debate is very much a case of a fight between two forms of determinism, two forms of reductionism, both of which ignore our capacity for transcendence. In the nature/nurture debate, we are often told that we are variously shaped by our genetics, and variously shaped by environment. But where is the person in any of this? Is the person merely a passive mechanism caught halfway between these two enslaving forces? Has the individual, or the group, no power at all to rise above both genetics and environment? Where is transcendence in the nature/nurture debate?

We will see much the same when we talk of the brain and its various parts. The person is continually dissolved into a conflict between (usually) two bits of the brain battling each other. Each of these brain parts is anthropomorphized, and "the person" becomes a mere by-product of the final outcome of the carnage wrought by these two opposing deterministic powers. Our morality, as we will see, will be cast as nothing more than a matter of which of these warring powers "in our brains" wins out. Moral enhancement, when we reduce the immense richness of moral living to such a convenient and tempting simplification, suddenly makes sense in biomedical terms. When understood against this conceptual backdrop, moral enhancement is a matter of properly arming the appropriate warring faction, the various bits of our brains that are on the side of the good. I invite the reader to keep in mind throughout how this reductionism is written pervasively beneath the surface in the assumptions, the narratives, and the speech patterns employed by so many commentators. There is no person, there is no moral choice, there is only gray matter, this matter is divided into "factions," these factions are at war, and moral enhancement is about "hacking in" to the infrastructure used by these parties—neuro-cyber warfare, as it were—and disabling the immoral enemies' powers to invade and win. This reductive, deterministic conflict/domination model will be encountered again and again in this book. The person is nowhere to be seen, transcendence is forgotten, and the relationship between constraints and freedom dissolved.

Throughout, then, let us respect this balancing act. Let us not discredit biological accounts of moral functioning as utterly without value just on principle (though it may be that much of the extant empirical research so happens to be of a particularly low grade), but let us never forget transcendence either, nor yet those profoundly human elements of the moral life which arise out of their balancing.

*

One very clear illustration of this need to find the complex balance between biological influences and the various broader nonbiological considerations can be seen in the recent phenomenon of "neuropsychiatry"—the idea that clinical psychiatric practice can be augmented by grasping the neurobiological and genetic factors which contribute to mental health issues, which cause psychiatric problems by their effects on the brain's anatomy and biochemistry. Now, understanding the neurological bases of certain psychiatric conditions in order to assist in treating the unfortunate persons subject to such conditions is a good thing, on the whole (again, things do have to be looked at on a case-by-case basis, for some disorders may be more or less influenced by such factors). The problem here, though, is exactly that which we face with respect to moral enhancement: as soon as one starts to think that one's object can be even partially understood in biological or neurological terms, or both, then this mode of analysis starts to consume all other aspects, either totally, or by assuming a de facto primacy of these biological dimensions.

There are many reasons why the biological mode invites its proponents to use it in this way, to eat up the other nonbiological elements. For a start, since the biological aspects are more amenable to being controlled and manipulated, this hope that biological interventions offer some greater purchase and control on the issue at hand is extremely seductive. It is also true that the biological aspects of such phenomena are much more readily subjected to rigorous empirical means of study, and so the results of such biological analyses carry all the weight and authority of the hard science that biology is (as opposed to the more contextual "soft science" one finds in the human sciences like psychology or sociology). Perhaps most significantly, biological and, in particular, neurobiological modes of analysis tend to consume other modes of analysis, for the very simple reason that they are *sexy*. That is, there is a strong cultural interest in the brain, and such research produces a wealth of colorful brain-scan imagery and the like, which, certainly in the recent past, has had a strange and hypnotic effect on the general public's perception of the quality of the research. So the biological or neurobiological aspects of an investigation end up taking on a de facto primacy, and we get a de facto reductionism of moral functioning to the biological level of analysis, to the detriment of the other many significant influences involved, and the nature of these causes as intermingled, as mélange.

Suddenly we start looking for biological causes everywhere, and for everything, even if they are but a minuscule part of the larger picture. In neuropsychiatry, for example, we look for the chemical bases of something like depression or melancholia and attempt to deal with the issue on these terms. There is nothing wrong with this in principle, but in practice, when the biological aspects have consumed the larger dimensions of the problem at hand, our understanding of the nature of that problem becomes oversimplified to the point of being travestied. This biological primacy then feeds subtly into entire worldviews and is used to support particular constructs of the human person, in particular that persons are merely passive slaves to the chemicals in their brains.[4] There is no question that there are biological dimensions to depression, but we trivialize the problem by looking at it predominantly in that way. Psychiatric problems, like serious moral problems (and there is not always a sharp distinction to be had between them), are extraordinarily complex. To resolve these problems requires hard work, analysis, and the changing of lives. Treating complex problems as exclusively or even as predominantly biological problems is a way of sweeping the larger causes of the matter under the carpet, a means of ignoring them, a means of looking for easy answers so that we can carry on exactly as we have been, when it is precisely the way that we have been living, our personal choices, and the larger social-environmental situation we inhabit, that so often dominate in causing the problem at hand. As such, looking for easy biomedical answers serves the interests of the general public also, who are hungry for such easy answers, rather than having to admit that significant changes must be made to our lives. The parallels with the biologization of moral functioning here abound.

Of course, context is incredibly important, and some shortcuts are more or less problematic than others. Yet it is in order to undo and prevent the similarly problematic distortions happening with respect to biomedical moral enhancement (which is certainly prey to this general wish for shortcuts and easy answers) and the biological study of moral functioning more generally that such a big deal is going to be made here in emphasizing the nonprimacy of biological influences on moral functioning. We will be disputing in no uncertain terms the idea that such biological mediation can be sufficiently well atomized to proffer reliable and straightforward relationships between the aspect of moral functioning in question[5] and the way it is biologically mediated. We shall see that in most if not all cases, there is no easy or direct cause-and-effect relationship to be had between biology and moral functioning—and this is precisely because of what I will be calling "the mélange problem," which is nothing other than the categorical ambiguity of consequence produced by the extensive range of influences on moral functioning, which do include biological influences, but which are simply too complex, too numerous, and too interwoven to think that biological means alone can, in most cases, be sufficient to deal with the moral problems at hand. The biological and the plethora of psychological, social-environmental, political, economic, religious/spiritual, and other influences on moral functioning are thoroughly interwoven.

While it is certainly true that a degree of atomization may be necessary for biological study to be conducted at all, this can never be an end point, nor sufficient to give a satisfying picture of the situation. In all cases, any conclusions to the empirical study of these biological influences must in some way be *retranslated* such that these influences can be properly understood as part of the massively complex, interwoven mélange that moral functioning is. The present book might be understood as an attempt at such retranslation.

As for behavioral control, certainly it is true that behavior can be manipulated, but the point here regards the idea of some direct cause-effect relationship between biology and moral behavior or judgment—an idea which also happens to carry a distorting misrepresentation of the human being as a passive nonagent, a "biological machine" which can be controlled, directly, by simply flipping a few biological, genetic, or neurological switches. None of this means that moral enhancement is impossible—quite the contrary, in fact. But it does mean that moral enhancement discourse has to transcend any hidden "mechanical" metaphors that are covertly embedded within it (we will see that such metaphors are disturbingly prevalent), and deal also with the implication that biomedical moral enhancement will inevitably be limited in its potential powers, at best only ever understood as a support mechanism for larger moral formative means (though even this idea of moral enhancement working "alongside" traditional formative means, is, as we shall see, also extremely problematic)—not as an equal partner, and certainly not as some "magic wand" which can turn bad people into saints by popping a pill or by means of some neurological implant.

Examples of this way of ignoring the complexity of moral functioning, and the way it can lead to trivialization of serious, real-life situations abound. Dealing with problems like inner-city violence on the biological level, for example (as various contemporary government initiatives in fact do—for example, the 1997 US "Violence Initiative"), be that on the genetic or neuroanatomical level, results in a very real avoidance of the actual social-political and economic situation of the inner city. Yes, we can take "those genetically predisposed toward violence" and monitor them, even sedate them, and this may well serve a number of lazy interests, including the patients themselves, but it misdirects attention away from the hard answers and the real embracing problems—poverty, lack of education, and poor parenting being toward to the top of the rather long list of contributing factors.[6] Looking only at the genetic level and thinking that we can then preemptively segregate or sedate persons who have what is deemed to be a problematic "genetic profile" or "underdeveloped amygdalae," when one is aware of the tremendous scope of the problem, is obviously a travesty. It is radically superficial and dissatisfying.

As such, we end up trivializing very important problems if we start to think that they can be dealt with primarily by biomedical means. Indeed, it is only on the grounds that we perpetrate such trivializations of the complexity of the human situation that we can even begin to think that biomedical moral enhancement proffers some world-salvatory potential. If a complex problem has profound social causes, then we have to be very careful and very modest when talking about biomedical moral enhancement potential. When Paul Zak indicates that the London riots of 2011 can in some significant sense be attributed to a shortfall in oxytocin (Honigsbaum 2011); or that the prosperity and happiness of entire nations can be predicted by their basal levels of oxytocin (Zak and Fakhar 2006); or, as others have suggested, that unconscious racial biases may be assuaged by taking beta blockers (DeGrazia 2013, 228); or that alcoholism might be cured by tampering with the addict's dopamine-reward circuits (Bass et al. 2013, in Clark 2014); or, as Joshua Greene's words have been presented in the public press, that criminality is a matter of "faulty machinery" (Goldberg 2014)[7]; we end up with pictures that are much worse than superficial—they are outright misleading, if not offensive.

In such cases, study of "the moral brain," and the collusion between neuroscientists and neurophilosophers, is not deepening our understanding. It is in fact diminishing it. This is a very strong claim, but a crucial one that will be made throughout—that the exaggerated primacy given to the biological dimensions of moral functioning generates empirical studies, which, by interpretation carried out through inappropriate metaphors, actually lessen our understanding (this is not to mention the very dubious methodologies applied in such neurostudies, a point which will be emphasized later on). We understand the human creature less because of this mode of empirical study; we are presented with a false picture of human judgment and behavior based on superficial assumptions and metaphors of the human creature, which, because presented in objective scientific terms, pushes the debate onward in dubious directions. Yet this distortion is what must happen when we overexaggerate the potential of biological intervention, or place too much faith in neurobiology as a means for understanding and manipulating complex and contextually driven human behavior.

The bigger concern, then, is that of primacy, and the more covert de facto reductionism that it produces. The question of primacy is a matter of which order of explanation one privileges. This is, in part, a social question. As noted above, for those with the eugenicist's faith, it was the genetic order of explanation that mattered above all. For us today, the eugenic dogma is less appealing. Today it is neuroscientific explanation—the idea that "we are our brains"—that more readily captures public conviction. Simply because this is the current popular mode, the lingua franca of the larger social discourse, even those aware of reductionist tendencies can fall prey to it in more subtle forms. While it is true that there is a backlash coming from within the biological and medical sciences themselves against the superficialities of bioreductionism, this larger, more embracing form of de facto reductionism—neuroprimacy—that comes from retaining the privilege given to these sorts of biological accounts threatens to undermine this pushback. We end up at the beginning, with reductionism once again, but a more insidious kind because it does not even see itself as reductive. To the contrary, it perceives itself to be the opposite.

How Inappropriate Metaphors Distort Our Understanding: A Case in Point

The need for a more thoroughly integral approach is at least as pressing with respect to the study of moral functioning as it is anywhere.[8] The ambiguity of behavior and judgment is a significant problem here. To give an example of how subtending narratives and metaphors can distort the very way we go about investigating the subject matter, coloring our objective scientific studies, we can return to Molly Crockett's (2008) use of the "Jekyll and Hyde" analogy for our two "brain systems in conflict." With this Jekyll and Hyde frame, we are provided with an account of the human being as subject to and caught between two very different kinds of brain processes, "compelling" him or her either toward cool-headed self-control and goal-centered thinking, or toward an emotional reactivity that disregards all sense of consequence. While Crockett uses this "Jekyll and Hyde" frame as an explanatory tool in a "popular science" context, the reality is that this framework runs implicitly and pervasively throughout the neurological study of human agency. Indeed, one can observe that there is a narrative even behind this narrative, which is arguably a boiled-down reincarnation of the Freudian drama of the ego caught between the moral authority of the superego and the powerful, unthinking, desirous id—except described now in neurological, organic terms, using the top-down model of brain function instead of the hypostases that Freud elaborated in his topographical account.

In either case, man is caught between his animal reactivity, which is presented as automatic and natural, though savage, and his restrained, forward-thinking capacities, which are characterized as taking effort and being slower and weaker more generally. Yet, continuing with this Jekyll and Hyde narrative, problems arise because, within this value-laden frame, the "Mr. Hyde"—that is, the pre-civilized, emotionally reactive bases within our brain matter—is clearly being presented in distinctly negative moral terms. But look at how context and ambiguity threatens this narrative. Very little reflection should be required to see that sometimes it is precisely our emotionally reactive selves that are exactly what is required to perform acts of tremendous moral worth.

Indeed, there are times when decisive, emotionally reactive responses are the necessary condition for enacting the good. What happens to the value-laden frame of Jekyll and Hyde when we discover that, in fact, sometimes we need to be our Mr. Hydes in order to be our best selves in that circumstance? The reactive, emotionally driven response required for acts of great moral heroism (say, intervening in a street robbery, diving into rough waters to save someone from drowning, jumping onto a grenade to save one's comrades, running in front of a bus to push a pram to safety, happening upon a person being raped and intervening, and so on, and so on), that is, the most extraordinary moral actions, are primarily the domain of our so-called nefarious "Mr. Hydes."

Where our "morally problematic" Mr. Hyde reacts immediately (and we forget that we have inborn emotional urges that manifest in potentially morally beneficial manners too—recall Midgley's (2002) point mentioned earlier: if we did not have cooperative and self-sacrificial impulses as part of our instinctual animal nature, we would not have survived to this level of evolutionary advancement to begin with), when it comes to circumstances where we need to be acting impulsively, irrationally (or nonrationally, nonreflectively), self-sacrificially, in order to perform the good, our morally civilized Dr. Jekylls, hesitant, reflective, goal-oriented, cool-headed, will likely only stand at the sidelines and watch. Our cool-headed Dr. Jekylls are utterly useless in such circumstances—standing by reflectively, refraining from what they quite sensibly perceive to be "egregious" and imprudent reactions. In this case at the very least, then, things are as John Harris (2010, 104) observed: "the sorts of traits or dispositions that seem to lead to wickedness or immorality are also the very same ones required not only for virtue but for any sort of moral life at all."

Clearly then, as analogues for our "two brain systems in conflict," the "Jekyll and Hyde" narrative is superficial, inappropriate, and misleading. Both so-called "Jekyll" and so-called "Hyde" are absolutely required for a person to express a satisfying moral range—each has its time and place, unrealistic as it may be to expect any given single person to have such self-mastery as to be both able to switch between such systems at will, and able to discern which system is most appropriate to the situation he is in. Perhaps what this really points to is that such a satisfying moral range is better sought out in *groups*, whose members might embody different moral strengths, rather than looking purely to individuals. A shift toward the inclusion of the importance of group moral functioning, where differing members of the group work as a "moral team," as it were, is certainly one development that moral enhancement discourse could benefit from considering (though how one would biologically fine-tune the moral functioning of a group so as to encourage a diversity of moral powers across the group is anyone's guess).

The problem is that the morally laden narrative of Jekyll and Hyde gets projected, albeit implicitly, onto the brain itself. The brain gets talked about as if it has bits of its organic matter which are themselves good, and bits which are evil, and that these bits of gray matter, which persons are compelled by, are in a state of war. This is the myth of the moral brain in action. Notice the moral bioreductive framework forwarded by applying such a narrative:

Dr. Jekyll =
Sufficient serotonin levels =
Goal-oriented thinking =
Appropriate amygdala response and activity in the prefrontal control network =
Properly functioning brain =
Morally good

Versus:

Mr. Hyde =
Low serotonin levels =
Emotionally reactive =
Malfunctioning frontostriatal dopamine system =
Faulty machinery =
Crime =
Morally bad

The difficulties with Crockett's use of this analogy are clearest because she has been at least self-aware enough to make her metaphor explicit; most contributors are under the sway of such narratives without even realizing it. Indeed, the very narratives through which study is framed are in many cases symptomatic of significant simplifications of the nature of the problem being investigated, and thus, right from the get-go, the framing of the study leads to distorting results. While this "Jekyll and Hyde" analogy is a particularly clear example of the problem, the worry should be that the scientific study and the philosophical reflections which result from them are replete with such narratives, covertly shaping study from beneath the surface and then being presented as objectively gleaned results rather than as the means for supporting particular views about the nature of the human person that they ultimately are. The inclusion of such narratives in scientific research is an inevitability, particularly with respect to the moral domain, but at the very least we need to disclose such narratives and interrogate the powerful way they have of shaping what is taken as indicating purely objective facts about the objects of inquiry.

Proceed with Caution—Final Words

Given all these remarks, it should not seem too controversial to posit the claim that there can be no direct, unmediated link between biology and complex, encultured human behavior. By definition, results which depend upon a confluence, a mélange of factors—psychological, cultural, conceptual, historical, political, economic, and so forth—and behaviors which depend upon all of these shaping variables interacting in some complex way will by definition be falsified by attempting to grasp them only, or even primarily, on the biological level. If influences are only influences in relationship, in the mingling, then no direct line of sight exists therefore wherein some biological "button" can be pressed, or lever pulled, which will produce absolutely predictable results with respect to such complexly intermingled behavior.

Of course, certainty is too exacting a standard to which to hold such interventions. The proper question is rather how reasonable or probable it is that any given biomedical intervention will have a positive impact on moral performance, all things considered. It is highly likely that some moral goods or evils are more or less amenable to direct biological intervention than others. Once again, therefore, a case-by-case analysis is necessary. Something like aggression is much more readily altered by using biological agents, sexual impulses too, though not without cost. It is already possible to increase or diminish a person's aggression or sexual drive using biomedical interventions, and has been for a long time. A range of antipsychotics and hormone therapies exist, and have been used for modulating such impulses and drives. Once again, there can be a tremendous cost to using such agents, and where such biomedical interventions are curative rather than punitive (say, treating the pathologically aggressive, versus punishing a self-possessed, predatory sex offender) the costs to the personality, well-being, and agency of the individual in question are tremendous. Whether such interventions would be worthwhile anyway is another question.

While certain very basic, impulse-driven moral evils might be potential targets for moral enhancement (though in this case the term "moral enhancement" would then be little more than a euphemism for chemical castration and psychiatric grade sedation), as soon as we get into moral goods which have the least complexity, the lines of predictability diminish radically. Even with aggression and sexual drive, one could very well make the case that such biomedical interventions, if taken in isolation, involve unduly simplifying the matter at hand. In actual practice it is never a case of simply dispensing the drugs and then letting the individual in question alone—such drugs are always applied as part of a larger psychiatric or state-monitoring package. Thus, even in these cases where biomedical sledge-hammers like antipsychotic sedatives and chemical castration are applied, there is a recognition that things are more complex than the purely biological angle allows, that the individual in question needs care and attention along a number of lines, and thus that there is an inner complexity to issues of aggression and sexual drive such that biomedical approaches are not adequate on their own.[9] There is some sense here among those that use such biomedical methods in real-life instances that simply using the biological approach in isolation will end up being self-defeating.

And these are the simplest, most direct cases. The more complex the sort of behavior we are looking at, the more social, the more scaffolded, the more cognitive-dependent, the more ambiguous, interpretive, conceptually formed (and so on), that a behavior is, the less we are going to be able to draw that reasonable or probable line between biological intervention and the desired moral improvement. Even forgoing an expectation of certainty, the predictability of such interventions is diminished more and more by the number of intermingling factors essential in shaping such goods.

We shall see that virtually every refined moral good has a tremendous degree of complexity attached to it. While aggression and sexual impulses can be clubbed down wholesale using biology, the idea that something like moral discernment, or generosity, conscientiousness, perseverance, gratitude, social conscience—the finer moral powers—can be likewise biologically reduced is unrealistic in the extreme, and any interventions aimed thereto will always come with serious strings attached.

Once more we return to this need for balancing a respect for what biomedical intervention can offer with a healthy skepticism. We are biological beings, and biological avenues are something that have to be considered. But we do have to avoid the radical overconfidence of the possibilities for biological and neurological enhancement so rife in the moral enhancement domain. We have to be wary about the extent to which moral functioning (or any given facet thereof) can, or should, be understood in biomedical terms. For this is the underlying thesis of this book: that in the vast majority of cases, the biological aspects of moral functioning have been massively overexaggerated in their potential significance. The biological approach to moral functioning, while certainly valuable and enlightening when viewed cautiously, is not the most appropriate lens through which moral functioning should be looked at. If moral enhancement is truly desired, and desirable, it may very well turn out that the biomedical angle of approach is actually the least powerful means of encouraging moral growth and transformation. That does not mean such biomedical explanations should be avoided. But this only heightens the need to assess, in each case, the extent to which biological interventions can be of benefit, to have a clear grasp on what their limits are, and to look at as many various influences as are relevant, such that a multidimensional and integral approach might be envisaged, which can then be brought to bear for positive benefit in an appropriately integral bio-psycho-social manner.

We will see some examples of biologically based interventions which are beneficial, whose desirability is very hard to argue with, and which decisively challenge those who think that biomedical moral enhancement is an all-out "bad idea" to be thrown out without consideration. Yet attempts to augment moral functioning by biological means must always be understood, even at their most optimal, as partial efforts, nudges, in the service of more traditional morally formative means, and certainly not as containing any kind of world-salvatory power. Attempting to go further would be an abuse of biology, and like all the efforts in moral bioenhancement's prehistory, because of the superficial picture of the human creature that they rely upon, self-defeating and doomed to failure anyway.

I Philosophy

2 The Philosophy of Moral Enhancement: Why Morally Enhance?

We need more understanding of human nature because the only real danger that exists is man himself ... he is the great danger and we are pitifully unaware of it ... we know nothing of man ... far too little ... his psyche should be studied ... because we are the origin of all coming evil.

—C. Jung (1959)

Even those with the most optimistic views of human nature will admit that there is much that is undesirable in human behavior. Even those who think that "underneath" or "in his heart" the human creature is "basically good" will admit that, for whatever reasons, human beings go astray, and have behaved and continue to behave in appalling and hideous ways. As such, if there were any human endeavor at all that could benefit from being given a helping hand, it would seem natural to think that *moral* behavior, the endeavor to become morally better persons, should be somewhere toward the top of that list. If only there were some way of making persons more altruistic, more compassionate, more cooperative. If only there were ways of making persons less bigoted, less territorial, less violent, less sadistic and cruel to one another. Yet there are those who believe that advancing technology and increasing scientific understanding of the nature of the human being and "the moral brain" will come to offer (and to an extent already do offer) some means of obtaining these grand aspirations. The belief is that human moral functioning is in no small part biologically conditioned in nature, and thus, if the right biological levers can be discovered, significant changes to man's moral nature can be affected. Given that even the most optimistic view of the human creature is forced to admit that terrible, terrible evils have issued, and continue to be issued, from none other than man's own hand (and let us not say any more then of what the pessimists think of human nature), should some way be found to alter the biological impulses involved in moral functioning, some way of strengthening the good impulses and of weakening the evil impulses, of developing our powers of moral reasoning, judgment, and action, who then could possibly be against such moral enhancement?

Nothing is so easy. While on the surface the idea of giving moral functioning a helping hand seems not only desirable, but laudable, and perhaps even necessary given the world we live in and the sort of beings that we are, in fact there are a range of obstacles and complexities which make the idea of moral enhancement much more challenging than it first seems. As is so often the case, how one defines one's terms can make all the difference between the sort of vision of moral enhancement that is laudable, that which is merely impractical, and that which is outright grotesque. The aim of the present chapter is to articulate and explore the rationales for moral enhancement as proposed by its three primary philosophical enthusiasts. We will explore a range of issues, many of which we saw in abstract form in chapter 1, in the context of the rationales that have been proposed. We will encounter first the thesis of Ingmar Persson and Julian Savulescu, whose apocalyptic vision presents compulsory moral enhancement as a means for preventing all-out annihilation of mankind by his own hand. We will encounter James Hughes, the arch-transhumanist, perhaps the most intellectually credible of all transhumanists, and his account of voluntary "virtue engineering" as a means for facilitating the development of a democratic citizenry's self-chosen moral codes. We will finish with an encounter with Tom Douglas, whose vision of voluntary noncognitive moral enhancement for dealing with "counter-moral emotions" offers a more realistic and presently attainable account of how moral enhancement might be operationalized. While each of these accounts has its problems (as we shall see, some more than others), each of the accounts has valuable lessons to teach us about how to go about thinking through moral enhancement, but also valuable lessons in how *not* to go about it. Thus, before moving onto the various gnarly issues the subject poses, and the science upon which such claims are based, it is important to explore critically the more influential rationales for moral enhancement that have been variously proposed.

Rationales

Persson and Savulescu: Monkeys with Guns

If safe moral enhancements are ever developed, there are strong reasons to believe that their use should be *obligatory*, like education or fluoride in the water, since those who should take them are least likely to be inclined to use them. ... That is, safe, effective moral enhancement would be *compulsory*.

—I. Persson and J. Savulescu (2008, 174; emphases added)

Persson and Savulescu's (2012) work on moral enhancement, culminating in their book *Unfit for the Future*, offers one of the most talked-about philosophical contributions to moral enhancement discourse, and this book outlines in nothing less than apocalyptic terms the threat to man's survival from himself by virtue of advancing technology, liberal democratic freedom, and various elements of his innate psychology. As such, they say, there is a need for society-wide, compulsory moral enhancement as the cure and salvation for our future, as a means to fulfill "the urgent imperative to improve the moral character of humanity" (Persson and Savulescu 2008, 162).

Be Afraid

In order to grasp the rationale for moral enhancement presented by Persson and Savulescu, we have to look at what the pair refer to as "the Bermuda Triangle of extinction" (Savulescu 2009); that is:

(a) radical technological power,
(b) liberal democracy, and
(c) mankind's "myopic" moral psychology.

Let us begin with the question of man's incredible technological progress. The essence of Persson and Savulescu's argument here is that the greatest threat to humanity is none other than humanity itself. Nowhere is this threat clearer than in the world-destroying potential of the technology that we have developed. Ever-increasing technological advancement, particularly in light of the ongoing interest in the development of cognitive enhancement, which only aims to make the human being more clever (without making him wiser, or more moral), implies a worsening of the odds that world-destroying technology will at some stage fall into the hands of a rogue state, or a lone, crazed individual, or some set of such individuals. Such a person, or group of persons, bent either on the overt annihilation of mankind, or as an indirect result of attacks made using such world-destroying technologies, would then have sufficient destructive power to reduce the earth to a completely uninhabitable condition. Such a condition Persson and Savulescu refer to as a state of "ultimate harm" (Persson and Savulescu 2011, 441): the idea that conditions on earth can be made so inhospitable that no *worthwhile* human life, no life worth living, would be possible. Such a state of ultimate harm is a very realistic possibility, one made ever more realistic as technology advances and the human population increases. It is rather hard to argue with the idea that mankind's technological progress far outstrips his wisdom, that mankind is in a precarious position, threatened by his own inventiveness. But before we get to the proposed solution—statewide, obligatory moral enhancement for everyone—it seems that things are even worse for us than this picture suggests, for liberal democratic freedom and man's myopic moral psychology only compound the risk that this world-destroying technology presents. Let us look closer.

Big Brother

The second concern that Persson and Savulescu raise as a threat to human existence is liberal democracy, with its emphasis on personal freedom, moral pluralism, and privacy. Why is this so terrible? For Persson and Savulescu liberal democracy creates a "heightened risk of destruction" (Savulescu 2009), compounding problems by promoting ideas such as neutrality, tolerance, freedom, multiculturalism, ideals which can make for a poor basis in coordinating the sorts of programs required to secure mankind's future welfare. The goal of maximizing the freedom of individuals and offering tolerance to violent, hateful minorities is a formula which, it is suggested, can only serve to compound the dangers technology poses in the future.

Certainly liberal democracy has its problems, as any regime will. However, whether the solution to the presence of violent extremism and intolerance in society is really the curtailment of freedom along the implicitly totalitarian lines that the pair describe—increased surveillance, increased encroachment into privacy, restraining or segregating those "at greatest risk to the community," and so forth (Savulescu 2009)—is quite another question (indeed, such measures are precisely those advocated by the violent extremists Persson and Savulescu are so concerned about). Nor has it been made particularly clear how reducing freedoms and privacy in a Western state can possibly assist in preventing a rogue state or extremist group in another continent from availing themselves of world-destroying power and using it.[1] It is surely the case that unless one is talking about moral enhancement in terms of *globalist totalitarian enforcement*, such reductions in the liberal democratic freedoms of but one or a few states can in no way be proposed as a solution to the world-destroying danger that certain groups and individuals might pose. For unless the obligatory scheme Persson and

Savulescu have in mind were enacted in a comprehensive worldwide way, there would be no point in enacting it at all—not if preventing ultimate harm is the rationale for so enacting such a scheme anyway. For as they themselves observe, those most needing such enhancement are those least likely to elect to undergo it. As such, some kind of global totalitarian political measure is the most natural foundation for their proposal and implicitly advocated for thereby.

What is helpful in thinking about this proposal, however, is the insight that moral enhancement has tacit and unavoidable political dimensions— the idea that a hard vision of moral enhancement has to be understood against certain worldviews, ideologies, and political philosophies. It is a mistake to think of moral enhancement (however it might be envisaged) as something purely technological or pharmacological in nature—rather, these means are prongs of an inevitably larger set of programs (benign or otherwise) which will involve distinct views on which particular values are worthwhile and which involve broader political, medical, mental health, or public policy efforts to bolster the technologies promoted. Persson and Savulescu's embedding their moral enhancement discourse within reference to such overt political foundations serves as an important indicator of the manner in which moral enhancement refers not just to technologies but to entire worldviews and, to varying extents, world-shaping aspirations. Moral enhancement cannot be talked about as if in an apolitical vacuum, and the validity or extensiveness of a particular vision of moral enhancement must be understood against a political or ideological backdrop in which such a vision might realistically be advanced. It is hard to imagine how a vision of compulsory moral enhancement of the sort that Persson and Savulescu have in mind could be justified and enacted in a liberal democratic state, or a world organized such as ours in fact is. Even if such technologies could be devised to make such a prospect possible, enactment of such a vision of enhancement would require a huge political-economic revolution. And not necessarily one for the better. In any case, no such political set-up presently exists in the developed world.

Cavemen in the Modern World: Our Myopic Moral Psychology

The final prong of man's threat to himself comes, it is argued, from his innate moral psychology. The basic idea here is that man's moral nature is a product of evolution, and thus at least in part biologically grounded. This all seems reasonable enough to assert. Humans are biological beings, and it would be rather surprising to find that moral functioning is not in some way mediated by his biology. Human beings, as the pair have argued, have a "myopic" moral psychology by nature. The claim is that we have evolved to thrive in small communities and have thus developed a strong territorial and in-group preference system.[2] This is particularly dangerous in a mutually interdependent world where various different conflicting groups have contact with each other, and it makes global problems difficult to remedy because humans are not "hardwired" to care enough about persons they cannot see or are not emotionally invested in by proximity (for example, starving children in far-off lands, or members of other groups, clans, tribes, and so forth). It is suggested that this evolved "parochial altruistic" moral makeup of the human being is highly threatening to mankind's survival in the kind of interconnected world that we live in today, and that what is necessary now is a set of biologically derived enhancements which can produce in us a disposition toward a more universalist way of making moral judgments and acting (i.e., focusing on the welfare of all equally).

But to what extent is it true, as Savulescu (2010) puts it, that "the *fate of humanity* may lie in its own biology"? To pose a question that will continue to resound throughout the rest of the book: is our biology so decisive a factor when it comes to our moral functioning? One needs to ask whether human biology and moral functioning map onto each other so readily, in such an uncomplicated way as is required for a reasonable hope of success in applying biological agents for moral improvement. It might turn out that humanity's fate does lie to some extent in its biology—no doubt this is so. But as has been suggested, the interwoven nature of such a fate must by necessity be spread across factors both biological and non-biological wherein no easy separation can be discerned. Thus it would be as crude to say "man's fate lies in his biology," as it would be to say "man is a product of his environment, and only as good as his environment allows," or, indeed, to say that "man makes of himself whatever he wills." None of these reductive explanations is adequate in isolation. The question will always be about how realistic it is to expect any given biological intervention, in combination with any given moral scaffolding, to be able to enrich a particular moral characteristic in a human person. And for this question, no generalized or abstract answer will do. In every case, we must appeal to the facts on the ground.

That being so, within Persson and Savulescu's framework, reliant as it is upon the rationale of ultimate harm, we have to constantly add the further questions regarding the likelihood of any such interventions (a) having even the slightest impact on the probability of self-annihilation, and (b) whether that justifies compulsory, state-sponsored programs to enforce them (assuming the political revolution required to provide the context through which such a program could be enacted has been effectively carried out). Given the inner complexity of moral goods, and the complex interwoven nature of the influences on moral functioning, we have good reason for thinking all this is simply too unwieldy to assert as plausible the idea that biological manipulation can provide a sufficiently stable fulcrum against which to shift those wheels of man's fate in a way that would justify what Persson and Savulescu are proposing. This will not so much constitute an argument against moral enhancement, but rather says more about the notion of ultimate harm as a rationale for the exhortation to so enhance.

Ultimate Harm

While both the ideas of ultimate harm and moral enhancement, when considered independently, can be construed in ways that are perfectly coherent, when the former is used to justify the latter, we get into highly questionable territory. The practical reality is that we cannot at present envisage any sufficiently potent intervention to (a) prevent malevolence-based ultimate harm, nor even (b) diminish the possibility of ultimate harm in any significant *real-world* way (the purely formal, abstract possibility will always exist)—and this is so regardless of whether one present such enhancement as compulsory or voluntary.[3] The reality is quite simple—workable moral enhancement interventions are not nearly strong enough to do the job that preventing ultimate harm would require them to (and we will have to confront this truth regarding the very limited efficacy of proposed interventions over and over again throughout the chapters to come). Given such a lack of efficacy, discussion of moral enhancement with respect to preventing ultimate harm can begin and end right there. If preventing world annihilation is one's concern, moral enhancement cannot be the answer. For if the practical reality is that no such intervention can stop a person decisively committed to carrying out a terrible harm anyway, then we had better start looking to alternative rationales for justifying such enhancement. Reality, as we shall see throughout, is the severely limiting factor which is the ruin of these more fanciful sorts of thinking. No moral enhancement that we can realistically envisage can turn a person into a "moral robot," thankfully, thus ensuring that free choice will always remain, no matter how compulsory or powerful the moral enhancement intervention will be.[4]

But what about *indirect* harms as threats to human existence? Can groups of persons doing petty wrongs create an aggregate of harm so devastating that it destroys the world? Might moral enhancement thus be capable of reducing the probability of ultimate harm in some meaningful enough, if indirect, fashion to yet justify its use? Some have suggested manmade climate change as an example of such a prospect. For example, it has been suggested that if the population of China were to have the same level of consumption of cars and petrol that we enjoy in the West, a smog so thick would be created that the entire world would choke. Currently, the choking smog in Beijing, a combination of coal emissions, dirty diesel, and industrial gases, is so thick that outdoor school sports have been banned, planes routinely require diversion, and the city has a color-coded system for measuring pollution which, when "red," requires factories to be shut down and half of the city's cars to be garaged. The capital's pollution is "10 times the level considered safe by the World Health Organisation" (Grammaticas 2013, MacLeod 2014). If this is true, then it is possible that indirect harms, involving deeds that are not necessarily explicitly recognized as immoral,[5] could compound together to create a world-destroying situation.

Does biomedical moral enhancement have the power to prevent or diminish the likelihood of such world destruction? I would argue not. Taking the example of pollution above, one has to ask what possible biomedical intervention, compulsory or voluntary, foisted upon the general population could be so potent as to stop the developing world from pumping out such absurd levels of poisonous emissions, or so powerful as to stop Western governments from paying such countries to pollute on their behalf? To suggest biomedical moral enhancement as even but a partial solution to global pollution is to be looking in the wrong direction right from the outset. It suggests that the very framework through which one is going about conceptualizing such problems is mistaken. How could such pervasive pollution be even *fractionally* managed by individual persons taking something like oxytocin or serotonin, or going through a process of genetic moral engineering? As we look at more and more cases throughout the book, we will be brought unassailably home to the vast complexity and multidimensional nature of these world-threatening problems, as well as their absolute imperviousness to biomedical resolution and their impenetrability with respect to individuals taking a tablet or having some biological intervention nudge them in a more generally compassionate direction. Even the suggestion that moral enhancement might produce some minimal resolution to these problems involves a simplification of the complexity of the problems that verges on the offensive. To think that the prospect of state-sponsored, compulsory programs of moral enhancement are being argued for on these grounds. Genuine apocalyptic potential is used to justify a course, compulsory moral enhancement, which could not possibly remedy the problem invoked to justify it.

What is it about enhancement enthusiasm which constantly seeks to reduce the immense complexities of human problems to the level of that which can be medicated with pharmacological or biotechnological interventions? Such complex problems have causes at both individual, national, and international levels. These problems are political, economic, historical, cultural, psychological, and social in nature. Such mass social harms are a combination of all these varying forces and more, and without specific and highly coordinated mass organized action (of the sort that moral enhancement is not realistically fine-grained enough to motivate, as we will later argue), the idea that such moral enhancement can have some palpable effect on the problems in question stretches credulity to its limits. Indeed, Persson and Savulescu offer us little to hope for here, for as they themselves state:

A moral enhancement of the magnitude required to ensure that this [ultimate harm] will not happen is not scientifically possible at present and is not likely to be possible in the near future. (Persson and Savulescu 2008, 174)

Then why did they suggest it? It is their contention, after all, that it is in the immediate present that the greatest threat looms over us. If no realistic potential can be imagined to exist that would even dent these massively complex world-destroying problems, what was the point in even suggesting any of it? We are simply told that humans do bad things, are too clever for their own good, prize individual freedom too much, and are hardwired for xenophobic aggression. Even if all of this were true (and no doubt, there is more than a grain of truth to this highly selective portrait), as soon as one even begins to grasp the complexity of the biology of moral functioning, the variety of influences that impact on such functioning, and the complexity of the global problems such enhancement is invoked as being the solution for, then the idea that biomedical intervention can be held up as some means for preventing such an apocalyptic endpoint holds no water. It may well be possible that moral enhancement, *construed differently*, can be applied to some positive effect, but we must manage our expectations.

As such, while Savulescu and Persson's proposal may be apocalyptic in focus, it is instead only their argument that is so catastrophic. Going from the propositions "the world is in a precarious technological position, liberal freedom is a threat, and humanity has evolved for close-knit moral functioning" to the implication "we need a compulsory system of state sponsored moral enhancement that all persons are to be technologically compelled to obey," involves just too great a leap. All practical and present political realities are simply against it. In the end, we have to agree that the world may very well be in precisely the dire circumstances that Persson and Savulescu suggest—yet even so, moral enhancement cannot be the solution, because it is simply not powerful or nuanced enough, and the problems themselves are just so tangled and multicausal that simply offering up a biomedical solution trivializes the problems to the point of absurdity. The very framework forwarded by Persson and Savulescu distorts and simplifies the nature of these absolutely serious problems to the point of ludicrousness, misdirecting us away from the very hard answers that need to be actioned by means of the false hope of biotechnological moral solutions. So let us put aside this idea of ultimate harm and seek for some alternative rationale for justifying moral enhancement.

James Hughes: Virtue Engineering

Civilization is moral enhancement.
—J. Hughes (2012b)

Transhumanist Optimism   For Hughes, neurotechnologies hold the power to create a society where we are happier, more intelligent, more self-aware, more socially minded, and—most significantly—more virtuous human beings. As Parens (2012) observes, Hughes is part of a tradition of thought which views enhancement as a natural expression of human nature, and which sees human beings as creators, all of nature as a mechanism, and technology as a morally neutral tool (Parens 2012). The underlying transhumanist proposition here is that the application of biomedical advances to human enhancement is a continuation of the project of civilization itself—that is, the human being is by nature the sort of being which seeks to enhance itself. This draws, I suppose, on the sorts of Nietzschean and existentialist ideas of man as his own creator, man as the being whose nature is to self-overcome, to design and redesign himself, to make and remake himself as he sees fit.[6] For transhumanist philosophers, advancing technology is envisaged as the means by which this self-making urge inherent in the human person can be realized to its fullest and taken to its most radical degree. Becoming a better human being in the *moral* sense through applying advancing biotechnology is thus understood as part and parcel of this larger need for enhancement, of "humanizing ourselves," a natural extension of our will to self-overcome and to better ourselves as we see fit. Hughes is very optimistic, he writes:

Our growing understanding of the brain and how it generates empathy, self-control, moral judgment and even spiritual transcendence suggests that we will increasingly be able to identify and *treat* not just psychopathy, but ordinary moral and spiritual weakness. Neuroscience will offer us all the possibility of becoming the better people that we want to be. (Hughes 2012a [emphasis added])

Choosing to Enhance   The key feature of Hughes' work, that which sets him apart so decisively from the sort of views presented by Persson and Savulescu above, is the staunchly and explicitly liberal democratic political framework within which his thesis is set. Hughes argues for a liberal framework as the only political set-up in which moral enhancement can safely work. For Hughes, a moral enhancement project needs to be elective, *explicitly self-chosen*—for that is precisely the sort of self-making project that Hughes has in mind—"becoming the better people that *we* want to be." So, moral enhancement is envisaged by Hughes as something which has the power to enhance human freedom, rather than detract from it. In this vision, what truly enslave us are our own passions and destructive urges, not technology, and indeed, if technology has the power to diminish these inner forces which enslave us, then the case can be made for saying that moral enhancement technology contains the power to liberate us to embody precisely the sorts of moral traits that we wish to so embody. As such, moral enhancement is very closely associated with, indeed it is the very extension of, Western freedom, which Hughes sees as its proper context (the only fear Hughes expresses is that of moral enhancement in authoritarian hands, where the project of "cognitive liberty" could be threatened and curtailed; Hughes 2012b).

In short, just so long as moral enhancement is a freely chosen, elective process, wherein citizens can choose what values they consider important, freely choose which moral traits they consider valuable, and then, as a purely self-chosen act, decide to engage with technologies which enhance those self-chosen traits, then moral enhancement is presented as being, in principle, a positive and desirable force, an empowering idea, one which has the capacity to increase both happiness and the human power of choice. Just as importantly, for Hughes, moral enhancement then increases and betters the *citizenry* potential of a given democratic state. Citizens are able to better pursue their self-chosen moral codes; this increases personal responsibility, as well as the very standards of responsibility against which persons can be held accountable. Moral enhancement is thus a way of "raising the bar" on the societal level in terms of responsibility, be that civic, moral, or political. This would then be better for democracy all around. Of course, this would only happen if all, or a sufficient proportion, of the citizenry were willing and able to take advantage of the enhancements available to them.

Virtue Engineering and the "Good Personality"   Just how wide of a choice is one supposed to have regarding the morals one can elect to enhance? Is one allowed to choose *any* value or disposition as a subject for moral enhancement? For example, if one were a member of the Taliban, would it be permissible to "enhance" one's daughter so that she better showed the virtues of submission and obedience? Are all values worthy of enhancement? Not according to Hughes—despite his strongly libertarian views, Hughes definitively argues for a limited range of moral goods that are worthy and conducive to the larger democratic good, a "liberal yet positively normative model of the good life" (Hughes 2012b) wherein some ways of thinking are discouraged and others encouraged. Since for Hughes, "civilization is moral enhancement," and moral enhancement is the extension of Western liberal democratic values—their very enactment—this commits Hughes to a vision of moral enhancement which is limited squarely by the extent to which it forwards the potential for liberal democratic society. You are free to morally enhance yourself in any way which encourages free society.

Having now considered Hughes' rationale for moral enhancement, we must now look at the manner in which he thinks such enhancements might be made manifest. Hughes' tremendous optimism starts to become a lot more problematic for him at this point. Taking what Hughes calls "the liberal personality" (and taking its exclusive moral superiority for granted from this point), he begins his proposal by associating liberal personality traits with elements from the classic Big 5 model of the personality.[7] A typical statement from Hughes, for example, would read as follows:

In fact variations in serotonin receptor genes have been linked to the liberal-leaning personality trait of "openness to experience," which makes you more likely to have spiritual experiences, ... and less likely to be a religious fundamentalist and political conservative. (Hughes 2012a)

Let us look a little closer at this statement, for Hughes goes on at great length making connections between liberal personality traits and genetic/neurobiological factors in precisely this way. Disregarding one's personal views on liberal values for a moment, let us take a look at the nature of the way Hughes goes about connecting moral characteristics and potential biomedical interventions—for this, in itself, raises important points about the indirectness and complexity of attempting to influence personality characteristics by means of biomedical approaches that need close inspection.

The general flow of Hughes' proposals seems to be as follows:

(a) genetic and/or neurobiological bases can be manipulated *to some extent*;
(b) these genetic and/or neurobiological bases influence the expression of positive liberal personality traits *to some extent*;
(c) these positive liberal personality traits predict certain positive moral traits *to some extent*.

Therefore, by a fairly loose chain of partial linkages and Chinese whispers, the conclusion is that:

(d) positive moral traits can be *to some extent* generated by manipulating genetic and/or neurobiological bases, which *to some extent* determine the positive liberal personality traits that *to some extent* predict the aforementioned moral traits.

How convincing is this chain linking of genetics and neurochemistry with "the liberal personality," and then to certain moral traits? Just how much accuracy do we think might be obtained by such means? And what kind of side effects might be created as the effects filter through this long and loose chain of linkages? The problem here is serious—there are just too many mediating factors to go reliably from manipulating one end of this chain to the desired effect at the other end.[8] Silviya Lechner (2014) makes this point well. She writes:

For many philosophers the intentional structure of consciousness is what distinguishes the mind from the brain. Consider the following chain, which takes us from the rudiments of the brain all the way to action: brain states, mental states (thoughts), intentions (beliefs, desires, fears), propositions (statements), actions, settled dispositions to act (character traits). The farther an element is located from the chain's origin, the more complex the account of this element becomes. Exponents of moral bioenhancement want to tell a story about the middle part of the chain, about (moral) motivation (Douglas 2008; Persson and Savulescu 2012), and sometimes about the end of the chain, about (moral) dispositions (Persson and Savulescu 2008; Sparrow 2014), but their explanans are drawn from the origin (from the category of brain states), leading them into the trap of reductionism. (Lechner 2014, 31–2)

Hughes is really talking about the creation of a chain reaction wherein, through a series of knock-on, indirect effects, we might have some kind of impact in engineering the particular moral trait desired. Looking at the complexity of the chain, as Lechner proposes, should help make clear why no fine-tuned way of influencing moral functioning can be achieved through this clumsy way of conceptualizing the operations of moral enhancement. This is so by definition: one cannot take explicitly broad-stroke influences, ignore all the many other mediating factors involved, and hope that this long chain reaction will have specific desirable results with respect to one's moral character, or not have unexpected side effects which may be just as morally problematic, if not worse, as the original condition. It is simply unrealistic. As with the concept of ultimate harm, while there is nothing in principle wrong with talking about liberal personality traits and moral enhancement independently, the problems arise when one uses the one as a rationale for justifying the other. Disregarding the question of whether liberal personalities have quite the moral superiority and weight that Hughes seems to think they have, realistic visions of moral enhancement are simply not fine-grained enough to plausibly assert that such interventions might consistently produce liberal-type persons anyway. This does not rule out moral enhancement, but it does mean that one has to be very careful with one's proposals and limit oneself with respect to what one plausibly expects such enhancement to be able to achieve.

Taking as Read

The biggest problem with Hughes' proposal is that he just refuses to think at all critically about the science he draws upon. This is a very important lesson that many moral enhancement enthusiasts need to take on board. Indeed, Hughes is the paradigmatic exemplar of the refusal to think critically about the quality of the relations between present science and the suggestions for enhancement he merely lists off. Hughes forgets that the scientific studies he draws upon are quite rudimentary in nature and extremely provisional. For example, Hughes makes the claim that we now have a growing understanding of "how the brain generates empathy"—certainly there is some truth to this, but if the science shows anything at all, it is, again, that there is no direct cause-effect relationship between empathy and the hormone (oxytocin) that supposedly causes it. The situation determines practically everything—for while increasing oxytocin in one situation can lead to increased empathy, in another situation the same hormone leads to far less pleasant behavior (we will go into this in greater detail in part II). Hughes also places too much emphasis on the supposedly genetic bases of character, which again relies on highly contested science. Similarly, regarding his claim that the genetic component of personality traits like conscientiousness is "*about* 50%": wonderfully vague as this proposition is, there is yet no way of sufficiently justifying even this, and the science used to defend genetically deterministic claims is notoriously fluid. Indeed, as soon as one begins to ask questions like *"which* 50% of my conscientiousness is genetic?"* the absurdity of such figures becomes transparent—if I spend an hour doing something conscientious, is the first half hour caused by my genes and the second half hour caused by my environment? Facetiousness aside, the idea that moral behavior is too complex to be divided into a neat dual-causal equation wherein "moral behavior = biology + moral education" should be prima facie obvious. Making moral goods is not like making a cake wherein the ingredients can be quantified as being *"about* 50% genetic"—yet this is precisely the level of thinking that pervades moral enhancement enthusiasm.[9]

One can go on in such a vein about the manner in which Hughes has simply taken the science he draws upon at face value without subjecting it to the least bit of critical scrutiny. Because Hughes has not thought even slightly critically about the sources he has used to justify his claims, but rather has simply taken them as "the final word," one ends up with a very poorly evidenced and massively overoptimistic account of moral enhancement possibilities based on highly provisional and contested research. This is actually a common feature of many of the moral enhancement enthusiasts who still continue talking about substances like oxytocin or SSRIs as if they had some magical world-salvatory power. And because one commentator talks about a particular intervention as viable, so many of the others simply assume the science behind it is sound, checking nothing for themselves, and disregarding any and all real-life limitations and the various qualifications raised by the scientists themselves who carry out the various empirical studies.

What we will be looking for in a realistic moral enhancement is in fact something far more limited, something present here and now. But just because such presently relevant moral enhancement may be limited, that does not necessarily make it any less useful for that. Since the biological bases of moral functioning are but one influence among a great number of others, a desirable moral enhancement should not seek to dominate these other influences, but to work alongside them in its appropriate place. *As such, a limited moral enhancement might be more ideal than a comprehensive one.*

Tom Douglas: Wanting to Be Good

Motives and Impulses   Tom Douglas' proposal for moral enhancement takes as its starting point the basic impulses and urges to do good and bad things. The basic rubric for moral enhancement here resides in attempting to strengthen one's "positive" moral impulses and diminishing those sorts of desires and impulses which lead us astray. What Douglas has in mind is the experience of inner moral dissonance—persons who want to "do the right thing," but for whatever reason feel forces in themselves pulling them otherwise. Everyone has experienced those times where they have been overwhelmed by emotions or desires to do things they know they really should not do. This is a universal experience wherein one knows, often with perfect clarity, what one should or should not do, and either does not care, or, with appropriate guilt, indulges anyway. There are times when everybody's good judgment is simply subverted—and, as such, Douglas' picture speaks to moral predicaments that most persons will face on a day-to-day basis. Moral enhancement for Douglas then is all about finding ways of altering the balance of the various conflicting inner forces such that what is left over is a less-hindered wish to do the good.

Now, the sorts of harms that can arise from this sort of moral conflict can be more or less terrible. They could be something as simple as having another slice of cake when one is on a diet, or it could be the source of the lusty adulterer's problems or the sudden urge to murder a man where he stands. Such impulses are not always evil or wrong—when a soldier throws himself on a grenade to save his comrades, when a bystander intervenes in a street robbery, quite often we are talking about immediate emotional impulses (the tendency is to forget that humans do actually have morally laudable desires and urges, too), things not necessarily produced by a long chain of rational reflection and dialectical thought but rather sourced in basic desires and immediate impulses. Without reducing moral functioning entirely to such immediacy,[10] it should be clear that the importance of such emotions and impulses in motivating moral (or immoral) functioning needs to be recognized and respected.

We can provide an example of how the attenuation of such "counter-moral emotions" might be applied to improve moral functioning or reduce potential harm. The connection between low serotonin and explosive aggression is one particularly prominent example given throughout the literature. Herein, if it is true that low serotonin levels are a causal factor creating an increased propensity for explosive aggression (the science here is more than a little ambiguous, as we shall see), then insofar as serotonin is a factor in these cases of explosive aggression, the possibility that the problem can be partially helped by offering some small increase in that person's serotonin levels should, by this reasoning, exist. Without reducing explosive aggression to low levels of serotonin only, one can see how tackling this one influence on the biological level, supposing a worthy intervention could be created to do so, might help tackle one dimension of the problem.

As long as we are very clear about the limitations of this sort of approach, which is very blunt, very brute, and lacking in any fine-grained targeted specificity, then, it is argued, we have some workable rationale for treatment here. Unlike Hughes' long chain effect proposal going between genes, personality traits, and virtues, or Persson and Savulescu's world-salvatory hopes, Douglas's account is more modest. Douglas makes clear that such methods offer no guarantees and recognizes their brute nature, so at least we have something more constrained and focused with respect to what we can realistically expect such interventions to be able to achieve.[11]

The main issue with this sort of approach is that it seems to presume a rather *atomistic* vision of our emotions and impulses. It does presume that there is some realistic way of disentangling "good" impulses from "bad" ones (or, worse, that impulses can even be thought of as "good" or "bad" in and of themselves outside of context), and it is hard to imagine how something so embodied and contextual as emotions or impulses are can be so readily severable as Douglas would need them to be for the project to be maximally effective. To illustrate, it may well be that a certain pharmacological agent could be devised which, say, reduced one's violent impulses— in fact, there are plenty of psychiatric grade pharmaceuticals which do precisely this—but there can be a tremendous cost to using such medication. Removing a person's aggressive impulses is only going to result in an overall net benefit of moral goods in circumstances where that aggression is being applied in comprehensively morally problematic ways. The fact is that aggression informs a huge number of various activities we perform on a daily basis, many of which are morally neutral (e.g., sports), and indeed in certain circumstances aggression can be a necessary evil, and in extremis a righteous moral good. So, an impulse to do a good thing may well actually benefit by being enriched with a certain amount of aggression, again, such as intervening to prevent a hostile individual doing a tremendous harm. Simply taking a substance which makes one less aggressive proffers no way of differentiating between morally appropriate expressions and the nonmoral or immoral manifestations of aggression.

Now it may well be that certain persons are simply aggressive and violent in purely harmful ways; for example, those with antisocial or borderline personality disorders (though it should be noted, as anyone who has encountered such unfortunate individuals will attest, that such mindlessly aggressive persons can in no way be talked about as having "moral" issues— extreme pathological violence is a very sad, but a very, very clear case of mental illness). In cases where we are dealing with an aggression unleavened by any orientation to a greater good, such persons could well benefit from anti-aggression medication, even if the cost is their ability to do the good things that aggression might facilitate. This being so, apart from very extreme cases, the point should be clear, good and bad impulses cannot be readily severed, since they are ambiguous. Our embodied nature defies this sort of atomic analysis, and knock-on effects from trying to remove "bad" impulses, even without bringing in the sort of long chain that Hughes has in mind, may yet result in the loss of good too.

Even so, Douglas' proposal is circumspect and modest enough to accommodate this problem to some extent. Douglas has no illusions that moral enhancement used in this way is likely to be anything other than crude— the question is one of cost–benefit analysis. Douglas can make the case that such moral enhancement could be worthwhile, just so long as there is a sufficiently attractive ratio of benefits against the various costs of using such interventions, including the various side effects they may have. Whether or not moral enhancement here makes one a "better person" or contributes to personal growth, his argument is to do with outcomes and desirability of preventing violent and destructive persons doing harms they would otherwise have been disposed toward perpetrating. There is some desirable, remedial potential in such a rubric, particularly in more pathological or borderline cases where organic deficits may be the predominant cause of the issue at hand. Whether it turns out that medicating such pathological cases are instances of moral improvement or not, they might yet provide in some cases an overall net benefit, and this might be sufficient for a bald consequentialist to want to advocate their use as moral.

It should be noted that Douglas' proposal is for *voluntary* moral enhancement, and so efficacy of such moral enhancement would rely on the relevant cases electing to engage with the moral enhancement in question, and then remaining disciplined enough to continue its usage. That means that only the already strongly morally inclined are likely to use moral enhancement in this way. Limiting as this is, Douglas' proposal is, again, modest enough to accommodate this, though, in essence, what we are saying here is that impulse-driven moral enhancement to correct one's impulses, in Douglas' accounting, must likewise be driven by a sustained impulse to enhance. How realistic an expectation this is will depend on the individual in question, the moral problem at hand, and the means of enhancement proposed for its augmentation, at the very least.

Noncognitive Moral Enhancement

There are more significant issues raised by this focus on the "noncognitive" approach to moral enhancement, so let us look a little deeper into the "motives-based defense" of moral enhancement. Douglas defines moral enhancement as follows:

I will understand moral enhancements to be interventions that will expectably leave an individual with more moral (*viz.*, morally better) motives or behaviour than she would otherwise have had. I will use "non-cognitive moral enhancement" to refer to moral enhancement achieved through (a) modulating emotions, and (b) doing so directly, that is, not by improving (*viz.*, increasing the accuracy of) cognition. (Douglas 162, 2013)

Focusing on noncognitive enhancement is a fundamental part of Douglas's proposal here. The noncognitive focus signifies that we are not talking about moral enhancement in terms of making persons more intelligent, more morally discerning, more capable of moral deliberation and moral judgment. There are those who have criticized Douglas rather heavily for excluding these factors from his account (e.g., Harris 2011). And there is more than a little validity to these critiques—a noncognitive account does not target the powers of moral discernment and moral deliberation that are crucial for any comprehensive vision of a person's moral growth, which requires reflection on the good rather than just an immediate emotional response. Rather, a noncognitive approach would be attempting to go beneath the rational elements of moral judgment and target instead the very impulses which are causing mischief in the first place. But how severe a limitation is this neglect of the cognitive dimensions of moral functioning?

Now, it should be relatively obvious that a considerable range of morally significant functions and situations rely rather heavily on the need for deliberation and judgment. Any time one faces a complex moral difficulty, one is confronted with the need for reflection, for weighing competing claims against each other. Reducing moral enhancement to a matter of unthinking impulses would do a disservice to the rich complexity of factors involved in moral functioning. But this is not really what Douglas has done, for he is not saying that *all* moral functioning is noncognitive in nature, nor is he looking for something comprehensive, but is simply looking for a "way in"—an entry point, as it were, some limited and workable rubric for moral enhancement, some means to at least get the ball rolling with respect to a constructive proposal for moral enhancement. Insofar as some moral functioning has some noncognitive dimensions, then to that extent Douglas's approach should have some purchase on improving the situation in question. Just so long as one does not reduce all of moral functioning to the level of pure immediacy, one can yet focus on the noncognitive dimensions as a partial and limited way of engaging with moral functioning. Again, Douglas is more than aware that such interventions are likely to be crude, to have side effects, but might yet be worth considering if an overall ratio of costs to benefits can be produced. This more limited and modest position is hard to take too much issue with.

While there is no question that a thorough picture of moral function-ing must make some significant reference to the cognitive dimensions that are applied as part of that moral functioning—discernment, practical rea-soning, figuring out the right thing to do—just so long as one does not reduce *all* moral functioning to the level of the noncognitive there is little weight to the critique here. What this shows, above all, is that Douglas's approach should not be taken as a complete package, but as a rationale for justifying one particular kind of approach to moral enhancement, one limited prong of a larger project. But this is exactly what we should be looking for anyway. In light of this, we can see that Douglas has managed to carve out a very limited but more realistic prospect for moral enhance-ment. It has its problems, certainly: it proffers a brute way of going about moral enhancement, and would work best with respect to moral problems that are predominantly or totally impulse-based rather than those requir-ing moral reflection and discernment (the examples raised in the previous chapter, chemical castration and psychiatric-grade sedation would be most fitting here, though, as we saw, not even these can be appropriately reduced to the brute biological level).[12] We are not talking about turning persons into "moral robots," but rather just giving persons who feel themselves to be in need of assistance a "nudge" in the right direction. The question then becomes one of looking at interventions on a case-by-case basis and deter-mining which of these interventions is liable to be effective and workable for a given morally related issue.

Why Morally Enhance? Keeping One's Feet on the Ground

Enthusiasm for moral enhancement has produced a very mixed bag of insights and considerations. We have utilized the various thinkers and their rationales as a vehicle to make a number of more general observations and claims about moral enhancement. We have seen that in the search for a rationale for moral enhancement, it is essential to keep one's feet securely on the ground and not ask of moral enhancement what it cannot realistically provide. What good does it do to defend moral enhancement on grounds that it cannot hope to remedy? We have seen moral enhance-ment defended as a means to stopping humans destroying themselves, moral enhancement as a means to forwarding liberal democratic values and creating a more democratic citizenry. We have rejected such rationales as utterly implausible. The ideals may be worthwhile, but these are simply not things that moral enhancement can be expected to contribute to, so they should not be invoked as reasons for encouraging such enhancement. This has helped us bring to the fore a particularly important point that it will serve us greatly to keep very closely in mind throughout the rest of the book:

We Must Manage Our Expectations about What Can Plausibly Be Realized through Biological Moral Enhancement

As we saw with Tom Douglas, some modest and brute vision of moral enhancement may be workable in principle, and potentially even desirable in certain contexts (an example will form the substance of chapter 9, treating alcoholism). If the context is appropriate, then certain ways of construing moral enhancement might help contribute to lessening overall harm (which should be attractive to consequentialists), as well as having potential application for persons to whom character development or the development of a stronger moral identity is important (which should be attractive to more deontological and virtue-based thinkers). Context is a massively significant factor here, and there is no reason why the idea of "moral enhancement" should be thrown out without serious reflection, or at least without attempting to find some context in which pharmacological or technological means might not be brought to bear for some greater good. The big problem is in expecting, or worrying, that moral enhancement will be forced upon persons to turn them into "moral robots" who do good things not because they want to, not because they feel it is right, but only because they have been drugged or technologically compelled into doing good things. This is the nightmare scenario that need not worry us at all—and we will continue to expand throughout upon all the many, many reasons why no such "moral robot" situation is going to be possible through biological means.[13]

But this all comes back to our main point, that we need to reign in our expectations of what moral enhancement can provide. We need to base our rationales for moral enhancement upon this foundation of what is realistically possible. We need to keep strongly in mind the manner in which realistic moral enhancement interventions will likely be quite crude and incapable of fine-grained adjustments to moral functioning (we will argue that this can be an attractive feature of moral enhancement, not a drawback, necessarily). We cannot expect something like moral enhancement, which will inevitably be limited in nature, to have anything like the world-salvatory power that some contributors seem to think it has. While there is no doubt that the many various influences on moral functioning will have biological or neurological correlates, the interwoven complexity of such influences on moral functioning, which are simply too finely calibrated, make it hard to imagine any way for a realistic pharmacological or technological intervention to gain sharp enough control over them. We have to keep in mind a healthy skepticism with respect to the idea that human biological makeup can be mapped in a sufficiently clear way onto his moral functioning to make fine-grained moral enhancement realistically possible.

Keeping moral enhancement as a limited hope in this way reduces the possibility of reductionism, of reducing all moral functioning to merely biological terms. At the very best, moral enhancement must be understood as working *alongside* its various influences—though what this "alongside" should look like is a vexing question that we shall deal with in parts III and IV. As such, in a theme to be continued throughout the book, it is the more modest answers to the question: "why morally enhance?" that will be seen to fare best of all. This is because it is the limited approach *only* which can leave sufficient room for such techniques to be comfortably assimilated within a more broadly embracing, integral approach. And, as will be argued, it is these more broadly embracing, integral approaches that are likely to be the most powerful and effective moral motivators of all. Insofar as any given case or technique for biomedical moral enhancement might realistically contribute toward the efficacy of such broader, more integral approaches for moral education, then we have something that may very well be worth talking about.

3 Conceptual Issues and Practical Realities in Moral Enhancement

Which Morality Should We Enhance? The Challenge of Moral Relativism

Having explored the philosophical cases that have been made regarding the idea of moral enhancement and how it might best be understood, it is time to take a step back and explore some of the more significant issues raised by the very idea of moral enhancement. The first issue that gets raised, one very indicative of our contemporary Western postmodern context, surrounds the question of moral relativism. One very popular idea is that "morality is purely subjective," and that the very notion of there being "objective moral standards" is merely an artifact of archaic religious thinking. Indeed, most persons that I converse with regarding this subject of moral enhancement tend to begin by articulating the same gut-level concern: "well, what does *good* mean anyway?" The general sense seems to be as follows: before one even gets to the question of whether moral enhancement is possible, must there not be some prior judgment as to what precisely constitutes the good? Moral enhancement, after all, necessitates that we have a sense of what is moral.

It is a valid concern, for what one person considers morally permissible (e.g., abortion, euthanasia) or laudable (e.g., campaigning for the rights to do such and the like), may be considered by others to be obscene and morally despicable in the extreme. So again, if one cannot even agree on what the good is, how can one even get started with the question of moral enhancement? For some, this concern alone is enough to dismiss the entire conversation. However, while it is a valid concern, it is not at all a decisive one. In real terms, people dispute over the ethicality of many laws, but that does not stop the state from legislating. The fact that universal moral consensus cannot be reached on all issues is never going to be enough to prevent a state taking action on matters that have moral weight—voluntary euthanasia is a very clear example of a moral issue that involves (in the

United Kingdom, and many other countries) the constraint of individual freedom on a matter of conscience that is highly contested.

Again, we need to be more specific and look at the particular context in which a particular program of moral enhancement is being instantiated. For example, a voluntary set of moral enhancement interventions might be more immune to the problem of moral relativism than a state-compelled program. If moral enhancement were voluntary (granted there are problems with the voluntary/compulsory distinction), something one could simply purchase from a local dispensing pharmacy, then, in principle, one could choose for oneself the desired moral objects and effects. Whatever problems may arise with this view, at least one can see that differences in opinion with respect to what makes for the good does not in and of itself negate the possibility of moral enhancement outright. This is not a question of permissibility, but rather about showing that the mere fact people disagree over what is good and what is not good is not the insurmountable stumbling block to moral enhancement that it appears to be at first sight. There is at least some wiggle room here—leaving it to the individual in question to determine for themselves which particular values they consider to be morally worthwhile and making the related interventions as voluntary as such things can be.

Another way of responding to the problem of moral relativity is to reject the idea that moral judgments are quite so fractured as a postmodern outlook tends to assume. Thinkers like DeGrazia, for example, have suggested that there is a realm of overlap, a bedrock of "reasonable consensus" with respect to the variability of moral valuations. In a postmodern world, one which often takes the standard of perspectivism as common sense and basic truth—that is, there is no objective truth or value, everyone is coming from "somewhere," and all expressions of value judgments must be understood as coming from the particular place that writer has in fact come from, as unassailably relative to their place in history, their culture, their psychology, their community, and so forth—the idea that there might be some kind of universal basis for moral judgment can be a shocking thought. Yet I think this idea that there is some kind of overlap in values, thus some manner of (at the very least human-relative) objectivity to certain human values is defensible, and one is right to reject the radical subjectivism which says there is only "my truth" and "your truth" and that we are absolutely free to choose whatever "truth" we wish to adhere to. Humans have a common ancestry, have lived in societies, have common bonds, and none of these things could have been sustained unless there be some kind of shared system of valuations, some basic commonalities to ground such common bonds. Thus despite the fact that, yes, there is massive disagreement on many particular issues (and even where there is agreement these shared values will inevitably be shaped by idiosyncratic and culture-relative formulations of those values), there has to be at least some semblance of commonality too—a "common law," if you will, wherein, for example, murdering and doing harm to members of one's in-group for no good reason is universally considered to be a bad thing to do.[1] There is at least some manner of a fragile universality between moral judgments and moral codes—and this universality, this consensus, fragile and narrow as it might be, is sufficient in and of itself to ground a similarly universal platform from which agreed values for moral enhancement might at least begin to be proposed. This being so, the question is not one of whether we can agree on the good, though this is significant. Rather, it is about how to actually go about encouraging those goods through biomedical means. It is this practical reality that needs to be explored.

Many enthusiasts for moral enhancement (in particular, Douglas, DeGrazia, Persson, and Savulescu) have attempted to delineate some expression of this commonality—a noncontroversial basis for agreement regarding which values are worth enhancing and which tendencies are worth diminishing. DeGrazia proposes the following: "as far as justified state policy goes, the idea is to locate points of overlapping consensus among *reasonable visions*" (DeGrazia 2013, 228; emphasis added). Savulescu and Persson (2012) offer the following noncontroversial proposal for commonly agreed-upon moral powers that might be useful objects for enhancement:

Moral enhancements which [can] increase altruism, including empathetic imagination of the suffering and interests of others, coupled with sympathetic response to this, together with greater preparedness to sacrifice one's own interests, greater willingness to co-operate, and better impulse control. (Savulescu and Persson 2012, 13)

Now, the problem here is not so much that Savulescu and Persson are wrong about whether these increases are desirable, or whether they would constitute an improvement in man's character. Rather, once again, the problem lies in how exactly a technology is to be developed which might actually make such improvements possible, and how they are to be manifested for public and private use.

DeGrazia provides a more ambitious "moral anatomy" of faculties and features that could benefit from enhancement. In fact, DeGrazia suggests there is a "fairly broad" area of overlapping consensus with respect to "any reasonable view of morality" (DeGrazia 2013, 228). He suggests such a range, including conditions like narcissistic and antisocial personality disorders; sadistic and malevolent pleasures; moral cynicism; defective empathy; significant prejudices and xenophobia; unwillingness to confront unpleasant realities (e.g., starving children); weakness of will and impulsivity; unwillingness to negotiate, compromise, and find common ground on certain issues; inability to find creative solutions to complex moral and social issues; and inability to deal with underlying issues with respect to complex moral problems.

As with Persson and Savulescu's list, it is not so much that DeGrazia is wrong about the tendencies and capacities that would benefit from being enhanced. Instead, the question is about how realistic it is to think that technology will be able to enhance these faculties in ways that traditional moral formation has not been able to. What DeGrazia is asking for is just too much to expect some realistic piece of technology or pharmacology to achieve—narcissism, moral cynicism, unwillingness to reflect upon the plight of those who are suffering, and the like are highly complicated phenomena; none of these concerns are monocausal or even predominantly biologically based issues to begin with. What sort of technology might be advanced that would help persons be more inclined to think about issues of poverty and inequality? This requires effort, will, discipline, imagination, and vision, at the very least.

DeGrazia is right that the concerns he mentions are all significant—very significant, in fact—but are there not more realistic and readily available techniques for tackling moral cynicism? For example, if one wishes to get others to reflect upon the suffering and injustice in the world, one of the better means may in fact be something much more mundane: something as simple as storytelling (Bloom 2007b; O'Connell 1998, 117). Narratives and stories can be very powerful means for evoking the emotions and empathetic response; such means are more dynamic, contextual, and engaging, too. For when one follows a story and becomes invested in the plight of the participants, one is much more inclined to be touched by the situation that is being referred to. Perhaps certain readers may scoff at such a suggestion, but the power of narratives, particularly when given through the media, have the power to convince nations to go to war. In the right context, an appropriate "human interest story" from the media can and often does evoke a tremendous public reaction—for better or worse, by creating public outpourings of outrage, of compassion, or atmospheres of fear, xenophobia, and hostility. Storytelling is the perennial and ancient power for motivating moral (and immoral) action.[2] Achieving this has never been particularly difficult, and the various propagandists throughout history have been more than adept at motivating swells of such solidarity of expression. For this no pills or neuroimplants are required.

In light of this, what then would make one even think that such problems—moral cynicism, apathy, refusal to even think about the problems of others in far-off lands—are biological problems to begin with? One might even go so far as to suggest that, by the very act of producing such a sophisticated moral anatomy, DeGrazia has ironically managed only to do the opposite of his intent, pricing moral enhancement out of the market by underscoring how unrealistic it is to expect technology to be able to offer sophisticated solutions to the genuinely sophisticated and multidimensional problems that undergird the moral weaknesses of our nature.

More helpful than some blanket biomedical intervention to be applied to everyone may very well be the presence and actions of persons of imagination, great speakers and persons of vision whose words and actions can touch the general public, evoking responses that motivate us to apply the tremendous capacities and powers we all already do have. That is, what might be more morally productive is the activity of persons who can be creative in conveying to large numbers of persons, and in touching ways, the nature of the moral issue at stake and how they might best respond. The reality is that most people have their own problems and their own lives to lead, regardless of how insignificant most of those problems may be in comparison to the extreme poverty that the majority of human beings alive have to bear each and every day. Giving the average Western person an enhancement to remedy their moral cynicism, should one be devised (and it won't), will not change the manner in which their lives are tied up with their own problems. What would be far more effective, again, is persons of vision and imagination who can take up leadership positions, inspire the public, perhaps create institutions, and make clear to persons living their everyday lives how a summation of small contributions from sufficient quantities of persons can generate a significant change: this is perhaps one way that moral cynicism might be overcome. Indeed, this seems much more realistic to me than the idea of technologically or pharmacologically based interventions at the biological level having the capacity to resolve something so complex and multicausal as moral apathy and its related moral shortcomings.

Are Not Institutional Means More Effective in the Long Run?

If, as we have just suggested, the sophisticated nature of the most serious moral problems are not predominantly biological in nature to begin with (and very inappropriately cast in such terms), then might moral enhancement be better served not by looking to technology and pharmacology, but simply by resorting to the various means of "paternalism" that are available to us now? That is, would we be better off simply offering various incentives, social advantages, and opportunities for those conforming to the desired moral characteristics prized by the society in question? This is a vexing question. The immediate response by moral enhancement enthusiasts (for example, Rakić 2014b, 249) is to assert that there is no reason why the two means might not be used alongside each other—institutional and technological means for encouraging moral behavior are not at all inconsistent, and even if it is the case that institutional or social-environmental means are more effective than technological moral enhancement, then this is still no reason in itself to deny the enactment of moral enhancement through technological means. It matters that moral enhancement on its own might very well be ineffective when coming up against deeply entrenched cultural imperatives to the contrary—but ultimately this means only that moral enhancement, if it is to be understood as some sort of comprehensive project, must be enacted as but a single prong in a larger project of institutional and social change; that is, it is still, in principle, a potentially desirable thing to pursue, if construed in an appropriate way.

Returning to DeGrazia (2013), we see a far more bleak analysis is possible. DeGrazia seems to be quite skeptical about the efficacy of traditional means of moral formation and gives examples of the many terrible moral evils perpetrated by persons and governments the world over to this day. For DeGrazia, traditional moral formation has not been sufficient to do an adequate job of preventing harm and evil in the world, which is why the need for moral enhancement is presented as a more pressing concern. DeGrazia makes the following observation:

The status quo is deeply problematic because there is such an abundance of immoral behavior, with devastating consequences, and serious risk of worse to come. Consider examples. In the 1990s, genocides occurred in Rwanda and Bosnia as world powers looked on. Slavery still exists in some parts of the world. Forced prostitution and participation in pornography, often involving children, is a reality on nearly every continent. Around the world, violence and other forms of oppression are committed against girls and women, religious and ethnic minorities, and others who are considered "outside" the group committing the violence.[3] (DeGrazia 2013, 228)

As far as I can tell, DeGrazia could have extended this list indefinitely, and as he rightly states, the moral status quo is deeply troubling—things may look good in our backyards, in our very fragile bubble of Western security, but in fact there are terrible crimes being committed at home and abroad on a daily basis. And while moral enhancement technology may be, in DeGrazia's own words, like using "an axe to mend a watch" (DeGrazia 2013, 228), the point is that not only have institutional and traditional measures failed to deal with the terrible moral ills that plague our world (and in fact rather support them), but that the very people we are relying upon to enact these institutional measures for an improved moral citizenry are themselves participants in the very evils that need to be remedied. The fact is that the moral evils of the world offer too many benefits to the powerful, and the unfortunate reality is that the average Western individual's very lifestyle is in many ways contingent upon the exploitation of individuals in countries too far away to be of concern. What we give in charitable donations with the one hand (usually to charities that draw profits therefrom), is taken away from the very same persons by way of cheap labor and goods bought at prices that no Western competitive market could match using humane working conditions. Thus the very institutions which are being relied upon for the means of moral enhancement are the very institutions that benefit from the continuation of the moral evil in question, and likewise the very individuals who are to be morally enhanced (i.e., you and I), are the very same persons whose lifestyles are contingent upon such society-wide moral enhancement never occurring. This Savulescu calls the "bootstrapping problem." No satisfying answer to this issue has been proposed.

DeGrazia is right about the problems in the world, but is wrong in suggesting moral enhancement as a remedy over and above traditional moral formation. The problem here is that *the very same thing that prevents traditional moral formative methods from working is the very same thing which will ultimately prevent moral enhancement from being the exultant solution certain enthusiasts hope it will be.* What is required above all for moral formation to take root is a predisposition of will. We can talk of the biology of morality, of social-environmental dimensions, and all the other manifold influences on moral functioning, but without the entirely personal predisposition of will on the part of each and every moral agent, without active participation in the process of moral development on the part of the individual, there is no sustainable moral growth to be had. The fact is that "traditional means" of moral formation fail to do their work, not because they are ineffective, but because they require active engagement if they are to flower into a strong moral foundation. Traditional means of formation, as we will see in part III, are numerous, wide-ranging, and very powerful indeed. What they cannot achieve, however, is a miraculous transformation of the individual or group without willing cooperation. Moral character, moral identity, and virtue are just not the sorts of things that can be produced without long effort and the willing engagement of the person so forming. If one thinks that moral enhancement—or, to use Hughes's term, "virtue engineering"—is going to be able to produce a virtue from scratch, then one has simply misunderstood what a virtue is. Virtues require practice, and sufficient practice, that they become to some not inconsiderable extent ingrained. The idea that any sustainable moral enhancement is possible without a preexisting disposition of will is simply unrealistic. While thinkers like Douglas hope that generating the motivation to be moral is precisely what moral enhancement will achieve, it would still require a person to take it, which requires a will to do so. As for the prospect of forced intervention, the reality at present is that the general public in Europe and America do not even trust their governments to be sufficiently competent to manage the yearly flu shot correctly. In the absence of something powerful enough to make us into moral drones, compulsory intervention is likely to result in mass unrest, to say the least. Rightly so; the irony with the prospect of powerful compulsory moral enhancement is that, since the very idea of turning persons into moral drones is itself evil, the first and only response for those who have been morally enhanced in this way would be to take action against those forcing them to be morally enhanced. *Powerful, compulsory moral enhancement only has itself for a target.*

So it may well be that traditional means of moral development have failed to remedy the moral evils of the world. This seems indisputable, and it may well be that the purveyors of such traditional means have failed to sufficiently motivate individuals and groups to take seriously the disciplines of moral formation and the various positive values inherent in their cultures and faiths. But this is not because such traditional means are not powerful and effective. Rather, it is because persons have refused to engage with them—the predisposition of will has been lacking. And for exactly the same reason, it is hard to imagine how moral enhancement technology will fare any better in this situation. We will continue to see throughout this book that moral enhancement is simply not potent enough to turn persons into moral robots—and this is a good thing. Freedom of choice will always be present, and moral enhancement can offer, at best, a nudge in the right direction. Thus, if skeptics with regard to traditional means of moral formation wish to appeal to the failure of such traditional moral formative means as a reason to encourage biomedical moral enhancement, they are in for profound disappointment. No enhancement can compel a person to become moral. Therefore, there must be willing cooperation on the part of the individual or group, and therefore the very same problem which impedes traditional moral formation—*the lack of such will*—will thereby become the very same limitation which prevents moral enhancement from becoming the superior alternative and "magic wand" that its many enthusiasts hope it will be.[4]

However, once again, it should also be noted that none of these observations constitute a reason for saying that moral enhancement should not be enacted at all, at least not where worthy means have been presented. It is simply a question of having a reality check, of reducing one's vision and getting realistic about what moral enhancement is likely to be able to achieve. There can be no perfect solution to dealing with the moral evils of the world—this is something that we need to be resigned to. It is a fact that must be put at the very foundation of a meaningful moral enhancement discourse. There is no solution to the possibility of malevolence-caused ultimate harm, and there is no final solution to the existence of evil, destruction, suffering, and harm. Realizing that moral enhancement will by necessity require compromise and limitation is a good first step in avoiding the more fantastical strains of thinking that we have encountered and will continue to encounter throughout.

Whether one's proposed solution to moral evil is institutional, technological, or a combination of both, we will always be dealing with imperfect means. This is the best one can hope for. Science and power are constantly abused by certain individuals and groups to forward their own ends. None of this can be a reason to not continue with the projects of moving forward with moral enhancement, since such corrupt tendencies exist independently of the question of moral enhancement. The moral evil in the world is something to be managed rather than resolved, and if moral enhancement can find some humane and desirable way of contributing to this management, then this can only be a good thing. This is a big "if." However, for those persons in whom precisely a predisposition of will and the sincere intent to become more moral are present, then—in principle at least—moral enhancement should be possible, and the confluence of biological moral enhancement with traditional social-environmental and psychological means for moral formation (a bio-psycho-social approach) has potential to be something encouraging and worthy of pursuit.

Beyond the Individual? Moral Enhancement and Groups

Where the moral enhancement proposal gets particularly interesting comes with the insight, one not often raised in the literature, that morality has social and collective dimensions. By this I mean that moral enhancement is treated, on the whole, as a purely individually focused phenomenon, focusing only upon the alteration of the traits of individuals within the context of smaller, more individually focused problems—for example, angry people who cannot control their rage, or psychopathic individuals who have deficits in the organic matter of their individual brains, deficits which might be mitigated on the purely individual level. In contrast, one must take note of the possibility that many of the most important moral issues facing human persons are not of a purely individual nature at all, but rather relate to concerns that require entire groups and societies to take collective action. This poses an entirely different set of problems to those usually considered in moral enhancement discourse. Seen in this light, moral enhancement cannot only be about "giving persons oxytocin to make them more charitable," or "taking SSRIs to be less aggressive." Moral enhancement, if it is to be effective on a grand scale and deal seriously with issues of global concern, must focus upon powers which facilitate higher-order social interaction and all the various elements which contribute to its fluid expression—the motivation required to come together, negotiate, discuss, reason, deliberate, conclude, take action, organize, divide labor, and make change.

I suggested earlier that DeGrazia's observations managed only to price moral enhancement out of the market by showing how unrealistic it is to expect technological means to be able to carry out what we need a sophisticated moral enhancement to achieve and the incommensurability of technology to the nature of the problems at hand. These observations regarding the social-interactive nature of moral functioning only serve to underscore that lack of realism and further price a comprehensive moral enhancement out of the market. The development of a morally evolved, discerning public, subject to trained and sensitive moral intuitions and adept at the socially interactive powers enumerated above, is, of course, a worthy ideal to pursue—but is moral enhancement capable of bringing us toward that goal?

The most significant moral problems that face us at present—poverty, lack of education, pollution, wars of aggression, wars for territory, to name but a few—are essentially immune to being resolved by targeting pettier individual evils. Rather, they require collective social action on a massive scale. They require wise leadership and a meeting of hearts and minds, such that nothing short of a coordination of various groups' powers of discernment, communication, and negotiation, as well as firm motivation and the appropriate affective responses, will be sufficient to even get the ball rolling in dealing with such issues. The most pressing moral issues of today require collective action, social interaction, and a complex range of powers.

Large groups of persons need to be motivated to engage with each other, reflect, learn, work together, organize, petition government, legislate—to take action wherever necessary, and toward the specifically moral goods in question.

If moral enhancement, realistically, is not even fine-grained enough to reliably encourage mere individual traits, how then is it going to be possible that moral enhancement offer any meaningful contribution toward developing all the faculties required for such complex collective action, and to motivating their effective utilization toward the desired goods? If individual morally related traits, capacities, and emotions are so ambiguous that it is context and intent that supervene in telling us whether their applications are moral or immoral or neutral, how much more ambiguous are these capacities when they are required to motivate and enhance such a long chain of intricate social-interactive powers?

Moral enhancement applied to foster mass social action—that is, precisely the sort of action that is required for our greatest, global moral ills—is the least realistic potentiality we have considered thus far. If this is so, then we had really better reign in our enthusiasm. At this point we must see moral enhancement as having a very limited efficacy at best, inasmuch as its most realistic prospect, regardless of how it is construed, applies almost exclusively with respect to individuals and their individual moral ills. For if one is thinking that simply giving oxytocin to a nation of people is going suddenly to motivate them to gather round, figure out what is required to solve the world's ills, and then go ahead in their various teams and enact it, then one is being terribly naïve.

As far as the social dimensions of moral functioning go, one might suggest that the best one can hope for here would be akin to the economic idea of "invisible hands," wherein the accrual of mass individual behavior and affect might at some level aggregate in some positive way at the larger societal level.[5] In any case, we must continue to be realistic with our expectations. The potential for such indirect aggregate benefits to solve something as complex as, say, poverty, lack of education, or sectarian conflicts that have calcified over generations is not only highly questionable, but one could suggest that it belittles the evils of the world and the suffering of those subjected to these evils in a way that is terribly inappropriate. Undue hope for aggregate effects might also draw attention away from the extended, focused and deliberate activity required for managing such ills.

Again, such solutions require highly organized and efficient action along numerous levels of power (political, economic, social, and so on)—and, as of now, no one has even begun to articulate any even half-plausible means by which technologically based moral enhancement can hope to motivate or sustain this level of complex, hands-on engagement on the grand scale. This is not to say moral enhancement might not have some overall net benefit and be worth pursuing anyway, be that on the individual level or with respect to some hope toward aggregate benefit. But let us be clear, and not pretend to ourselves that such moral enhancement is going to provide some magical cure for the global moral evils that plague mankind. The bootstrapping problem comes back to us again: there are simply too many powerful interests, including those of the general public, which are premised on such global ills never being remedied.

An Investigation of Means

It is said that to will the end is to will the means. Thus far we have been discussing the various rationales regarding the need for moral enhancement and some general issues surrounding the idea. But realistically, what prospects are there for credible moral enhancement intervention? And are these prospects something toward which we ought to be welcoming or afraid? Before concluding the discussion of the philosophical contributions to moral enhancement, there are three classes of interventions that have been proposed throughout the literature that need to be considered. We shall be asking how helpful these proposals are as ways of going about thinking through how moral enhancement might be enacted.

(a) Pharmacological approaches—the use of pharmacological agents to either increase certain moral or morally related tendencies, or to remove others.
(b) Neurostimulation for emotional retraining—activation of certain brain regions implicated in the emotional responsiveness requisite for certain morally appropriate responses.
(c) Genetic approaches—the attempt to locate moral or morally related traits in terms of their genetic precursors and thus encourage certain moral traits by means of selecting embryos or by artificially creating gene clusters which favor such traits.[6]

Pharmacology

The pharmacological approach to moral enhancement represents a category of interventions which is, on the whole, best characterized under the label "not at all well thought through." There are practical realities which pose tremendous difficulties for the idea of purposefully developing pharmaceuticals for moral improvement or moral treatments. This does leave over the possibilities of existing pharmaceuticals being applied as crude moral enhancements, or of future pharmaceuticals developed for other purposes then being found to have some manner of moral remedial effect. Thus such approaches are worth considering on that account, at least. Proposed pharmaceutical moral enhancement interventions include, for example, oxytocin to make persons more generous, charitable, compassionate, and trustworthy; SSRIs to make persons less explosively aggressive; or propranolol to reduce unconscious racial bias (DeGrazia 2013, 228; Persson and Savulescu 2008, 162).

However, it is important to note that pharmacological interventions for moral enhancement share a number of family problems. Let us begin with the most general such problem—for the first thing to be considered is the state of despair coming from within the pharmaceutical industry itself regarding the riskiness of developing new psychopharmaceutical substances. Ed Boyden (2011a, 2011b), for example, reminds us that the FDA process for developing and getting approved any given pharmaceutical agent takes about a decade to complete and costs in the region of a billion dollars. Such practical realities have been seen to severely disincentivize risk-taking with respect to developing and testing new compounds. And as patents expire on lucrative agents, like the newer generation of SSRIs that have been produced, the pharmaceutical industry has shown a tremendous unwillingness to invest in and develop any more (Hyman 2013, 2–3). Steve Hyman, of Harvard University and the Stanley Centre for Psychiatric Research, has long bemoaned the pharmaceutical industry's unwillingness to invest in new product development for serious psychiatric problems. He writes:

During the past three years the global pharmaceutical industry has significantly decreased its investment in new treatments for depression, bipolar disorder, schizophrenia, and other psychiatric disorders. Some large companies, such as GlaxoSmithKline, have closed their psychiatric laboratories entirely. ... This retreat has occurred despite the fact that mental disorders are not only common worldwide but ... there is, moreover, vast unmet medical need. ... This retreat has happened despite the fact that different classes of psychiatric drugs have been among the industry's most profitable products during the last several decades. ... This withdrawal reflects a widely shared view that the underlying science remains immature and that therapeutic development in psychiatry is simply too difficult and too risky. (Hyman 2013, 2)

Hyman has hit the nail on the head here, "the underlying science remains immature," and thus the risks to the companies are perceived to be too great. If this is so of psychiatric science, which is so much further advanced than that science which looks at "the moral brain," and with which much of the moral enhancement discourse overlaps (particularly with regard to addiction and aggression), how much more likely is this concern for the immature nature of the science to be relevant to something as contextual as moral functioning?

The basic reality is quite simple: the pharmaceutical industry tends to invest in products which are most likely to produce the greatest profit margin. This is just basic economic sense. When one understands the costs to the pharmaceutical industry to invest in a product and the nature of competition within its market, it is hardly surprising that they take the least risky and most profitable routes available.[7] This is an important point with respect to moral enhancement through pharmacological means: aside from all the other practical impediments that will be raised throughout the book, there is likely never to be enough demand for most moral enhancement drugs to motivate the pharmaceutical industry to invest in extensive and time-consuming clinical testing of pharmacological agents for such purposes. Becoming more moral just isn't sexy. On the one hand, we have a culture which prizes a "you only live once" party mentality, the cult of youth and youthfulness, and on the other hand we have an institutional backdrop which constantly shies away from explicitly prescribing moral courses, which is territory that many seek to avoid in contemporary Western culture for fear of being associated with undesirable religious fundamentalism. If governments are (rightly or wrongly) too afraid to touch moral values, and the general commercialized segment of the population is uninterested in the subject of morality as an explicit project, where would moral enhancement gain a profitable-enough foothold to motivate its sustained investment? And which pharmaceutical company would risk the time and vast monies that it would cost venturing into that territory?

For example, Hyman has made a very strong case for the impracticability of using SSRIs as moral enhancement drugs. He observes that there are important issues to be considered relating to getting ethical approval for the testing of such compounds on healthy persons. Disregarding the cost of the clinical trials, gaining ethical approval for such testing of toxic chemicals on persons who have no clear illness is highly unlikely to get the go-ahead in our current climate (Hyman 2014, 1). This is a good thing. The toxicity of most pharmaceuticals is another important practical impediment that enthusiasts simply have not thought through. While treating a serious medical complaint with such substances is acceptable, because pharmacological treatment is essentially "the lesser of two evils"—indeed, the goal for finding a good medicine, according to Hyman, is for medical developers to find workable classes of chemicals which are, in his words, "*not too toxic*" (2013; emphasis added). In clinical cases, this balance of evils becomes a matter for patient and doctor to monitor and assess whether any side effects of such treatment are worth the benefits that might be provided. The idea of clinical trials here, of subjecting persons without serious medical problems to such substances, is as Hyman rightly says highly unlikely to gain ethical approval. This all adds up, then, to a rather long list of serious practical impediments which, taken in combination, amount to a rather decisive account of why pharmaceutically based moral enhancement is not a sustainable project or one that will find the deep-pocketed investors that such prospects would require to become a credible reality—in terms of "positive" moral enhancement, anyway; that is, for the treatment of those without clear medically grounded, morally related conditions.

On the other hand, "remedial" moral enhancement—that is, treating morally related issues which are more pathological in nature—offers a little more promise for investment. And where it can be shown that there is a significant market share to be exploited, the pharmaceutical companies might be a little more likely to act against their current reticence. For example, alcoholism and drug addiction are massive morally related public health problems of arguably epidemic proportions. One agent that might be useful in moral enhancement (and we will return to this in part IV), nalmefene, an anti-opioid drug for use against alcoholism, has been tested and produced and is currently being applied as one means for helping alcoholics treat themselves. So while it is possible for morally related products to find investment (though most likely not using the word morality, but rather "public health" or "mental health" intervention—thus constituting "soft" rather than "hard" moral enhancement), everything hinges upon the nature of the market for the particular product. Just how marketable do we think a product would be that promised to increase compassion in a world driven by competition? Marketing deep kindness is not going to be easy. Indeed, if motivating such behavior were easy, there would be no call for moral enhancement to begin with. The whole thing is going to be a tough sell.

Thus, these questions regarding development costs, approval time, and clinical tests, weighed against the need for a large-enough market to justify taking a risk on such a product, compound to create a practical reality that I would ask readers to keep in mind whenever the topic of pharmacologically based moral enhancement is raised. This is something that needs to be thought through more thoroughly by many enthusiasts. In a world such as ours, there might not be a huge amount of room for moral enhancement drugs. It is likely for reasons such as these that Persson and Savulescu have advocated compulsory moral enhancement, since all such concerns about profitability would be immediately put asunder. However, this would still rely on the expectation that such drugs could be effective. How effective do we think such drugs could be? We noted in the previous chapter that Douglas' proposal was most realistic because it was most modest and appreciated the nature of such interventions as crude and having numerous limitations.

Let us look closer at how such moral enhancement drugs might or might not work. Let us take the example of propranolol for reducing unconscious racial bias (DeGrazia 2013, 228). Propranolol is a beta-blocker which interferes with the effects of noradrenaline and disrupts the regulation of memory encoding. The suggestion is that it can be used prophylactically in the treatment of post-traumatic stress disorder by interfering with the encoding of emotional memories and their reconsolidation (Wallach 2012[8]). No doubt there could be a whole range of clinical uses in which propranolol might well be very useful. But what about in a moral context, as used by otherwise healthy persons? Let us ask precisely how much of a market one thinks there might be for persons wanting to "reduce their unconscious racial bias," or how often during the day of the average person their "unconscious racial bias" causes them sufficient mischief to make them cry out for a cure.

Once again, let us ask the core questions: who would realistically elect to take such a substance? In what context could the side effects be justified for mandating it? And to whom would it be mandated? Are we to mandate the use of such agents for the police, for judges? That would require an institutional admission that the police and judiciary is pervasively racist (and this is too crude a characterization of these institutions anyway). One could doubt whether such an admission would ever, or should ever, be forthcoming. These are important practical realities that the mere abstract contemplation of such compounds do not raise. Moreover, it is to be remembered, again and again, that such drugs cannot make a person do anything, a predisposition of will is required, and none of the science comes even close to suggesting that propranolol, or any other substance, might be potent enough to prevent racial profiling in real-life situations (or in any context outside the extremely limited and unrealistic test conditions that the propranolol studies applied), or aggression by police against minorities, or harsher sentencing of racial minorities by judges, nor yet that it has any effect when applied in combination with a willed effort to become aware of and overcome racial biases. If such substances are used universally among the police and judges then one has to know whether they have different effects on persons who are not racist at all as they do, say, on persons who have quite potent unconscious racist attitudes. If such substances are not used universally that would mean outing particular individuals in the police or judiciary as racist, after which such persons would not have a particularly attractive set of career prospects ahead of them, whether they take the enhancers or not. Just a moment of critical thinking unveils the entire prospect as the absurdity that it is.

Herein lies the crux of the matter, as we have seen before: these are not primarily neurobiological problems to begin with. They are predominantly social and institutional issues, and suggesting that propranolol might have any serious impact on such issues demeans the true nature of the problems and reduces to an absurdity what is in fact a very serious social problem. An example of this might be found in Douglas's (2013) schematic of "the biased judge." Douglas describes a potential moral enhancement situation as follows:

> James is a district court judge in a multi-ethnic area. He was brought up in a racist environment and is aware that emotional responses introduced during his childhood still have a biasing influence on his moral and legal thinking. For example, they make him more inclined to counsel jurors in a way that suggests a guilty verdict, or to recommend harsher sentencing, when the defendant is African-American. (Douglas 2013, 161)

One can take the point that someone taking a drug (Douglas is clear that he is not necessarily talking about propranolol here) to ameliorate such a bias might constitute an example of moral enhancement, but the danger here is that such an example *only works in conceptual space*, for it provides only the most superficial simulacrum of a real-world scenario. Sadly, this image of James, the well-meaning racist judge, taking some sort of medication to prevent his or her biases is a ridiculous mischaracterization of the reality of racism in the judiciary and police force. Such suggestions radically trivialize very important social problems by touting poorly evidenced and inadvertent effects of extant pharmaceuticals as solutions to immensely complex historical and cultural difficulties. As such, I would encourage enthusiasts to think very carefully before getting too excited about these sorts of possibilities, and to consider first of all the ground-level practical realities and the real complexity of the contexts through which enhancement is being proposed for enactment. We will continue to inspect the potential for pharmaceutically based moral enhancement throughout the course of the book, but for now we have had more than a glimpse at the sorts of issues such drug use raises in an enhancement context.

Neurostimulation

The idea that changes in moral functioning can be produced, or augmented, through direct stimulation of the brain is very enticing. Some evidence exists, for example, that disruption of the right temporoparietal junction, an area involved in mental state reasoning, with transcranial magnetic stimulation alters the way persons go about judging attempted harms. By interfering with the brain's capacity to infer mental states, moral judgments based on assessing a person's intentions is likewise disrupted (Young et al. 2010, 6753). While this particular result may not be especially surprising, the mere fact that the way persons go about making moral judgments can be influenced by neurostimulation has sparked the imaginations of proponents of such neurostimulation. In this neurostimulation category we can also include proposed use of neurofeedback as potential means for treating psychopathy and antisocial personality disorder.

The rationale behind such proposals is that psychopathic individuals (and this has been found to be the case particularly with respect to criminal psychopathic individuals) may have deficits in the metabolic activity of their brains' "fear circuits." Since fear conditioning is understood to be essential for behavioral development (Ranganatha et al. 2007, in Jotterand 2011), training such individuals how to reactivate the respective brain regions through "brain-computer interfaces" or neurofeedback techniques is hoped to offer a means of training or retraining the psychopathic individual, providing a means for encouraging the restoration of the appropriate affective responses which usually check violent behavior. Indeed, even moral enhancement skeptic Fabrice Jotterand (2011) can see a place for such technologies. He writes:

According to Sitaram and colleagues, researchers are developing a real time fMRI (rt-fMRI) system for the treatment of criminal psychopathy in which criminal psychopaths are trained to control localized brain regions involved in the disorder. ... Such technologies would require the stimulation of brain regions involved in moral decisions such as the ventromedial prefrontal cortex or the amygdala—associated with affect and decision-making capacity. .... These emerging neurotechnologies could enable us to ... reform "morally deficient people" (e.g., psychopaths or incarcerated criminals).[9] (Jotterand 2011, 5)

While these possibilities are certainly exciting, there are a number of cautionary remarks one ought to make about the potential efficacy of such techniques on criminally psychopathic individuals. For example,

(a) psychopaths are notoriously manipulative and adept at faking socially appropriate affective states. Even if the treatment were able to reactivate the respective brain regions, when it comes to applying such responses in real life, it may be difficult to know for sure whether the treatment has taken and the changes are being embraced, or whether the individual in question is simply using the intervention as a way of becoming even more adept at deception and simulating the appropriate affective states. This is a particularly vexing possibility when our subject group is composed of violent, fearless psychopaths;

(b) by implication, for the therapy to work, it would require a preexisting disposition of will on the part of the patient to so improve. Since such techniques are incapable of compelling the subject to do anything, there must be a genuine will on the part of the individual to be rehabilitated—and since remorse is precisely the quality so often lacking in such persons, one wonders how far such a treatment can go in helping criminal psychopaths who may have no will, desire, or intention to be reformed.

As Jotterand observes with respect to the prospect of treating psychopathic individuals with SSRIs (2012), such interventions may even serve to make matters worse by enhancing adaptive psychopathic personality traits such as social charm and interpersonal boldness. In other words, we have to be vigilant that our very treatments for psychopathy do not end up being techniques that actually make criminally inclined psychopaths more effective at pursuing their goals (ibid.). Indeed, if such therapies offer the chance to reduce incarceration times or to allow greater freedoms to institutionalized individuals, the incentive to deceive is as strong as it is obvious.

Another significant concern here is the possibility that such methods could be used to train persons to have controversial emotional responses which are deemed appropriate or inappropriate merely on social or cultural grounds. We will return in part IV to this question of how far our definitions of "normalcy" and "health" are socially constructed and socially relative, and whether this is a problem. For the time being, we can make ourselves aware of the more general issue that "emotional retraining" with the aim of bringing persons to experience "normal" or "appropriate" affective responses will always be to some extent a socially relative business, one which brings with it controversial realms of application.

For example, one possible manifestation of affective retraining might be the use of such neurofeedback technologies for purposes of augmenting the "sexual reorientation therapies" so popular within predominantly homophobic cultures, fundamentalist groups, and Putin's Russia. Indeed, feedback and reconditioning methods have long been a staple of such therapies, and one might suggest that it is yet more likely for such neurofeedback and real-time fMRI technology to be applied by such parties to "rehabilitate" socially unacceptable homosexual preferences than for them to get ethical approval for use in treating criminal psychopaths.[10]

As was hinted at in the previous chapter, the crudeness of such methods can be of benefit. It is true that such methods would not really be able to train (on their own) specific fine-grained responses to a comprehensive range of morally demanding situations. The near-infinite possible range of contexts in which emotions might play a role in guiding moral functioning makes things such that generalized techniques are our only realistic option here. This is good. What would be much worse than a crude or generalized kind of intervention would be the possibility of exceptionally finely grained capacity to induce powerful affective responses at will, which would be a threat to moral freedom. The capacity to induce a very sharply defined set of responses to a range of significant morally demanding situations, one which allowed for no reflection, no self-directed personal moral growth, no potential for social critique, or no acknowledgment of all the many other factors that are important contributors to moral functioning, would be concerning—much more concerning than a more generalized approach demanding the patient be an active participant in his or her own moral development, biomedically augmented though such development may be.

So, while the above limitations do need to be kept closely in mind, it is completely feasible in principle that such technologies be applied to morally improving effect in the right kind of person, and the right kind of context. It is true that the emotions can lead one astray; however, the solution is not to remove the emotions, but rather to retrain them. Precisely this is what the neurofeedback Jotterand references aims to assist in. For persons in whom a predisposition of will toward moral development does exist (say, someone who has an organic deficit created by a later-life brain injury or accident which has impaired their capacities for emotional regulation and response), then this technique could indeed constitute a valid and helpful intervention. How far it would be useful with regard to treating criminal psychopaths is another question. Insofar as moral discernment requires a certain emotional responsiveness, and insofar as certain individuals are lacking in capacities for such emotional responsiveness, neurofeedback or real-time fMRI techniques offer a partial and limited technological potential for augmenting these elements of moral functioning. As Jotterand observes, this does not "make people moral"—it does not even guarantee a moral improvement. All the same, such techniques, applied appropriately and combined with a preexisting disposition of will, might be beneficially directed, in principle, for morally developmental purposes.

Genetics

What an extraordinary effect might be produced on our race, if its object was to unite in marriage those who possessed the finest and most suitable natures, mental, moral, and physical!

—F. Galton (1869, in Larson 2010, 170)

Yet more problematic than the interventions hazarded by certain commentators without properly thinking them through are those which fall more into the domain of science fantasy. In this category I would include, for example, most of the proposed interventions relating to gene therapy or genetic selection for moral traits. Superficially, the idea seems to have potential, but as soon as one grasps the complexity of moral functioning and the responsive competencies that moral living requires, the idea that morality and genes can be linked in anything other than the crudest possible ways becomes less and less plausible.

Let us look closely at the more rudimentary attempts to link moral character (and desirable characteristics more generally) to genetic markers. The linking of moral traits with genetics has a long and ugly history, and the suggestion that the complexities of what is good and desirable in human behavior is describable in simplistic genetic terms is a highly suspect strain of thinking not only applied by Nazi doctors. In fact, eugenics has been pursued in both the United Kingdom and USA (most explicitly between 1895 and 1945), as evidenced by the documented sterilization of the mentally ill (and we must be aware of how easy it is to describe anyone whose behavior we do not like as "mentally ill"), sterilization of the disabled, of the homeless, of the unintelligent, and so on (Larson 2010).

The idea that genes can somehow be linked to the complexities of human behavior has been a long-standing hope—a lazy way of displacing blame for social problems onto biological causes and attempting to excise everything socially unacceptable, not by putting in the hard work of analyzing and rectifying the institutional causes to present problems, but by looking for easy biological answers in "eugenic remedies for perceived social diseases" (Larson 2010, 172). In this respect, one might suggest that the eugenic project is a very clear mirror for what is now being proposed as moral enhancement. I would suggest that the introduction of genetic discourse into the moral enhancement conversation is nothing more than a continuation of this sort of false hope and lazy thinking, which seeks out easy answers in order to displace blame from where it belongs and to save the hard work of meeting moral challenges head on.[11] Indeed, we shall encounter below numerous contemporary examples of how the linking of genes to crime and other undesirable behavior continues to this day.

Even the least fantastical of such genetic links to morally related issues is deeply problematic. Let us look at the MAO-A gene variant, sometimes called the "warrior gene," a much-publicized gene variant that engenders in those who have had abusive upbringings a propensity for antisocial behavior, lack of empathy, and explosive aggression. The MAO gene codes for an enzyme which is critical for catalyzing serotonin. The MAO-A variant disrupts this process of breaking serotonin down, resulting in a pooling within the synapse, a build-up of which can result in various pathologies, including violent aggressive outbursts (Casebeer 2012). All this being so, linking such a gene to violent behavior in a direct causal fashion is a highly dubious strategy. The gene, as has gradually come to be accepted, is not a destiny which compels violent behavior; rather, the gene interacts with various environmental influences across the entire lifespan of the individual in question. It is the combination of the MAO-A variant with deprivation, mistreatment, and abuse that makes it most likely to find expression. However, it should be noted that deprivation, mistreatment, and abuse make violent and psychopathic behavior much more likely anyway, regardless of one's genetic make-up.[12] Where a loving, nurturing upbringing is found, the gene variant is much less likely to result in destructive, antisocial behavior. In such cases, possessors of such gene variants can remain perfectly anonymous to others as well as to themselves. Certainly Savulescu's (2010) proposal that bioprediction and "targeted social interventions" for "those at higher risk" such as increased surveillance, segregation, and gene selection may work to ameliorate potential problems with carriers seems a little severe (such solutions create more than an echo of ghettoization and nightmarish state control). That being so, if less ethically objectionable interventions were available then a case could be made for a balancing of consent with the concern over means.

What conclusion can we draw from this other than the idea that upbringing should be our primary focus with respect to this dimension of moral functioning? But this is the hard answer, certainly not what enhancement enthusiasts want to hear. What would be more desirable for such thinkers would be some magical thing in our DNA that we can identify—that we can control, manipulate, and make everything all right thereby. That would be a great relief. But when it comes to complex and contextual human behaviors, genetic predisposition offers very little predictive potential. As with Douglas' work on emotional modulation, in which emotions themselves are thoroughly ambiguous with respect to their moral effects, much the same can be said of the supposedly genetic bases for moral functioning, too.

If the MAO-A variant, the strongest case for positing a link between genetics and violent crime, is but one influence, one cause interacting with a complex constellation of many other causes, how much explanatory power can we really expect to garner by isolating genetic causes?[13] It is precisely as interwoven, and not as atomic, influences that genes are best understood as having effects here. Even if, for example, a gene for the diminution of empathy could be identified, such a predisposition could yet be shaped into something positive—there are professions where a lack of compassion, an unflappability with respect to making horrible but necessary decisions in morally challenging situations, is precisely the sort of trait that is required. As Wasserman (2013) put it in his article "When Bad People Do Good Things," losing psychopaths may, in fact, leave humanity poorer rather than richer. Just because one lacks empathy does not mean that one will turn into a serial killer. One could just as well become a businessman, a politician, a sportsman, an academic, a member of the security services. Many of these professions offer positive contributions to human society.

It is important to be clear here. We are not disputing that genes predispose us in a range of ways. Rather, we are talking about moral enhancement in particular. There is no question that other, more basic human characteristics might be enhanced through genetic manipulation. The point is not to say that genetic enhancement is impossible or impractical in general—no doubt elements of man's various physical powers and some of his cognitive capacities can, and inevitably will, be objects for genetic enhancement. But such objects are much less complicated, relatively speaking. There is a categorical line to be drawn between something like man's physical capacity and something so fundamentally contextual as complex human behaviors, such as the virtues. Metabolic powers may well be enhanced through genetic means, but altruism, justice, *wisdom*? Whenever one tries to atomize complex and contextual phenomena, one ends up with ambiguity. The "constituents" of the phenomena in question are tied together with an indefinable "glue" which can only be determined when its particular manifestation arises in context. This is why we need to look at practical reality first.

The reality is that we do not educate our children for practical wisdom. What would be much more valuable, and effective in even the short- to mid-term, would be an effort in educating persons in how to think imaginatively about applying their already existing capacities in constructive ways. It is *responsiveness*, more than generalized dispositions and traits, which are required in sophisticated moral functioning. Thus, our focus should surround educating persons regarding how to shape their own dispositions in dynamic and responsive ways, rather than attempting to reduce persons to a genetic mean where everybody's morally related genetics are "averaged out." But only practical wisdom is capable of the fine-tuned, dynamic responsiveness so integral to sophisticated moral functioning. Thinking that there is going to be some gene or genes which code for some virtuous character trait in an uncomplicated fashion, some manner which is somehow not governed by the intricate and manifold nature of contexts in which action is made manifest, represents a particularly crude and fantastical kind of thinking better off set aside.[14] As such, it is much more likely that the genetic dimensions of morality constitute a red herring, and that even if some rudimentary link between such complexities of moral living and genetic predisposition could be demonstrated, actually is it the shaping of disposition that should occupy our primary focus.[15]

Conclusion: Moral Enhancement, Hard and Soft

Moral enhancement has raised a number of vexing issues. Problems arise with respect to the actual development of such technologies, all of which require considerable financial investment without a reasonable expectation of any attractive financial return, assuming any such means would be effective anyway. And this leads us to ask ourselves: "Who would elect to morally enhance?" Of course, no generalized answer will do here, and it depends on which particular intervention we are talking about, and in what context. An intervention to assist in treating alcoholism may be much more popular than something to turn one into "a generous person," imagining such a thing were possible. Again, it is presumably in recognition of the possibility that pervasive moral enhancement is unlikely to have mass appeal that commentators like Persson and Savulescu suggest compulsory moral enhancement, and why those like Rakić try to find a mediating position wherein the state offers various incentives, carrots and sticks, in order to encourage the use of moral enhancement technologies by the general public.

Yet something very important is missing from these ways of thinking about moral enhancement's public enactment. Actually, the realism of enacting moral enhancement is much more likely to turn on whether we are talking about making persons moral in some explicitly moral sense ("hard" moral enhancement), or whether we are taking some nonexplicit means toward manipulating behavior in order to create more "desirable" social patterns. It is a simple practical reality right now that no citizenry that we can imagine in the Western world would give their political assent to a project of moral enhancement. No politician would ride to electoral victory with the slogan "compulsory technological moral enhancement for all." It would not matter if moral enhancement were proposed as voluntary or compulsory, there is no politician in his or her right mind who would even mention the words moral enhancement in their appeals to the general public. It would be political suicide. This fact alone should make it abundantly clear that there is no way—no way at all—that explicit moral enhancement would become an overt reality. At least, not in any of the ways that fit within Persson and Savulescu's framework, and not in any kind of political setup that we are presently used to. This is an incredibly important practical reality that must be made absolutely clear. Thought experiments aside, if we are talking about explicit, overt moral enhancement, legislated for by compulsory means or by utilizing carrot-and-stick social incentives, we are simply wasting our breath. An explicit project of state-sponsored moral improvement of the general public is *unthinkable* in liberal states.

That being so, there are ways of conceptualizing moral enhancements, the enactment of which do not rely on the public's consent, which do not rely on even bringing the word "morality" into the equation at all, and which do not require the public to realize they are being "nudged" toward certain behaviors and goods. This leads us to an insight for the moral enhancement domain that will be developed throughout the remainder of the book:

If moral enhancement is ever to be made manifest, it will most likely never actually go by the name "moral enhancement." Moral enhancement, in any present Western climate, will always be an indirect, subtle, covert affair that might not be recognized as having any relation to moral matters even by those who advocate it.

It is this latter case which is of particular interest. The relation between moral enhancement and matters of behavioral control or social engineering have not really been examined in the philosophical literature, yet it seems clear that a substantial overlap can be discerned between the two concerns. While the practical reality may very well be that "hard" moral enhancement will never get off the ground in a liberal society, a "softer" approach, quite to the contrary, can be said to be much more plausible. Indeed, as we will continue to argue throughout the book, it might be said that such moral enhancement is already upon us and has been for quite some time, for "soft" moral enhancement, concerned with behaviors which have strong moral dimensions (though not necessarily dealt with as explicitly moral concerns), can be snuck in covertly by other, more acceptable names—for example, mental health interventions for antisocial behavior; public health interventions for addictions; or public safety interventions for preemptive incarceration of known recurrent violent offenders (a paper for such a bill has been proposed by Lamparello 2011). So-called violence initiatives for screening inner-city youths concerning their potential for antisocial behavior in later life are a present reality in many US states (this will be discussed at greater length in chapter 8). All these existing moves signal to me that covert moral enhancement, "soft" moral enhancement, or moral enhancement by another name is likely to find itself being applied in subtle and piecemeal ways under the radar, without any reference to morality whatsoever or even knowledge on the part of its advocates that subtle moral enhancement is being enacted thereby.

That is why the point made in the introduction regarding "hard" and "soft" forms of moral enhancement is so crucial to the debate, much more so than the "voluntary/compulsory" question. If moral enhancement commentators are thinking that a given intervention is moral enhancement if and only if it is explicitly directed at the moral faculties, then these more subtle and covert forms of moral enhancement—which are moral enhancement in effect (though without using the word "morality" in its rationale)—get swept under the carpet and removed from view. Yet these are not only the most interesting cases, they are the cases that are most practically viable.

In my view, it is this more diffuse and covert form of moral enhancement that is the most realistic shape through which the various prospects involved are likely to be made manifest. Among certain moral enhancement commentators, it seems as if there is a notion that moral formation, or the moral dimensions of life, are somehow separable from the various other dimensions of living. The reality is that the moral dimensions of living overlap with everything else we do. Thinking that moral enhancement has to be a separate enterprise, distinct from the other dimensions of living, relationships, practices, politics, economics, institutions, and the whole gamut of elements that fill the lived world, is an error.

Moral living and development are not always about reflecting upon the nature of the good; most often, it merely involves being embedded within the habits and practices and value ideals of the group in which one is living and aiming to conform thereto. One does not simply put aside half an hour a day for being moral and then leave it at that. Moral development is not a separate activity in the way that learning to play a musical instrument is, or learning to drive a car; it is something that is diffused within daily living. Moral living is embedded and ad hoc. We will go much deeper into this idea in part III. For now, it is important to grasp that the moral dimensions of life are not separate from the whole of the various elements that interweave to make up daily living. By reframing moral enhancement in terms of mental health, public safety and public health interventions, persons might not even realize that they are getting "soft" moral enhancement at all—which, because it overlaps with all these other already existing institutional concerns and is inseparable from them, makes such enhancement almost as inevitable as the prospect of hard moral enhancement is politically untenable.

Then we see that the debate changes completely. Using this "soft" sense of moral enhancement, one can discern that the seeds of moral enhancement are not only present now, but have in fact been applied for quite some time. The eugenics project was in part a project of moral bioenhancement; pseudoscience or no; the goal was the breeding of persons of good character and preventing the breeding of persons with poor character, criminals and so forth. So many clinical mental health interventions, as we shall later see, traverse a very thin line between "badness" and "madness," with much debate as to whether such interventions are dealing with people who are actually quite ill and lacking in responsibility, or whether they merely have behavioral problems in need of conventional punitive action. As we shall see, moral enhancement actually bleeds quite naturally into a number of different domains, from governmental paternalism and behavioral control to mental health intervention, medical practice, and public safety—all of which draw upon the various biological sciences as means for justifying their interventions. Eugenics drew upon the genetics of the time; in the 1970s lobotomization drew upon what little brain science was available to control "undesirable behaviors" such as senseless aggression; and today we draw upon, as we have seen, a range of genetic, biochemical, and (as is particularly modish at present) various neuroscientific sources, such that we can better grasp the workings of "the moral brain" and potentially generate some change. This is no speculation. Paul Zak, a prominent "neuroeconomist" has written about the goal of understanding how the brain "causes" our various behaviors, and talk of such neurobiological "causality" is disturbingly prevalent across both the neurobiological literature and its expression in the popular media. For Zak, neuroeconomics seeks "proximate causes" of choice behavior (which, apparently, give policymakers "more leverage when seeking to affect behavior through policy"; Zak 2004, 1,737). As such, the rubric is as follows:

Introducing laws that seek to influence individual behavior can be done more effectively and precisely when the proximate mechanisms producing the behavior are known. (Zak 2004, 1,737)

Whether the grand vision of neuroeconomics is a pipedream based on an incredibly shallow misrepresentation of how neural factors interact with the world with respect to influencing human behavior (Tallis 2012) or not, the *aspiration* is indisputably there. The essence of the matter is to understand which neural substrates associate with which kinds of choices and outcomes, and to use that understanding for the creation of laws and policy which influence individual behavior in the desired directions.

Precisely this is the sort of way we can expect moral enhancement to be introduced—in "soft" form, as a series of "nudges," and most importantly, without the words "moral enhancement" ever being uttered. That is, the use of psychological and neuroscientific understanding, unconscious motivators, cognitive biases, and other such means to gradually steer behavior toward more desirable ends are already being applied, and all this without saying a word, and certainly not mentioning morality. In other words, moral enhancement's most realistic and practicable prospect will likely be enacted through softer, nonexplicit attempts to shape behavior and opinion. It may well be that the hopes placed in "hard" moral enhancement— explicit attempts to use the technology of pharmacology to develop the moral functioning of persons—have been massively overstated. We have seen and will continue to see throughout the various ways in which too high a set of expectations have been placed on hard moral enhancement. Soft moral enhancement is not like this. To the contrary, soft moral enhancement is more than an inevitability, it is already a reality, and we will give numerous illustrations in part IV of how such soft moral enhancement is already in play. We will find that moral enhancement in this softer sense is readily assimilated within such soft paternalism as a means for managing "undesirable behavior"—for the line between moral behavior and undesirable behavior is never so clear. Insofar as such interventions involve technological or pharmacological means, we can say that we have de facto biomedical moral enhancement. Moral functioning—and, by extension, moral bioenhancement—does and will continue to interweave imperceptibly with daily living and our everyday interactions. So far from moral enhancement being futuristic, so far from it being inevitable in some form or other, construed in this way, moral bioenhancement is something that has already begun.

II Science

4 The Biochemical Bases for Moral Enhancement: How the Brain Does Morality

In part I, we explored the various philosophical rationales for moral bioenhancement, and we encountered a number of ways that such enhancement projects might be brought to bear. But what does the science tell us? Up till this point we have dealt with the concepts involved in moral enhancement discourse on their own terms. What do the empirical studies have to tell us about what may or may not be possible in terms of biomedical moral enhancement? It is important to be able to grasp what empirical supports, if any, there may be to ground and justify the hopes and expectations that surround the idea. To that end, the present chapter will focus on three exemplars pointed to by the philosophers in part I as having moral enhancement potential—oxytocin, serotonin, and dopamine—and the various studies conducted regarding their connection to moral judgment and behavior.

As we have maintained throughout, one can suggest without any controversy at all that moral behavior is to some greater or lesser extent influenced by individual biology. Such a claim is impossible to dispute—we are embodied, biological beings, and to some extent at least, moral functioning can be described in terms of its various biological correlates. While there are many possible influences for moral behavior, it has to be said that biology is certainly one among them, though disagreement exists as to the precise extent to which moral behavior is controlled by biology. Yet the various studies conducted in this area do seem to agree that there are certain chemicals, hormones, and neurotransmitters, the presence or absence of which set a biological context, and out from this context certain kinds of behaviors and judgments may be made more or less likely to manifest.

In exploring how realistic moral bioenhancement might turn out to be, one important step is generating an understanding of the various ways in which neurobiological substances can dispose persons to behave and think in certain ways. Returning to our exemplars, it has been suggested that oxytocin plays a role in encouraging empathy, generosity, trust, and trust-worthiness. According to Paul Zak, the goal here is to promote the "natural goodness" of human beings, improving human relations and happiness by reinforcing empathic and trustworthy behavior. Molly Crockett's work on serotonin suggests that this neurotransmitter has a profound role to play in regulating emotions, promoting harm aversion, and influencing our judgments about fairness and justice. Herein, serotonin helps down-regulate the less evolved "Pavlovian" parts of brain, as we have seen, our "inner Hydes," which govern basic responses to signals of pleasure and pain in favor of more recently evolved capacities for "goal-centered" kinds of thought. Finally, the neurotransmitter dopamine will be examined with respect to its role in reward mechanisms and behavior reinforcement. It has been suggested that new technologies for ultra-precise control of dopami-nergic pathways might be effective in driving the learning of new behaviors and the dissipation of destructive learned or addictive behaviors.

Oxytocin—Empathy, Trustworthiness, Generosity—Paul Zak

So, I discovered this chemical in the brain that essentially makes people good.
—P. Zak (2011a)

Oxytocin and Trust

Paul Zak has made many grand and very well-publicized claims about oxytocin, calling it "the love hormone" (Honigsbaum 2011), "the moral molecule" (Zak 2012) and "the source of love and prosperity" (Zak 2013). Indeed, according to Zak, oxytocin is no less than "the *ultimate* moral mol-ecule" (Honigsbaum 2011; emphasis added). So grand and effusive are Zak's claims, in fact, that even the Royal Society for the Arts (2012) were suffi-ciently awed into introducing his team's research with the following words: "Paul J. Zak reveals how a single chemical governs all of our morality and behavior."

This proposition, that "all of our morality and behavior" is "governed" by one single chemical is a bold claim, to say the least. What is the empirical basis for these declarations regarding this simple neuropeptide, whose pri-mary functions are the facilitation of childbirth, breastfeeding, and orgasm in women? Zak and colleagues have conducted a range of experiments exploring the causal role of oxytocin with respect to trust, trustworthiness, and generosity, even going so far as attempting to plot the economic suc-cess of entire nations in terms of oxytocin levels.[1] For present purposes we will focus upon the more localized, interpersonal studies which Zak describes as being discovered using the "vampire economics" (Zak 2012) methodology—an exciting title to describe the rather more mundane reality, the process of doing blood tests on participants before and after the test conditions were explored.

We begin with Zak's work on trust and trustworthiness. To explore the role of oxytocin in facilitating expressions of trust, Zak appropriated a standard "trust game" from economic science. This game involves giving one set of participants (156 UCLA college students; Zak and Kurzban 2005) a ten-dollar sum, and then giving them the choice to entrust any proportion of that ten dollars to another person (the trustee). Whatever monies are trusted are then trebled, and the trustee is given the choice to either keep all of that trebled money for themselves, or of returning a portion of it to the person who had entrusted them with their cash. This study, which was conducted blind (that is, the person doing the trusting and the person being trusted were anonymous to each other), involved two potential transfers of cash. The first transfer, giving the money away, was regarded as a measure of trust, whereas any return movement of monies given back to that person was taken to be a measure of trustworthiness.

A number of variations of this study were conducted. In the first of these, blood samples were taken before and after the game from both parties to correlate oxytocin levels with the proportion of trust given and trustworthiness returned (Zak and Kurzban 2005). The results appeared strong: when persons were trusted with higher amounts of money, their oxytocin spiked and they tended to return more of the money, being more trustworthy in return. Zak's conclusion was that the very act of being trusted is a signal to the brain to produce oxytocin, and this facilitates reciprocal trustworthy behavior. Thus, the increased levels of oxytocin in the receiver correlated with increased generosity in splitting the money with the trusting party. The conclusion drawn: oxytocin is a "trust drug" released when we feel trusted, and it facilitates reciprocal and cooperative behavior.

Zak and colleagues took the study to the next stage, the direct introduction of synthetically produced oxytocin into the bodies of the participants (via a nasal spray)—as he puts it, "packing their neurons with oxytocin" (Zak 2010)—in order to see what effects this had. Redoing the trust game accordingly, this time dosing one group of persons with oxytocin and keeping a control group of persons who were not given oxytocin, Zak again found favorable results. Levels of trust (the proportion of money given away to the trustee) were increased in persons given intranasal oxytocin, and far more candidates were willing to trust the other party with all of their money (Zak 2011b). Moreover, the intranasal introduction of oxytocin likewise increased the return of the money back to them (Zak 2011b). Zak's conclusion then was that oxytocin can be seen to have a direct, causative relationship between trust and trustworthiness, such that, according to Zak, "when oxytocin levels are high we trust others in a tangible, objectively measurable way" (Zak 2010).

There were some notable exceptions to the above conclusions. In addition to the many positive findings, Zak also discovered that approximately 2% to 5% of the populations tested were utterly unresponsive to being trusted or stimulation by administered oxytocin (Zak 2010). They either did not release oxytocin upon being trusted with the other's money, or they did not reciprocate when oxytocin was present (Zak 2011a). Zak has come to refer to this category of persons as "bastards": these are people who "play like a bastard," keeping all of the money entrusted to them for themselves (Zak 2010). Zak asserts that this tiny minority of "unconditional non-reciprocators" (Zak 2010) have "some of the traits of psychopaths" (Honigsbaum 2011); he describes them as advantage-takers who are, for whatever reason, untrustworthy in spite of the chemicals in their brain disposing them to reciprocate. Nonetheless, given how proportionately small the size of this population is, Zak takes the result overall to confirm his belief that people are "basically good," and much more trustworthy than they are given credit for.

While Zak asserts that a small proportion of persons are "hardwired" in this way, there are other factors which influence the extent to which persons produce and respond to oxytocin. One key finding is that persons' "oxytocin systems" require a degree of nurturing if they are to develop properly. It has been found that women who have suffered childhood abuse tend not to release oxytocin when prompted by trust signals in the games (IJzendoorn et al. 2011, 7). It seems that attachments with parents and within one's social group are tremendously important in allowing the oxytocin system to fully develop. Women with histories of abuse have been found to have higher base-line levels of cortisol (a stress hormone), and the increased presence of cortisol usually means a decrease in the levels of oxytocin. This seems to be borne out in a number of studies exploring differential effects of oxytocin on trusting and generous behavior, where it was found that oxytocin's effects of donating to charity were moderated by "experiences of parental love-withdrawal," and thus that "the positive effect of oxytocin administration on prosocial behavior may be limited to individuals with supportive backgrounds" (IJzendoorn et al. 2011, 1). And again, in a study by Madelon et al. (2013), the finding was that the "oxytocin effects on complex brain networks are moderated by experiences of maternal love withdrawal ... as a disciplinary strategy involving with-holding love and affection after a failure or misbehavior" (Madelon et al. 2013, 1288).

However, unlike Patricia Churchland, who emphasizes cortisol and oxy-tocin as natural antagonists (Churchland 2012), Zak emphasizes testos-terone as oxytocin's opposite number. If oxytocin is the moral molecule, representing the better aspects of our nature, the suggestion is that tes-tosterone represents the worst. Zak (2009) writes that testosterone encour-ages "greater punishing and selfish behavior." According to his team's 2009 study, Zak writes that:

High (testosterone) males are more likely to have physical altercations, divorce more often, spend less time with their children, engage in competitions of all types, have more sexual partners, face learning disabilities, and lose their jobs more often ... suggesting that high (testosterone) men may behave differently than other men. ... We conclude that elevated testosterone causes men to behave antisocially. (Zak et al. 2009, 1)

Zak goes even further, for where oxytocin is "the ultimate moral molecule," testosterone, by extension, must be the cause of its opposite. Zak continues:

At times of stress ... we are physiologically in "survival mode," prompting the release of testosterone and its bioactive metabolite, DHT. These stress hormones prevent oxytocin from binding to brain receptors, tipping the balance towards distrust and away from pro-social behaviour. This process, he says, explains "the petty evils nor-mally virtuous people exhibit." In some cases, this bad behaviour may also be exac-erbated by genetic and environmental factors. (Honigsbaum 2011)

Angels whisper to us when we are high on oxytocin, and the devil dwells in testosterone.[2] So, we begin to see a value-laden polarization of concepts arising from Zak's discourse. On the one hand, we find stress, aggression, selfishness (effects which Zak argues create distrust, as well as emotional and socioeconomic poverty); on the other hand, we have peace, trust, and kindness, as well as doing good unto others (which Zak argues is in accord with our true natures and creates happiness as well as emotional and socio-economic prosperity). Regardless of what one makes of Zak et al.'s various studies, it is interesting to see how oxytocin, presented as the Angel on our shoulder, comes to be a conceptual placeholder for a particular set of values which are expressive of an entire worldview.[3]

Oxytocin and Generosity

Let us move onto Zak's work on generosity. Zak and colleagues have con-ducted a range of studies investigating the relationship between oxytocin levels and giving. The first study (Zak, Stanton, and Ahmadi 2007), used an "ultimatum game" setup involving two persons who are given a pot of money, one of whom then splits that money in any ratio he likes, on the understanding that if the other player does not accept the split, both parties end up with nothing. Zak and colleagues found that those infused with oxytocin "were 80% more generous than those given a placebo" (ibid., 1). In another study (Barraza and Zak 2009), oxytocin's effects on generosity were tested with respect to charitable giving. Short video clips were shown to participants, some of a very emotional nature (a father talking about his young cancer-ridden child), some of a more neutral nature. Participants "rated the emotions they experienced" (Barraza and Zak 2009, 182), and then played a $40 ultimatum game to gauge their generosity. The results were that "empathy was associated with a 47% increase in oxytocin from baseline" (ibid., 182). Zak concluded as follows: "Our findings provide the first evidence that oxytocin is a physiologic signature for empathy and that empathy mediates generosity" (ibid., 182).

Finally, Barraza et al. (2011), exploring the prosocial effects of oxytocin, let various participants have the opportunity to earn money in a series of economic games, and then let them donate a portion of those earnings to charity. While oxytocin infusions did not significantly increase the number of persons who decided to donate, for those who did decide to donate, Barraza writes:

Among the 36% of participants who did donate, people infused with (oxytocin) were found to donate 48% more to charity than those given a placebo. The amount of money earned in the experiment had no effect on whether or not a donation was made or the size of a donation. This is the first study showing that [oxytocin] increases generosity in unilateral exchanges directed toward philanthropic social institutions, as opposed to immediate benefits directed at individuals or groups. (ibid., 148)

Oxytocin: "The Prosocial Neuropeptide"[4]

While one would certainly not get this impression from studying Zak's own publications, his team's work is actually part of a much larger body of research into the effects of oxytocin which both preceded him and continues into the present to put forth a great deal of output. Since this chapter is concerned with the science of oxytocin, it would be remiss to focus only on Zak and his colleagues. Indeed, it would appear that oxytocin is an incredibly versatile substance. According to Hurlemann et al.: "oxytocin ... is becoming increasingly established as a prosocial neuropeptide in humans with therapeutic potential in treatment of social, cognitive, and mood disorders" and has been found to enhance "amygdala-dependent, socially reinforced learning and emotional empathy in humans" (Hurlemann et al. 2010, 4999). Macdonald and Macdonald (2010) also write:

The oxytocin system is a promising target for therapeutic interventions in a variety of conditions, especially those characterized by anxiety and aberrations in social function. ... These effects include alterations in social decision making, processing of social stimuli, certain uniquely social behaviors (e.g., eye contact), and social memory. (Ibid., 2010, 1)

According to IsHak, Kahloon, and Fakhry (2011), oxytocin also has a role in enhancing well-being. Apparently, oxytocin

induces a general sense of well-being including calm, improved social interactions, increased trust, and reduced fear as well as endocrine and physiological changes. ... Just as [oxytocin] has widespread effects in factors encompassing well-being, its dysfunction is associated with morbidity and decreased quality of life as observed neuropsychiatric conditions such as autism, schizophrenia and social phobias. (IsHak, Kahloon, and Fakhry 2011, 1)

Oxytocin, then, has promise for a range of treatments in the treatment of schizophrenia and autism (Kéri, Kiss, and Kelemen 2009) by facilitating the decoding of faces expressing complex mental states and social emotions (Domes et al. 2007a, 2007b, 2010), increasing gaze to the eye regions of other people's faces (Guastella, Mitchell, and Dadds 2008), and modulating brain areas responsible for social cognition (amygdala and prefrontal cortex; Domes et al. 2007). However, the effects of oxytocin in these treatments seem to diverge considerably depending on what dose of intranasal oxytocin is applied (Goldman et al. 2011)

Conflicting Evidence

Versatile as oxytocin is in the range of effects it is purported to have, it is to be noted that the studies performed have not yielded uniformly positive results. In fact, there is a considerable amount of ambiguity and conflict in the results discerned. While I will return to these discrepancies and my interpretation of their significance in the next chapter, it is important to note at this point that such discrepancies do exist. For example, some of the prosocial effects of oxytocin have been famously disputed by De Dreu et al. (2010), who found that oxytocin, while promoting in-group cohesion, has also been seen to promote out-group aggression. This raises the question of whether oxytocin really has the morally affective power to bring together a divided world, or whether it might even exacerbate existing problems on the basis of those very divisions. Similarly, regarding oxytocin's anti-anxiety effects, Hoge writes:

Our finding of higher circulating oxytocin levels associated with greater social dis-satisfaction and increased severity among individuals with social anxiety disorder provide preliminary support for a link between social anxiety and plasma oxytocin level, albeit in the opposite direction to that originally hypothesized. (Hoge et al. 2008, 167)

Similar conflicts have been uncovered by Hoge et al. regarding oxytocin's effects on depressive patients and with respect to the bonding-inducing powers of oxytocin (the "cuddle hormone"). Quite contrary to Zak's well-publicized findings regarding massive oxytocin increases for couples on their wedding days and couples interacting over social media (up to 150% spike in blood oxytocin levels; Zak 2012), the Hoge et al. (2008) study found that there was also a massive increase in blood oxytocin levels for those expressing extremely low levels of relationship satisfaction. Indeed, interestingly given his claims regarding oxytocin as the moral molecule, Zak notes that several of his subjects had to be excluded from his stud-ies because of extremely high blood oxytocin levels: they had been found perusing on-line pornography just prior to their participation.

With reference to oxytocin's potential for moral enhancement in par-ticular, these studies (if we are to take them seriously), seem to present a very modest potential for such enhancement. However, a number of severe problems with the idea of using oxytocin in moral enhancement context have arisen:

(a) oxytocin tends not to be particularly efficacious with respect to those persons ("bastards") who would be the most pressing targets for oxytocin's desired effects;

(b) oxytocin's power to modulate trust and trustworthiness is to a consid-erable degree supervened upon by developmental concerns, particularly to do with upbringing and nurturing (which is significant here, given the importance that will be placed upon the social-environmental aspects of moral formation below);

(c) oxytocin increases out-group aggression, which is obviously a counter-productive effect if oxytocin is being presented as something that might cohere various disputing groups; and

(d) oxytocin seems to be raised in persons engaging in certain hedonisti-cally enjoyable though morally questionable deeds, so can hardly be associ-ated with improved moral functioning across the board.

If these studies are to be believed, we have good reason for thinking that oxytocin is not a particularly impressive basis for moral bioenhancement on the grand scale. That being so, oxytocin does seem to be effective in improving the empathetic response in those already disposed toward displays of empathy. Oxytocin then does have some potential as a "nudge"—it may help improve social functioning if only through its anti-anxiolytic qualities, heightening such moral goods where they already exist in the person in question, and perhaps further encouraging prosocial behavior in persons already so disposed toward such displays.[5]

Serotonin—Harm, Fairness, and Aggression—Molly Crockett

Serotonin and Judgments on Harm   Serotonin is known to play a role in regulating a large range of bodily functions, primarily sleep, mood, appetite, and digestion. It also functions in wound vasoconstriction, learning, temperature regulation, cardiovascular function, endocrine regulation, memory, sexual behavior, vision, and many other bodily operations. Indeed, when first isolated by Page in 1948, he commented that no other physiological substance known possesses such diverse actions in the body as does serotonin. With this long list of mechanisms influenced by serotonin, one might suspect that pinning down its potential for influencing moral judgment might be a particularly difficult task. This is in fact the case. However, one way of attempting to chart the possible relationship between serotonin and moral judgment has been provided by Crockett.

In describing the manner in which serotonin can influence moral judgments, Crockett, as we have seen, uses the story of Jekyll and Hyde to represent what she calls "two brains in conflict." Crockett writes:

Just as Robert Louis Stevenson wrote about multiple personalities within one man, our brains have multiple systems that *compete* to *control our behavior*. The Pavlovian system governs basic responses to signals of pleasure and pain. It is very old, in evolutionary terms ... and its actions evoke very predictable responses. ... For example, signals of danger provoke aggression, which probably evolved to protect us from danger. Free from conscience, the Pavlovian system is like the neural Hyde. It directs us to seek rewards and react aggressively to punishments, and is responsible for violent urges, like those unleashed by Jekyll's potion. ... Another system, called the goal-directed system, is responsible for more detailed representations of the world and helping us visualize our goals. The goal-directed system is evolutionarily newer than the Pavlovian system, and is thought to have reached its peak sophistication in humans. Like Jekyll, the goal-directed system considers social regulations. It allows us to override Pavlovian impulses in favor of long-term goals. ... When we face temptation, our primitive, emotional brain system appears to wrestle with a more recently evolved, goal-driven system. At any given time, temporary changes in our brain chemistry may determine which one wins. (Crockett 2008; emphases added)

For Crockett, serotonin plays a significant role in determining the balance of power that exists between these two neural mechanisms as they "compete to control our behavior." So, according to Crockett's studies, a person's serotonin levels will influence their levels of prosocial behavior and help reduce aggression by down-regulating the conflict between these two systems in favor of our more forward-thinking selves (Crockett 2008).

Crockett's key work has involved investigating the effects of manipulating serotonin availability on persons' judgments about two central kinds of values: fairness and harm. Let us first consider the studies attempting to relate serotonin levels to judgments about harm. In these tests, Crockett exposed candidates to a version of the common "trolley problem," a moral dilemma that confronts candidates with a situation in which they must make an imaginary moral choice between harming one person or saving a number of others. According to the setup used, participants could choose either to push "an imaginary fat man" in front of a train in order to save five other persons, or, refusing to harm the fat man by inactivity, allow the five to die. Again, the test is intended to examine persons' judgments about the permissibility of harm (rather than actual enacted behavior), and, according to Crockett at least, the setup offers a choice which corresponds broadly to deontological versus consequentialist kinds of reasoning (Crockett 2012b).

According to the experiment, the candidates are thus presented with two diametrically opposed schools of moral thought. Very crudely put:

(a) consequentialist thinking, which favors the greatest good for the greatest number—herein it is right to push the fat man to save the rest; and
(b) deontological or rule-based thinking, rather than the outcomes or actual consequences of the decision made—herein it is not okay to push the fat man, even if it saves the lives of more persons, simply because murdering the fat man is "just wrong."[6]

The results of Crockett's studies indicate that when serotonin availability is manipulated, persons' judgments about the acceptability of harming the fat man change in significant ways. Persons whose serotonin levels were increased tended to be far more averse to harming the fat man in order to save the others' lives (Crockett 2012b). Moreover, persons with stronger baseline tendencies against harm, or higher degrees of empathy, are even more affected by the increased serotonin, and even less likely to say the harmful act is acceptable regardless of the consequences to others (ibid.). Crockett's conclusion is that serotonin selectively influences moral judgment through affecting changes in persons' aversion to causing harm (ibid.). Increasing serotonin availability tended to encourage in persons a deontological style of thinking when it came to judgments about harm. This suggests that there is at least some role for serotonin in understanding moral judgments. Crockett is very careful to add that she herself is not making any judgment as to whether consequentialist or deontological viewpoints are better or worse moral standards; she is simply remarking that manipulating persons' levels of serotonin tends to affect judgments about the acceptability of harm.

Serotonin and Judgments on Fairness

In order to test persons' judgments about fairness and justice, Crockett conducted studies using the same ultimatum game setup we saw with Zak. This is a game which, as we saw earlier, involves two players and a pot of money—one of the players (the proposer) suggests a way of splitting the money, and the other player (the responder) then has to decide whether to accept this offer or not. If the responder accepts the split, both parties keep the money so divided; if the responder rejects the split, both parties leave with nothing. The assumption is that if the responder is acting from rational self-interest, he or she ought to accept any measure of split money simply because "some money is better than no money," regardless of how unfair the split, for at least then the responder walks away with something, which is more than he had to begin with (Crockett 2012b). Punishing the proposer for an unfair split is viewed here as an irrational, emotion-driven response, a pointless act of spite.[7] However, in reality, when offers are unfair, it has been found that people tend to predictably reject them, reliably choosing a "scorched earth" policy despite the fact that rejection means not getting paid. In short, responders would rather walk away with nothing than see the proposer walk away with an unfair gain. Crockett writes:

That behavior smacks of the Pavlovian brain. Accepting an unfair offer, however, means suppressing the *emotional* urge to reject, which researchers presume is the work of the goal-directed system. ... Unfair offers activated the Pavlovian system—and the greater the activity in this system, the more likely the volunteers were to angrily reject the offers. In contrast, when volunteers were willing to stomach the unfair offers, their brain scans revealed increased activity in their goal-directed systems and decreased activity in their Pavlovian systems. Their Dr. Jekylls were in charge. (Crockett 2008; emphasis added)

When participants' serotonin levels were manipulated, two clear trends were discerned which support the above picture. When responders were subjected to acute serotonin depletion, they were far more likely to angrily reject and punish unfair offers put to them. In complete contrast, increasing serotonin levels tended to produce an acceptance of the similarly unfair offers. As with harm aversion, judgments about fairness would appear to be subject to the influence of the availability of serotonin in persons' brains. Crockett's tentative conclusion drawn from the fairness test is that serotonin levels affect the brain's response to anger. She writes: "chemistry does play an important role in controlling aggressive impulses" (Crockett 2008). For those whose serotonin had been depleted, "Like ... brain-damaged patients, these volunteers seemed to be driven by their inner Hyde" (ibid.). The suggestion, then, is that serotonin levels in the brain have a powerful influence on the way persons make certain kinds of judgments, Crockett writes:

To study how decisions go wrong in the brain, researchers have looked at the brain chemistry of patients who have made particularly egregious decisions. ... They have found that alcoholics, suicide attempters, and violent criminals have unusually low levels of the neurotransmitter serotonin. Since this brain chemical is involved in *emotional* disorders like depression, researchers suspect that serotonin is also critical for regulating *emotions*. (Crockett 2008; emphasis added)

Crockett's central proposition, then, is that serotonin regulates emotional decision-making, promoting or inhibiting "rational versus irrational reasoning." Crockett continues:

Scientists who study decision-making have discovered that most human decisions are not exactly "rational," but do follow "predictably irrational" patterns. In particular, we stray from our long-term goals when *emotions* interfere with our brain's tools for self-control. ... So what does all of this mean for the everyday decision-maker? We're beginning to establish that artificially lowering serotonin levels lets our *emotions* influence more of our decision-making. ... If serotonin is necessary for making "rational" decisions, then lowering serotonin levels should unleash *emotions*, making people more "predictably irrational." Sure enough, ... by tinkering with a few grams of a single chemical, I reliably provoked irrational, *emotionally*-driven decisions. (Crockett 2008; emphases added)

There are a number of significant problems with Crockett's framing of the situation here and her interpretation of the various findings. What is most troubling is the very broad and crude manner in which Crockett uses the word "emotion" (as highlighted in the quote above[8]), which she seems to think is inherently irrational, directly contrary to and conflicting with reason, thus leading necessarily to poor decisions when "the emotions" win out. Ostensibly, her thesis is about "conscious control" versus "unrestrained, reactive animal responses," but a look at the deep grammar reveals a deeper, more simplistic metaphor at work, one between an "emotional brain system" versus a "rational brain system." In short, a conflict/domination model of the brain and human agency in terms of "reason versus emotion."

To repeat Crockett's point: "If serotonin is necessary for making 'rational' decisions, then lowering serotonin levels should unleash *emotions*, making people more 'predictably irrational'" (ibid.; emphasis added). I would suggest that the sharpness of this split is unfortunate. And not only the split between emotions and reason, but the split between emotions as being in a conflict relationship with goal-centered thinking is unwarranted, as if the goals we aspire to are not emotionally driven or emotionally nourished and sustained in a number of very important ways, as if emotions cannot in fact support rational decision-making rather than needing to be presented as antagonistic to such processes.

It is, of course, true that emotions can and often do interfere with our goal-directed thinking, or self-control—such an image is enshrined in the very popular science fiction trope of the enlightened "rational" alien being who has successfully suppressed its emotions. Indeed, one of the very few films dedicated to the idea of moral bioenhancement, *Equilibrium*, premises its entire dystopian moral-social order on the compulsory prescription of the emotion-suppressing drug "Prozium"—the rationale for this being that all of man's evil deeds, "man's inhumanity to man," arises out of his emotions, which need thus to be suppressed. All works of art, poetry, and music, any and all things that might inspire an emotion, are destroyed by the state, and all persons appreciating poetry and the like are taken for summary execution. Ironically, the punishment of these "sense offenders," incineration, can only be enforced because the soldiers involved have had their empathic emotions suppressed.

The implication here, and certain parallels with the Holocaust are obvious, is that some of the greatest evils are not caused by the emotions, but precisely by the lack of them. The true moral victory at the end of the film is motivated only when the protagonist has an extreme, explosive emotional reaction to the injustice of the situation. Thus it is only an overwhelming rage response (his emotionally reactive "inner Hyde") which motivates him to take the necessary moral course. It is somewhat worrying that the very basic, simplistic thinking of a Hollywood action flick has been able to treat of this subject with more finesse than some of our most esteemed neuroscientists.

I wonder whether the empirical researchers, "neurophilosophers," and psychologists that share Crockett's underlying paradigm have been subtly inspired by the many popular fantasy archetypes that pervade our culture, the enlightened rational being with absolute control over his or her emotions, or whether there is some deeper source for this assumption which seems to have worked its way into the objective neurobiological study of the brain, framing and shaping such study from the outset. The cultural stereotype of the unflappable Zen monk, the calm center of the universe, or the Stoic imperturbable promontory retaining his dignified aloofness regardless of life's vicissitudes all seem to contain this same harsh split and conflict between rationality and emotion that undergirds the very frame used by Crockett and many others (we will see that this paradigm is perennial).

In contrast, more Aristotelian-based conceptual frameworks, such as one finds in, say, virtue ethics theories (see Nancy Sherman, 2000, on the idea of "wise emotions"), take a very different view of the way in which emotions and rationality relate to each other. In these more integrative frameworks, the observation is that certain forms of emotion are absolutely essential for a refined moral functioning, that emotion and reason are never (and can never) be fully severed from each other. This is not just a case of saying "we need our Mr. Hydes where decisive moral heroism is warranted," as was suggested in chapter 1. Rather, much more comprehensively, we are asserting that our entire emotional range is required to nourish our capacities for refined rationality from the outset. The question would then become not one of reducing "emotionally driven responses" that are "in competition" with rational processes, but rather about refining one's emotional responses so that the two can best work together as an effective team.

Indeed, despite the fact that our most precious goals are always profoundly emotionalized, one might suggest that goal-centered thinking is a very limited way of construing moral action to begin with. This is because the moral value of one's goal-relating thinking depends on the moral value of one's goals. For where a person's goals are self-centered or even evil (e.g., planning out how to abduct and torture someone, designing Auschwitz), enhancing the capacities of the goal-centered workings of the brain at the cost of "the emotions" could well be the very worst thing one could do. As such, neither goal-centered thinking nor emotional reactivity, nor the brain structures involved in these processes, have anything at all to do with moral action or judgment in and of themselves. It may just be that whole "reason versus emotion" framework employed by Crockett, and by extension those who share these sorts of assumptions, is just too hopelessly superficial to do any real work regarding moral judgment from the get-go.

Crockett should not be isolated here, for such a conflict/domination framework appears in many guises, particularly in those applying "top-down" models of the brain which involve proffering a picture of brain function in terms of an:

oppositional tension between the higher-order thinking of the more recently evolved neo-cortex and the deeper limbic region that is the home of emotion and more basic, primitive thinking categories. (Savage 2013, 160)

Note once more, in this quote, the association of "emotion" with "primitive thinking," and its dissociation from "higher-order thinking." Certainly the public perception, with much justification, is that the "current wisdom" in psychology holds that there are two brain-systems in operation at all times, an "emotional" system, and a "rational" system, and that these two systems are in a constant state of conflict.

From Plato's charioteer and Hume's psychology of the passions to Nietzsche's description of the instincts as our primary rulers and Freud's topographical depiction of the unconscious, and onwards into neuroscience, Crockett's Jekyll and Hyde, Haidt's elephant and rider (Haidt 2006), Greene's dual-process theory of moral functioning, and so on, and so on, we find perennial variations on a very similar narrative. They recur, of course, because they contain more than a grain of truth—there is no doubt that they say something important about the human experience. But the big problem is that, like all narratives, they limit us as much as they describe us. Which is to say, they exclude as much as they reveal.

While these pictures certainly do capture our subjective experience well enough, it is the projection of the metaphor onto a set of empirical studies—which admit of a variety of possible metaphors when taken as a whole—in order to give this one metaphor the stamp of "objective scientific truth" that is so problematic. The metaphor has power, and when it is invested with the grand robes of scientific validity it takes on a weight which can be problematic if taken on board by policy makers. As we shall see in chapters 5 and 8, there are many bodies in the political sphere that are very much appealing to neuroscientific studies to justify public policies and generate legislative proposals. Conflict/domination narratives will then be the implicit guiding force in the enactment of various legislative and public measures (and, as we will argue in part IV, in "soft" moral enhancement also).

The difficulty is that this conflict/domination paradigm is everywhere, and it closes down alternative narratives which can be justified by appealing to much the same body of empirical evidence, mountainous and equivocal as this body is when seen from above. Such paradigms, like all grand-scale scientifically justified depictions of human reality, are necessarily selective about which particular facts they choose to emphasize and which particular facts they choose to ignore.

The Jekyll and Hyde story, then, is not merely a "simplification" for purposes of explaining "difficult science" to the general public. Rather, it is a particularly naked articulation of a guiding metaphor which is extremely influential in the discourse. Alternative metaphors are possible. Iain McGilchrist (2009), who is worth looking at if for no other reason than the manner in which he draws a completely different way of understanding human agency from the masses of empirical sources available, presents a much more balanced and cooperative ideal. Here, it is not a case of emotion against reason, where the conscious rational self must come to some arrangement with the beast he is tethered to, but rather of two complementary ways of approaching reality, two complementary ways of having emotion, two complementary ways of going about reasoning that arise from the brain's two hemispheres' very different performing of the same tasks.

McGilchrist's work is grounded in thousands of neuroscientific studies. Thus, when it comes to questions of empirical truth regarding the brain, we do have to note the tremendous ambiguity that exists when that research is looked at as a whole, and also the equally tremendous room for contrary interpretations and narratives that the data allows for. Not everyone buys into McGilchrist's narrative; it is after all, more than a little neuroreductionistic, inasmuch as it refers to the construction of the entire Western world and all its various epochs throughout history in terms of the interplay between the brain's two hemispheres. The point is not to say McGilchrist's account is "right," but to show the manner in which simple appeals to the empirical neuroscientific evidence are inadequate to ground a decisive account of human agency and moral functioning. Ultimately, the question is really about what kind of image you want to have of the human being and its moral nature. It seems both optimistic and pessimistic visions, that is, complement/partnership models and conflict/domination models, can be equally grounded in the empirical studies thus far performed.

In other words, I would caution the reader to not simply accept the empirical work provided here as the product of "objective science" which then provides an "objective account" of the brain, human agency, and our moral natures. Throughout the empirical study, the science is constructed in advance according to guiding metaphors. This is inevitable, but we must be very careful not to take such science at face value (as have the likes of Hughes 2013 and Savulescu 2013), and look deeper in order to grasp the extent to which the empirical conclusions drawn have been shaped by their guiding metaphors going in—metaphors which have worldviews, and values, and judgments regarding the nature of the human creature already formed before the empirical studies are even conducted.

Serotonin and Moral Enhancement

Despite these conceptual issues, Crockett has been quite careful about the sorts of conclusions she wants drawn from this research, particularly with respect to moral enhancement. She has herself raised a number challenges to the idea of using serotonin as a basis for a moral enhancement intervention, and has also herself taken to task at least one enthusiast, David DeGrazia, for misappropriating her work as indicating that SSRIs might be used, as we have already seen, "as a means to being less inclined to assault people" (DeGrazia 2013, 228). "In fact, our findings are a bit more subtle and nuanced than implied," she writes, continuing:

For the sake of argument, suppose that we were to amass a body of evidence that a single neurotransmitter ... reliably and substantially reduced people's propensity to physically harm others. Before we pull out the prescription pads, it will be important to consider the potential unintended consequences of altering the function of that neurotransmitter. ... Most neurotransmitters serve multiple functions and are found in many different brain regions ... serotonin plays a role in a variety of other processes including (but not limited to) learning, emotion, vision, sexual behavior, appetite, sleep, pain and memory, and there are at least 17 different types of serotonin receptors that produce distinct effects on neurotransmission. Thus, interventions ... may have undesirable side effects, and these should be considered when weighing the costs and benefits of the intervention. (Crockett 2013, 1)

These sorts of comments Crockett has made to disabuse commentators of the idea that serotonin, or any other single molecule for that matter, might at some point be used as the basis of some sort of "morality pill," should make it surprising that SSRIs have been so prominent among the theses of the various philosophical moral enhancement enthusiasts. However, it is quite easy to make the claim that Crockett has been very coquettish with her work. On the one hand she has been very cautious about what she is prepared to allow other persons to claim about her research. On the other hand, she has herself made a number of obviously seductive claims, and it is not so surprising that the various enthusiasts have become so excited by her work and drawn inappropriate conclusions therefrom. For example, serotonin's potential for moral enhancement, particularly in remedial contexts, has been clearly advocated for in her work. Writing on the moral enhancing potential of serotonin, Crockett et al. suggest that their findings: "provide unique evidence that serotonin could promote prosocial behavior by enhancing harm aversion, a prosocial sentiment that directly affects both moral judgment and moral behavior" (Crockett et al. 2010, 17433), and as such that these findings "have implications for the use of serotonin agents in the treatment of antisocial and aggressive behavior" (ibid., 17437). These are precisely the sorts of tantalizing claims that have allowed the less circumspect moral enhancement philosophers to get so excited about the possibility of using SSRIs for purposes of moral enhancement. As Crockett continues: "your sense of justice could depend on what you had for breakfast this morning" (2011). Certainly all this leaves wide open the sense that serotonin might at some point be appropriated as a means for affecting moral judgment and behavior.

However, the more fundamental point remains: all this neglects the basic appreciation of the value of aggression—and, more specifically, the moral value of aggression, even harm. It is not just the side effects and dosage that we need to be worried about, but the superficiality with which we fail to recognize the ambiguity of certain qualities like aggression and the immense plurality of ways in which something like aggressive impulses can be mediated, shaped, and transformed into such a wide variety of possible manifestations, some of which may be morally bad (attacking someone for no reason), some morally good (defending one's family), and some completely morally neutral (becoming a feisty sportsman). This gap between impulse and manifestation is everything—and the more complex the behavior we have in mind, the bigger the gap that exists between impulse and manifestation. Such manifestations have more to do with psychological, social, and cultural factors, factors which mediate and then help shape the mere biological impulse into the expression that it takes. It is then the expression which has moral worth or not, which again makes serotonin a third wheel to the moral debate. It is the shaping of the impulse, and not the impulse itself, which makes the manifestation morally worthy or not.

Conflicting Evidence

Just as with Zak et al.'s work with oxytocin, Crockett's work is but the tip of a very large iceberg, a mountain of research relating to serotonin's various effects on the body and behavior, and particularly with respect to aggression and encouraging various prosocial behaviors. And, as with oxytocin, the various findings throughout are rather ambiguous, as Carrillo et al.

(2009) note: "The role of serotonin (5-HT) on aggression has been exten-sively studied; nonetheless, the role of this neurotransmitter in aggression is still inconclusive" (Carrillo et al. 2009, 349).

Note, however, that Carrillo's survey of the various results are based mostly on rat studies—indeed, most of the research into serotonin is not performed on higher mammals, and the findings are not consistent even within the same species, but vary even between type of rat. So, for example, in one kind of rat (Long Evans), it was found that when increased serotonin was combined with a range of conditions (e.g., drug-induced aggression via injections of citalopram or L-tryptophan), those increased serotonin levels were found to reduce certain kinds of aggression. In contrast, Wistar rats were found to become more, not less, aggressive when serotonin was mixed with another range of conditions, such as social isolation, in short treatment spans (Carrillo et al. 2009, 349). Carrillo concludes: "Although 5-HT has an overall inhibitory effect on aggression, the animal's genetic background, drug, treatment time, aggression inducing paradigm, and aggression type are critical variables that influence and modify this effect" (ibid., 349).

It would seem, then, even in creatures as comparatively simple as rats, there are a tremendous number of factors which can modulate how sero-tonin has its effects and factors by which serotonin can increase rather than decrease aggressive behavior. Unsurprisingly, when tested on yet more complex creatures, human beings, the results are yet more ambiguous. Kamarck and colleagues (2009), in looking at the relation between sero-tonin, anger, and risk for cardiovascular disease, did find some encouraging results. Investigating persons "high in hostility" (i.e., bad-tempered people who are not quite angry enough to be described as psychotic), Kamarck et al. found:

Treated subjects showed larger reductions in state anger ..., hostile affect ..., and, among women only, physical and verbal aggression ... relative to placebo controls. Treatment was also associated with relative increases in perceived social support. (Kamarck et al. 2009, 174)

However, one key point to be kept in mind—something moral enthusi-asts have not adequately represented in their writings—is that aggression is not an easily defined, singular phenomenon. There are many, many kinds of aggression, and many reasons for becoming aggressive (some of which may in fact be morally laudable). Indeed, Prado-Lima (2009) points out that aggression can be measured in many different ways—by target (e.g., against persons or the self), by mode (verbal or physical aggression), or by intensity (from mild annoyance to apoplectic rage; Prado-Lima 2009, 59). Takahashi et al. (2011) explore the effects of serotonin on a range of different kinds of aggression, human and animal, including, for example, family aggression, frustration-heightened aggression, social provocation, maternally protective aggression, affective defense ("rage"), territorial aggression, and many other kinds, which, while having elements of overlap, are not necessarily reducible to a singular idea of "aggression" such that one can say serotonin does or does not ameliorate it (Takahashi et al. 2011, 185–6).

There is also a difference between what is called "reactive" aggression (i.e., impulsive, explosive, sudden) and premeditated aggression (a distinction actually enshrined in law, with premeditated homicide meriting a charge of "murder in the first," whereas murder done "in the heat of the moment" brings slightly lesser charges). Crockett's work seems to relate to reactive aggression only, and it should be noted that while citalopram may well reduce reactive aggression, it does not reduce premeditated aggression. Indeed, in some cases various SSRIs, including citalopram, have been found to increase premeditated aggression (Breggin 2003, 31).[9] Again, this is something to be borne in mind when considering SSRIs as a moral enhancement intervention: what SSRIs stand to improve with respect to one kind of aggression, they may in fact worsen with respect to another (Wiseman 2014b, 27).[10]

This general ambiguity in the serotonin findings is summed up neatly by Carrillo as follows:

Several studies have shown that treatment with SSRIs significantly reduces aggression, hostility, and violent behaviour associated with different psychiatric disorders. ... Conversely, various reports show increases in aggression and the emergence of violent suicidal ideation during treatment of various psychiatric disorders, including depression, anxiety, bipolar, and psychosis with SSRIs. ... Summarizing existing findings, it is difficult to establish a clear relationship between the serotonergic system and aggression. (Carrillo 2009, 350)

Dopamine—Rewarding Behavior—Ed Boyden

As with oxytocin and serotonin, dopamine has a range of important functions in the body. Dopamine is a neurotransmitter whose primary functions surround reward-driven learning, motor control, and the flow of information across the brain. For present purposes, it is dopamine's role in rewarding and reinforcing behavior that will take center place.

When a state is experienced as pleasurable or rewarding, dopamine is released, signaling that the behavior is to be repeated. This is good or bad, or morally relevant, depending on what it is that is found to be rewarding. On the one hand, many beneficial or morally laudable behaviors stimulate dopamine production and are found to be rewarding, making such activities more prone to being repeated. On the other hand, dopamine is implicated in many kinds of addictive behaviors too—many drugs, opiates, and activities (including computer games, gambling, and pornography) also stimulate dopamine production, creating a destructively reinforcing reward cycle. Thus, like serotonin, dopamine has an ambiguous character with respect to its implications for moral behavior. However, because of its power to encourage habit formation by means of reward and its implication in addictive behavior, dopamine is also worth considering as a biological factor that relates to moral enhancement.

In enhancement discourse, dopamine's chief advocate is James Hughes. Hughes is quick to implicate dopamine in a range of morally important functions. For example, Hughes asserts that dopamine receptivity is a key factor influencing the extent to which one is likely to have "self-control" and the personality trait of "conscientiousness" (Hughes 2013, 33). This trait has obvious relevance to certain elements of moral functioning, Hughes writes:

Our capacity for self-discipline is genetically determined and chemically malleable. The personality factor of conscientiousness is associated with carefulness, self-discipline, and thinking carefully before acting. Low conscientiousness is associated with substance abuse and criminality (Ozer and Benet-Martinez 2006). Conscientiousness appears to be associated in turn with genetic variations of the dopamine receptor (Blasi et al. 2009). Dopamine is the neurochemical that mediates the pleasure we receive from novel and risky activities such as gambling, drugs, alcohol, and sex (Zald et al. 2008). ... Drugs like Adderall or Ritalin increase the amount of dopamine signaling in response to ordinary events, at least for folks who aren't on the high end of the dopamine bell curve to begin with. They make whatever you are doing more interesting, giving you more ability to control your own behavior and perform your best. (Hughes 2013, 34; Hughes 2012a)

Apart from these already available pharmaceuticals for modulating dopamine receptivity, another exciting avenue through which dopamine's effects can be explored in their relevance to moral behavior is through the extraordinary work conducted by Ed Boyden, leader of the Synthetic Neurobiology Group at MIT. Boyden's team has achieved increasing success in inventing new tools for analyzing and engineering neural circuits and in devising technologies for controlling specific classes of brain cell. The methods developed by his team are capable of direct stimulation of very specific cell groups in the brain, as well as switching these circuits and cells off. Through such methods, various potent possibilities for engineering the brain and understanding its workings have opened up.

The specificity of Boyden's approach is significant. As Boyden comments, the main problem with pharmacological interventions for neurological disorders or augmentation is their current incapacity to manage the micro-scale nature of the problems (Boyden 2011a). Pharmacological intervention is a "blunt weapon"—it is highly indiscriminate, at best manages to treat only symptoms, and does so only at the price of producing many unwanted side effects (a necessary consequence of bathing such a complex organ as the brain with chemicals that affect it as a whole).

In contrast, Boyden's approach allows for highly localized alteration of very specific neural circuits. His method of "optogenetics" involves using gene therapy to alter the DNA of certain classes of brain cell in a such a way that they produce light-sensitive elements (like solar panels), which then either activate or deactivate that cell when a certain color of light is shone upon it. In short, Boyden has created an on/off switch capable of micro-scale targeting very specific parts of the brain. This, according to Boyden, was, "not so hard." To an increasing degree, Boyden can now "control the brain in two colors, driving information in at will" (Boyden 2011b). His hope is to build brain co-processors that work with the brain to augment functions in people with disabilities—and by modeling the brain's activities in binary strings of 1s and 0s, he hopes it will be possible to discover and control which neural codes can drive certain behaviors, thoughts, and feelings.

Currently testing has been very successful (in mice and rats, at least) in targeting and manipulating the reward centers of the brain. Again, reward responses drive learning, and they "go awry" in addiction. Knowing, as we do, which cells mediate the signals of reward, it becomes possible to impact upon learning systems and addictive forms of behavior by manipulating the brain's various signals of reward. Indeed, this is exactly what has been found. In Boyden's experiments, a mouse is placed in a box which has two buttons on the wall. Using the appropriate-colored light to stimulate the reward circuits in the mouse's brain when it reaches for one of the buttons, it has been possible to artificially generate repetitive and addictive behavior. Pressing the button creates a flash of light in the mouse's brain, signaling pleasure and causing the mouse to repeatedly select that button. In simple creatures, at least, reinforcing and motivating certain behaviors is a relatively easy business, wherein a brief activation of the dopamine neurons is enough to drive learning in a clear and measurable way.

What, then, of using reward circuitry to undo problematic conditions? With a possible treatment for post-traumatic stress disorder in mind, Boyden and his team conditioned a fear response into a group of mice using classic conditioning techniques (electric shocks and pulses). Once the conditioning had taken, an attempt was then made to recondition the mice brains such that they no longer feared the pulses when sounded. By stimulating the reward centers when the pulses sounded, the conditioned fear response quickly dissipated (within but ten minutes of the process). Boyden suggests a similar technique might be useful in treating destructive addictive behaviors. The conditioned part of addiction, which is often characterized by the repetition of damaging or meaningless behaviors, might be likewise amenable to reconditioning through affecting the dopamine-reward circuitry in the brain.

At the very least, Boyden indicates that optogenetics can facilitate an understanding of which neurons are active during the performance of a range of addictive behaviors. Such behavior tasks, with mice at least, indicate that brief activation of these specific cells can reinforce whatever the brain was doing before. Whether optogenetics is itself used for such treatments, or whether it is simply used as a preliminary technique for helping design drug and treatment protocols, the process does show that the neuron classes that are associated with the addiction in question can in principle be blocked, modulated, or stimulated in order to reinforce, or dissipate, a range of learned responses.

More recently, a set of experiments have been conducted using the ontogenetic methodology as a means for treating alcoholism (Bass et al. 2013). We will return to consider this study more deeply in chapter 9, but for now it is worth noting that optogenetics has been used in rats to condition a preference for alcohol, and then, by the reverse process, caused the rats in question to simply cease their drinking. This cessation was not temporary, either; Bass et al. conclude that this study offers strong initial evidence that the dopaminergic response which "governs" the behavior of alcoholics (Clark 2014) might be amenable to treatments which target specific dopaminergic pathways.

Such technology is both exciting and shocking. Yet no one will deny that optogenetics, as a procedure with humans in mind, is not very practicable. The process is extremely invasive on a number of levels, for a start, involving as it does drilling holes in the head of the experimental candidate in order that fiber optic tubes be inserted therein, as well as altering the expression of cells in that person's brain on the genetic level. This degree of invasiveness means there are problems on both the practical and bioethical levels, with some commentators (e.g., Gilbert, Harris, and Kapsa 2014) suggesting that optogenetics should not be sanctioned for human testing at all. This raises an interesting conflict for Boyden. On the one hand, Boyden is adamant that pharmaceutical approaches are deeply problematic in terms of their production, testing, and bluntness. On the other hand, given its impracticable nature, Boyden is forced into suggesting that the promise of optogenetics with respect to humans is precisely in the development of pharmaceuticals (albeit, hopefully, more targeted in nature). So, while optogenetics is itself a nonpharmaceutical approach, ultimately it is directed, to a large degree, toward producing approaches which are pharmaceutical in nature—all the problems with pharmaceuticals' development, testing, and regulatory approval notwithstanding.

All this being so, it is clear that using such technologies makes it possible, in principle, to target the brain's reward pathways with an ultra-precise level of specificity, thereby gaining a measure of control over what persons find to be rewarding. Certain kinds of addictive behavior, damaging to self and others, might then be undone, at least in part, by affecting the way dopamine is released in the brain. While the argument will be made later on that there is much more to addiction than mere "dopaminergic responses," this does not undermine at all the promising contribution such interventions might proffer. In principle, this sort of approach can serve both to help (to whatever extent) undo the effects of dopamine in rewarding destructive or immoral behavior, or the dopamine reward circuitry can be positively manipulated to reinforce certain classes of action as desirable. Again, whether used directly or as a means for testing new drugs, interventions, prosthetics, and therapies, Boyden's research on dopamine and reward-circuit manipulation is truly remarkable and has many implications for encouraging or diminishing the motivation to undertake certain kinds of learned behavior.

Conclusion

Even though the quantity of research relating to these three substrates is simply prodigious, relating them concretely to moral functioning is still very much at the embryonic stage. Moreover, while the coming chapter will focus on the many serious problems that exist with these studies, conceptually and methodologically, it is important to reiterate at this juncture that the impact of human biochemistry and neuroanatomy on moral judgment and behavior is profound. There can be little doubt about that. That being so, the radical ambiguity of the various findings is such that the impact of these biological agents remains every bit as unpredictable as it is profound. It can only be that context, and the larger psycho-social environmental conditions of the person involved, shape the influence of these biological factors to a tremendous degree. Disentangling the effects of these biological factors from one another and from the variety of other surrounding influences on moral functioning is going to be a weighty problem for those who wish to appropriate these chemicals as mechanisms by which moral functioning can be augmented.

5   Not Fit for Purpose? Methodological and Conceptual Problems

Neuromythology—which claims that neuroscience can explain far more than it can—seems halfway plausible only if it is predicated upon a desperately impoverished account of our many-layered, multi-agendaed, infinitely folded but wonderfully structured and organised selves.

—R. Tallis (2008, 160)

Before taking as read the conclusions drawn from the scientific work seen in chapter 4, it is important to think critically about the quality of the studies put forward. Just how much can we realistically hope to learn about moral functioning by looking at the brain? It is obvious, and should be assumed throughout, that neuroscience is an incredibly important and germane realm of study. There is no question about that. The problem arises when we start to look into very complex and contextual human realities, concerns which are predominantly philosophical in nature, and then attempt to analyze them through technologies, conceptual categories, and guiding metaphors which are too crude to accurately describe the subject matter involved. Moral functioning is precisely such a domain where neuroscience, marvelous as it is, is much more limited in its present capacities to enlighten us about the nature of what is going on.

In fact, the argument will be presented that—with respect to our present focus on moral functioning, at least—current neuroscientific research, far from developing our understanding, actually diminishes it. Indeed, the quote used at the head of this chapter encapsulates perfectly the premises upon which this chapter will build. The worries that we will be exploring with respect to the biological study of moral functioning are that (a) as Tallis has famously observed, such research often utilizes a picture of the human creature that is radically oversimplified and impoverished, and (b) that the conclusions drawn are tortured from experimental conditions in no way representative of the way in which moral functioning actually happens. From this foundation, it will be argued that there are a number of significant problems with the sorts of investigative methods encountered in the previous chapter—that much of the scientific methodology applied is, when it comes to understanding moral functioning, not at all fit for purpose.

Yet, as we shall see, the putative criticisms of the particular studies we shall be considering are but a means to an end—a device for helping articulate the more philosophical reflections on moral functioning that we have been considering hitherto and that we will continue to develop throughout the book, and a way of making more general claims about the very nature of neuroscientific investigation into "the moral brain"—insofar as it relates to the idea of manipulating judgment and behavior, at least. As the chapter proceeds, the various criticisms will be used to develop the suggestion that such brain-centric methods, as they stand, are inherently inappropriate means for investigating how something as complex, nuanced, multifaceted, and inherently contextual as moral functioning might be brought under biomedical control. This must be made very clear. We are not making general comments about neuroscience at large. We are talking specifically about the potential (or, rather, lack of potential) such present methodologies proffer with respect to moral enhancement.

The studies we considered in chapter 4 do have many decisive problems, problems which bear considering in order that the science may improve. Furthermore, given the contextual nature of enacted moral functioning, a solely biological approach to study should raise immediate suspicions for anyone wanting a rich and deep account here. Given the interwoven nature of the phenomena involved, it is very difficult to see how isolating biological causes can provide anything other than the most superficial (and false) account of this tremendously complex thing we call moral functioning. In other words, while I would contend that neurobiological investigation into "the moral brain" may be possible, in principle, not much of value can be produced until a more integral set of approaches can be discovered which do justice to the interwoven nature of such influences (such an approach has been suggested by the likes of James Giordano, 2012).

The conflicting findings that we have discerned throughout chapter 4 regarding serotonin and oxytocin are, I would suggest, a very clear symptom of the refusal of the neurobiological investigation to seek an integrated approach to study. It is worth noting that if moral behavior is a situationally embedded, profoundly embodied, and often deeply social or relational affair (and it will be argued that this is precisely how moral functioning is best characterized), then, actually, study into "the moral brain" will not be able to take a single step forward until it recognizes that this "moral brain" is something which exists inside a moral person who lives in a moral world populated with other moral people.

The larger concern of this chapter, however, will be to question and critique the idea of *neuroprimacy* in this moral domain. While this chapter is exclusively critical of the studies we have thus far encountered, and critical also of the explanatory power of the methods presently used, it is only toward those who assume a primacy of these accounts—that neurological accounts of moral functioning somehow have a special status and authority over and above other empirical and philosophical perspectives—that the critique is directed. Above all, I wish to question the implicit notion that present neurobiological approaches should be given "the final word" on the matter. To be absolutely clear then, while a skepticism with respect to the limits of biological investigation of morality will certainly be prevalent throughout, the main target is actually the attitudes toward the value of such approaches—and this constitutes far more of a social critique, a critique of the current mode of explanation, than one directed at the science or the many scientists who have so painstakingly pursued these lines of research.

Methodological Problems

Poor Ecological and External Validity

The idea of "ecological validity" refers to how well a particular empirical study adequately approximates the reality it is meant to be modeling and measuring (Brewer 2000). In contrast, "external validity" deals with the extent to which the results of a given study can be generalized, or taken as generally indicative of what is being studied. Thus, if a particular study has low ecological and external validity, this means that the results of the study in question cannot be safely taken as being indicative of the subject matter in question. In this context, it may be that a certain moral trait is related to a certain brain state or hormone in the laboratory, yet this does not make it so when such a moral trait is actually embodied and expressed in the real world. Now that we have looked carefully at a range of neurostudies which focus on moral functioning, or at least are morally related in focus, I would like to make the case that these studies considered have particularly low ecological and external validity. They do not really tell us much about the moral phenomena they purport to be describing, and thus are not a particularly strong foundation for justifying the sorts of claims that moral enhancement enthusiasts are apt to make.

This being so, a more fundamental question arises: is the low ecological validity, the low relevance to actual real-life moral functioning, merely a feature of the science applied, or is it actually something deeper, something more problematic? Is there something about the nature of moral functioning itself which defies analysis in this way? Are the studies used just poorly constructed, bad science—or is the nature of moral functioning simply too complex to be captured by the sorts of methodological paradigms that are routinely applied? Certainly these are tough questions, and important, too, for if turns out to be the case that even the most seemingly simple moral traits are too subtle, too opaque to this brain-centric mode of study to isolate in biological terms, then that would place severe limits on the potential for neurologically derived approaches to moral enhancement.

As has been contended, things need to be looked at on a case-by-case basis; it will turn out that some of the biological bases for moral functioning are more closely linked and amenable to manipulation than others. However, before entering into the critique of the studies thus far considered, let us have a very brief reminder of the investigative methods that have been used.

1. *Trust and trustworthiness* A variety of "trust games" were used, sitting persons anonymously at computers and seeing how willing they are to split $10 with another person, drawing blood before and after the experiment, usually with effective doses of oxytocin intranasally administered;

2. *Generosity* A variety of "ultimatum games" were applied, sometimes preceded by an emotional video, observing how generous persons were in splitting a pot of monies with another person; other studies involved administering oxytocin and observing how much experimentally earned monies candidates were willing to donate to charity;

3. *Aggression* A variety of "trolley problem" dilemmas were presented to persons, some studies involved fMRI machine analysis of neural blood flow, while controlling the levels of serotonin available in the various candidates' brains; a range of aggression studies have been performed on mice and rats in various conditions controlling for serotonin levels; and a number of in-group/out-group aggression studies have been applied focusing on oxytocin;

4. *Justice/fairness* A variety of "ultimatum games" were pursued examining how persons reacted to unfair monetary offers, with levels of serotonin or testosterone being controlled, respectively;

5. *Rewards and addiction* A range of mice and rat experiments have been performed, getting the creatures in question to either like or dislike alcohol

(or other conditions) by switching on or off the dopaminergic neurons in the creatures' brains using optogenetic techniques; this is taken as a model for human addiction.

I will be arguing that these techniques do not adequately model the moral trait being investigated. Perhaps it is because the investigators using such methods come from within a domain where these techniques are just assumed to be valid; that is, are part of the package of simply accepting the "false truths" the domain relies upon (every domain has them). But closer inspection should make clear—particularly to those not invested in such paradigms—that such methods are quite limited when it comes to making significant or durable claims about complex phenomena such as trust, generosity, altruism, and so forth. So what, then, is it about these studies that makes them such poor models of reality?

Rich Kids and Mice: The WEIRD Bias   One of the first things to notice about the studies performed is that the sample set is almost exclusively WEIRD (that is, white, educated, industrialized, rich, and democratic; Arnett 2008; Henrich, Heine, and Norenzayan 2010). In short, the results of the studies are derived from subjects taken from a very select and narrow demographic—usually, students at exclusive universities around 20 years of age. On the one hand, this narrowness can be a good thing: the results are likely to be more indicative of the group in question if there is not too much deviation in the subject pool, rather than reducing culturally and demographically different people into some large, homogenized muddle. On the other hand, the inevitable consequence of having a WEIRD subject pool is that the results of the studies involved are, for the most part, incredibly skewed toward a particularly narrow range of person. Even if the studies conducted were perfect in every regard, they would tell us less about moral functioning in general, so much as about how affluent college kids make moral judgments—arguably not an ideal foundation for grounding durable and serious claims about how the brain "does" morality, per se. Indeed, it does seem as if most of the science upon which moral enhancement enthusiasts draw is conducted either using Ivy League students, or mice. Such studies might be valid outside this range, but this is precisely what the science does not say. A diverse, culturally and demographically comparative literature is at present conspicuous by its absence—and, while this is a present problem easily overcome by further research, there are other problems more intrinsic to the investigative method used.

Fun and Games: Zak and Crockett   One such problem that needs to be considered is the *game* quality of so many of the studies we have encountered. This notion of the "game" in investigating moral functioning is problematic for a number of reasons. First, the sort of moral decisions and behaviors that are being inspected do not, in reality, occur in a game context—they are often very serious matters, decisions and behaviors which play into the very identity, self-perceptions, values, aspirations, and emotional make-up of the persons in question. Certainly they are not something to be "played" with a sense of fun or general disregard for any medium- or long-term consequences of the actions involved. Secondly, since it is clear that investigators know that the setups are games (they are called the "ultimatum game" and "trust game," after all), one does have to wonder why it is that these investigators with almost total unanimity treat such games as adequate models for actual real-life moral functioning. In a rare but significant deviation from the norm—a study conducted by IJzendoorn et al. (2011)—the following suggestion is made about the various game setups used in the oxytocin studies (though the same is true generally):

Oxytocin has been implicated in a variety of prosocial processes but most of this work has used social dilemma type tasks to evaluate oxytocin's prosocial effects. ... While these tasks offer a high degree of experimental control they might lack ecological validity. Furthermore, laboratory tasks (such as the ultimatum game or the dictator game) have a game-like dimension, in which empathic concern with the co-players is made difficult because they usually are anonymous or fictional. (IJzendoorn et al. 2011, 1)

This statement is tremendously significant, and the study conducted by IJzendoorn et al. abandons the game structure for a more realistic and embedded approach. IJzendoorn et al. write: "The first aim of the current study is to investigate the influence of oxytocin on real, high-cost donating to charity (UNICEF) without a game-like dimension" (ibid., 4). But what does this admission mean for the studies which continue to rely upon these game methods? The essence of the investigative rationale, as IJzendoorn rightly points out, is "experimental control." What these games offer is a very constrained, very controlled environment through which that element of moral functioning can be explored. This is a completely fair motivation. Science functions only by controlling its variables and constants; only in this way can the effects of the various components be measured and decisively linked. But at what point does experimental control actually strangle the moral phenomena in question, leading to an unreal scenario with decisively distorted results? I would agree with John Shook's analysis that where a theory's specified sense of morality does not resemble how humans generally do morality, this serves as a disconfirmation of that theory's applicability in moral enhancement context (Shook 2012). I would add that it disconfirms the theory's applicability to real-life contexts in general.

Let us take the case of the ultimatum game as used by Crockett for measuring fairness and the punishment of norms of unfair behavior. The first thing to notice is that this study is highly confused about what is actually being measured. All that is really measured is whether persons reject an offer; it makes no reference to *why* such offers are rejected or not. But let us look at the study as the game that it is, and let us look at the matter phenomenologically, that is, as experienced by the participants. Volunteers enter the laboratory; perhaps one is given a milkshake, or perhaps they have been fasting from the night before. They are greeted and the situation is explained to them; they sit down, and a small amount of money is given to one, who then proceeds to split it. Another person is then asked to keep the offer anonymously proposed (and the rather puny amount of money that split entails) or to reject it. Both parties get up and leave.

I would like you to put yourself into this scenario and imagine the process through the eyes of the participants. Is it not bizarre? Notice how clinical and detached the entire situation is. Compare this with any real-life instantiation of appreciating fairness or responding to injustice. Imagine a person whose house has been foreclosed by a bank that was bailed out by the government, but who was then refused financial assistance by that very same bank. Imagine a person wrongly accused of a crime, convicted in a court of law and handed a custodial sentence. Imagine now all the manifold and various ways, both extreme and petty, in which a kind of unfairness can be proposed. What have all these real-life instantiations to do with the game that has been played? *What do such instances even have in common with each other?* How can a small fairness infraction—for example, getting a smaller slice of pie at a diner than the pretty person sitting at the other end of bar—compare with a major fairness infraction, such as getting sentenced to death for a crime one did not commit? How can such crude ultimatum games be taken as even vaguely capable of forwarding credible general conclusions with respect to fairness? The idea that two anonymous persons splitting a small pot of cash can be taken as representative of anything beyond itself, let alone the "willingness to punish perceived infractions of fairness norms" in any general sense is hard to take seriously.

As we have suggested throughout, the context of the matter is a powerful consideration that cannot be simply bracketed out of the equation. Justice infractions, perceived or otherwise, can be large or small, and the cost of punishing those infractions varies just as much. Sometimes resolving an injustice, particularly petty ones, involves nothing more than standing up for oneself, and sometimes it requires massive action at the societal level. In tyranny, it can mean sacrificing one's life. Again, the variation is massive. It just seems strange that anyone could think that the ultimatum game intimates any useful information about anything at all beyond the very idiosyncratic and narrow context in which the game is conducted.

Much the same can be said of Zak's studies using trust games, and Crockett's use of the trolley problem dilemma to investigate willingness to harm. The trolley problem dilemma, again, involves simply getting random persons into the laboratory, giving them some milkshake or asking them to fast, and then simply posing them a question. Unless one is Oedipus standing before the sphinx, moral decisions simply do not present themselves in this way (and even that was not a moral dilemma). As for Zak's trust game—studies supposedly indicating a causal relationship between being trusted and oxytocin—we are expected to accept that gathering together a group of young and energetic Princeton students, drawing blood from them, placing them in front of a computer, and giving them a miniscule amount of cash to play with, is somehow meant to realistically model the plethora of situations in real life in which trust is given and received.

Imagine for a moment all the times during the day in which trust is given and trustworthiness is shown. Consider for a moment all the different kinds of trust there are, all the unspoken, taken-for-granted moments and all the societal, functional elements of daily life that must go smoothly if life is not to descend into chaos—say, trusting that the car will work, that others in their cars will stop when the lights go red or drive on the correct side of the road. Then, imagine all the different kinds of relationships one can have and the different kinds of trust that are engendered within those relationship contexts—the trust of one's spouse, perhaps to be faithful; the trust of one's children to not do something dangerous and very foolish; the trust of one's parents, trust which changes its nature across the lifespan; the trust of one's friends or of strangers: these are all very different kinds of trust. They are qualitatively different, experienced differently, manifested differently, and, most importantly, it is context and situation which play a key and decisive part in making that kind of trust precisely the kind of trust that it is. Zak and his colleagues seem to think their studies are capable of speaking for all of these qualitatively different kinds.

These are not elements which can be simply "controlled" for experimental purposes; rather, they are constitutive of the nature of the phenomena being examined. The trust game ignores all of these considerations, yet, because it involves using a scientific method, the results, which indicate nothing beyond the bounds of the narrow and unreal context in which the study is examined, are taken with the authority that science has. This is an abuse of that authority. Ultimately, Zak's studies are zero cost, zero gain. It is worth speculating whether, if the trust game were played with higher stakes—that is, candidates had to trust monies in the hundreds or even the thousands of dollars (rather than a maximal take-home of $30)— the college kids participating would have been quite so willing to trust and be trustworthy in the experiments conducted. The larger situation of the candidates' lives, whether they were in financial need, their attitudes toward money, and so forth, are completely excluded. Observe how Zak and his team label anyone who does not perform in the game as he hypothesized as a "bastard," as someone who possesses "some of the traits of psychopaths" (Zak 2010, Honigsbaum 2011). These are strong words to label someone with, and for no other reason than their not playing the game as Zak wanted. Could it be that the "non-reciprocators" were having a bad day? Could it be that they were in debt and needed that money to pay someone back? Could it be that they just were not taking the game seriously, it being a game after all? Could there not have been a multitude of reasons why the candidates in question did not behave in a "trustworthy" manner, that is, as defined within this strange and artificial game structure? Perhaps simply labeling these people "bastards" and "psychopaths" was a little rash?

The problem is that these game methods are merely accepted means of investigation. Their validity, particularly in this context, is simply a "false truth," one of those basic axioms of the discourse that insiders to the domain had better not question if they want to get on well with their careers. And, ironically, the more studies that get built on such false foundations, the more zealously those foundations need to be defended in order to prevent the edifice from collapsing. Game setups, certainly of this level of clinical simplicity, are completely inadequate means for investigating the complex phenomena which are their objects of study. They simplify to the point of absurdity what is being investigated and strangle the context out of the situation, sacrificing reality and representativeness of the results for experimental control—leaving not even the bare bones of the phenomenon under investigation left over.

Functional Magnetic Resonance Imaging and Disembodied Reflection
Functional magnetic resonance imaging (fMRI) is an exceptionally powerful and important technique when used for detecting the effects and

progress of strokes, brain injuries, and neurological diseases. When used in such contexts, the value of the technology is simply beyond question or measure. Its use in a moral investigative context is more dubious. There are many critiques of the way in which fMRI machines have been used inappropriately in neuroscience (Tallis, for example, provides a powerful meditation on the misapplication of fMRI studies; Tallis 2012, 80–2), and certainly these problems are entirely relevant here in attempting to associate brain activity with moral traits and the influence of neurotransmitters in modulating these moral traits. Yet there are plenty of deep problems with the way investigators have gone about utilizing this technique.

In essence, there are at least two issues at stake which are particularly relevant for our present discussion: (a) the use of such machinery in moral contexts at all, and (b) the strange effect that fMRI evidence, and the colorful brain-scan imagery that it produces, can have on the public imagination. Precisely this is the neuroprimacy mentioned earlier: the assumption that neurological studies have some kind of general special priority as explanatory accounts. Of course, neuroscience does have a special primacy in answering certain kinds of questions (neurological questions, for the most part). However, it is not always made clear where the strengths of neuroscientific explanations are such that it should be our primary recourse, and those areas which are more dubious when thought of in primarily neuroscientific terms.

This has filtered through to the public, in particular wherein there seems to be an unquestioned assumption that neuroscience has the seemingly magical power to offer ultimate explanations for all human things—that once something has been accounted for in neuroscientific terms, one has thereby reached "bedrock," and while other modes of explanation may be helpful in "filling out" the matter or offering some interesting and informative window-dressing, the matter has been essentially dealt with and closed once the neurological correlates of the object of enquiry have been observed.

Is such an attitude of neuroprimacy justified in the domain of moral functioning? I would suggest not. It was argued above that game setups hijack the authority of science by using its methods, even when the methods are not appropriate to the particular subject matter in question. It is hard to find a clearer case of the authority of science lending undeserved credibility to research than brain-scan imagery in moral contexts. These scans can be convincing, and much work has been done in revealing the persuasive power of such imagery. The classic study is McCabe and Castel's (2008) work on the rhetorical power of brain-scan images—the "seeing is believing" effects of brain images on public judgments about the credibility of scientific reasoning. They write:

Presenting brain images with articles summarizing cognitive neuroscience research resulted in higher ratings of scientific reasoning for arguments made in those articles, as compared to ... bar graphs, a topographical map of brain activation, or no image. These data lend support to the notion that part of the fascination, and the credibility, of brain imaging research lies in the persuasive power of the actual brain images themselves. We argue that brain images are influential because they provide a physical basis for abstract cognitive processes, appealing to people's affinity for reductionistic explanations of cognitive phenomena. (McCabe and Castel 2008, 343)

Indeed, Tallis' and McCabe and Castel's meditations on the subject are part of a slew of further studies investigating the curious effect brain-scan images have on the public mind. This is due in part to the growth in the public profile of neuroscience. Trout's (2008) article on "neurophilia" examines how "non-expert consumers" of behavioral explanations "assign greater standing to explanations that contain neuroscientific details, even if these details provide no additional explanatory power" (Trout 2008, 281). Such "placebic" information is described as offering "counterfeit cues" of scientific credibility (ibid., 281).

Brain scans are appealed to for making some unjustified, simplistic claims. O'Connor and Joffe (2013a) cite many examples of this phenomenon in the public press, for example that mass murderer Anders Behring Breivik's cruelty is to be found lying "in his brain"; scientists have found in the brain a "misery molecule"; there are brain scans "explaining" crime; racism is presented as "hardwired"; we have images of "the brain on love"; there can be discerned "dark patches" inside the brains of killers and rapists; and rioters have lower levels of serotonin and are thus less able to regulate aggression (O'Connor and Joffe 2013b). Meanwhile, fMRI images of babies' brains are being used to construct and defend childrearing policies and judgments about good or bad parenting (apparently, the "hardwiring" of the baby's brain for emotional and social interaction is settled before the child is three years old, and irreparable damage is done if the brain is not cared for in specific value-laden, neuroscientifically justifiable ways (O'Connor and Joffe 2012).

To illustrate, O'Connor and Joffe's (2012) work inspects how fMRI images of babies' brains are being used to construct and defend childrearing policies and judgments about good or bad parenting. Such rationales are "soft" and financial in nature. O'Connor and Joffe (2012, 2013a) illustrate this with the government report "Early Intervention: Smart Investment,

Massive Savings," independently conducted by British Parliament member Graham Allen (2011), the poster for which features two pictures of brains, one looking large and healthy, underwritten with the word "normal," the other looking rather shabby and shrunken, underwritten with the words "extreme neglect." Next to the two pictures of brains are stacks of gold bars labeled "costs to the taxpayer," with the neglected brain costing in terms of low attainment, benefits, teen pregnancy, violent crime, shorter life, poor mental health, failed relationships, and alcohol and drug abuse. In contrast, the cost for the "normal" brain is only "early intervention." We have here an instance of precisely what I am talking about with concealed soft moral enhancement. The upshot is that financial motives are being used to justify programs of "early intervention" in order to prevent outcomes which have a strong morally related foundation—the idea that without certain interventions or certain child-rearing practices, what is produced is a neurological degenerate, a creature incapable of having relationships and liable to violence, mental health problems, drinking, taking drugs, breeding, and being unemployed.

The clear message throughout is that "we are our brains," and if we do good or bad—or in the present context, if we are moral or immoral—the necessary implication is that "my brain made me do it." And colorful brain-scan images are prevalent throughout to evidence this view. However, there are suggestions this neuro-fad is coming to an end. Yet if things are as Struthers and Schuchdardt (2013) suggest, and the overloaded cultural emphasis on fMRI imagery has created a "neuro-fatigue" in a public no longer impressed by such discourse,[1] even then it is important to make clear—to those for whom the limitations of these studies have not entirely filtered through—the deceptive rhetorical power such images have, the various failings of such studies, and their inappropriateness as a basis for measuring and potentially manipulating moral phenomena.

*

With that background made clear, let us instantiate these claims by looking at the use of moral-dilemma testing conducted using fMRI technology. The point to be made is perfectly simple:

Putting someone in a noisy fMRI machine and getting them to mull over a moral dilemma is nothing at all like how moral decision making actually plays out in the embodied human reality of moral choosing in the real world.

One thing that needs to be borne in mind is that we are dealing specifically with moral cognition here—that is, making moral judgments. We are not talking about actually being moral, or moral behavior, or anything with any consequences at all. Important as it is to limit one's area of focus, let it be noted, we have already shaved off thereby the relevance of this sort of data from anything other than (a) explicitly dilemma forms of judgment, and (b) the choice between generally consequentialist and rule-based thinking (assuming that these are even adequately modeled by the dilemma in question). Let us not forget that there are actually other ways of envisaging and measuring the value of moral cognition and judgment (for example, a virtue-ethics framework which relies more on practical reasoning and trained perception is entirely opaque to the process Crockett has used, as are situationist or cultural-linguistic practice-based approaches to moral functioning). Indeed, the entire dilemma method implicitly assumes that there are only two possible responses to the given problem—but the fact is that more sophisticated moral reasoning has to cope with greater complexity than "either/or" type problems can capture, so right from the get-go we have a very, very superficial setup.

Let us look further into how limited this kind of trolley study is. First of all, the particular harm in question is murder, or at least, pushing a person to their death or not. As moral judgments go, it is pretty extreme. Thankfully, for most of us, the decision over whether to cause or allow fatal harm to a given person is not one which has to be faced or even seriously considered. This should make us question the range of applicability of the study from the start. A doctor, a soldier, or a fire fighter may have to face the sort of situation where the value of one life is measured against another or a number of others, but let it also be noted that these sorts of decisions tend to be, to a certain extent, readymade. Questions of triage are part of a doctor's training. Indeed, this is much more like the point made about virtue above: a doctor is trained in advance about how to make decisions of triage so that he or she does not have to waste precious time in the moment making complex moral decisions. The basic framework for how to make such moral decisions is already trained and habituated into his or her mind—it is already decided in advance that it is right to measure one person's life against another, and how to do so, usually in terms of their chances of survival or the number of persons who might be saved. This is much more similar to how moral decision making is made in real life. Formed habits (good or bad), ingrained assumptions, and practiced ways of being and thinking snap quickly into place so that the decision-maker is not left standing there contemplating his options.

Moreover, in real-life decision making things are usually done in situ; that is, in an embodied, situational way, in which intuitions and immediate impulses, experience and training, habitual mental patterns, concepts of one's identity, and emotional and behavioral patterns take the fore rather than abstract cognitive processes of reflection. The fMRI methodology for investigating how moral judgments are made, at the very best, measures only those sorts of times when we take the time to distance ourselves from situations and sit and reflect abstractly about moral concerns which are clearly defined into one of two distinct options which have no mid- or long-term consequences (for, in the trolley problem, we do not have to worry about the psychic scars that pushing a fat man off a bridge or watching others get run over by a train will have—for example, guilt, insomnia, and so forth). Setting aside that how we think we will react in a given situation and how we actually do react can diverge considerably, all that is being measured, therefore, is a very specific kind of thought process which very few persons actually engage in, which indeed persons are very rarely called upon to engage in at all.

Whatever it is that the brain is doing in dealing with such a dilemma, and questions about the acuity of fMRI machines aside, it is hard to imagine how the images generated by this machine are going to indicate anything significant at all about moral cognition more generally. The potential for generalizability of such findings does not seem particularly extensive. Indeed, as Crockett writes: "Laboratory studies of human morality usually employ highly simplified models aimed at measuring just one facet of a cognitive process that is relevant for morality" (Crockett 2013, 1). This is true enough. However, I reject the idea that once simplified to this extent the measurements have any generalizable value at all. Moreover, I disagree with Crockett's claim that "these studies have certainly deepened our understanding of the nature of moral behavior" (ibid., 1)—quite the contrary, in fact. Such studies have given us the illusion that our understanding has been deepened, when, arguably, the superficiality of the methods, combined with the authority the domain carries with it, have served to undermine and in fact reduce our understanding of the real-life phenomena these studies are supposed to be representing.

Nor can we find the excuse that "we are limited in our methodologies by ethical considerations" particularly satisfying. There are, of course, ethical problems here; we have to use machines because we cannot allow participants to literally throw a fat man in front of a train and measure the brain activity involved. Such methods may well be the best we have available, but that does not mean they are worth using. This is the big problem. We use the best methods we have available, and there is no question that the various investigators have been rigorous and painstaking in their experimentation, but the point remains: if these methods lack all capacity to offer anything other than the most superficial, and arguably misleading, representation of what they are investigating, perhaps they should not be used in moral contexts at all. If we are tasked with building a bridge across a river and all we have available is a hammer and a small plank of wood, it will not do to say "well, this is the best we have." Rather, it is better to simply admit that "the best we have" is so incommensurate to its task that applying such methods leaves us only worse off than when we began.[2] The problem comes not just from the studies themselves, but rather from the philosophers who then appropriate these studies as given fact without the least critical inspection.

In any case, there is a yet more fundamental limitation inherent to the fMRI study here: fMRI machines are completely incapable of indicating participants' reasons for action in making their moral decisions. That is, the motivations of the person in question are absolutely opaque. For even if we were not constrained by ethical considerations, even if we could perform the trolley dilemma in real life and accurately measure the brain activity involved, it can never be clear from the scan itself whether the person makes their judgment about pushing the fat man in front of the train for moral reasons (it is right), or for immoral reasons (e.g., they hate fat people, or just enjoy the thought of pushing someone in front of a train), or for completely nonmoral reasons altogether (perhaps they are just squeamish, or generally indecisive).

In short, there is no way of knowing whether what is being measured is moral judgment at all—or, more likely, given how intermixed persons motivations usually are, to what extent the judgments being made are a compound of moral, immoral, and nonmoral considerations. In other words, such fMRI studies tell us precisely nothing at all about moral judgments regarding harm. Or, worse, they tell us less than nothing because they purport to be indicating something significant, when in fact they say nothing realistic about moral cognition at all. We are led to believe that "these studies have certainly deepened our understanding of the nature of moral behavior" (Crockett 2013, 1) when nothing could be further from the truth. We cannot even be sure that what is being measured is even a case of moral cognition or behavior at all.

The Experimenter Effect and Tainted Sample Pools Before we put aside the methods used in Zak et al.'s oxytocin studies, there is one last bias which is worthy of mention as a possible distorting factor. A case could be made with respect to Zak's trust game studies that there has been a degree of *playacting* with respect to the various candidates' participation in the various games, an expectation of playacting implicitly built into the way the experimental environment was constructed. This problem, the "experimenter effect," has been defined as

any of a number of subtle cues or signals from an experimenter that affect the performance or response of the subjects in the experiment. The cues may be unconscious nonverbal cues, such as muscular tension or gestures. They may be vocal cues, such as tone of voice. Research has demonstrated that the expectations and biases of an experimenter can be communicated to experimental subjects in subtle, unintentional ways, and that these cues can significantly affect the outcome of the experiment. (Rosenthal 1998, in Carroll 2014)

Indeed, if ever there were a case to be made for the "experimenter effect" causing problems, it would be here. Zak is charismatic, the topic of oxytocin is widely known, and anyone doing a study on the subject will already have been primed with enough information to undo the blindness element of the tests—particularly as time went by and oxytocin was subject to the media frenzy that it was, it is hard to imagine that the participants did not have some previous notion of what was being examined and what kind of behavior that was expected of them. Moreover, Zak ("Doctor Love," as those who have attended his conferences will know him) favors giving these affairs a distinctly party atmosphere. At conferences, co-workers and the crowd are "spritzed" with oxytocin, and everyone cavorts together hugging and laughing and having fun. This is the sort of spirit with which the business is done.

Given that Zak is adamant that, to be effective, intranasal oxytocin requires two tablespoons worth directly inhaled (this is how Zak reassures concerned parties that oxytocin cannot be used by nefarious tricksters to manipulate the general public into trusting them), I do not see how this "spritzing" and the consequent joyous and celebratory effects it has on the audience and participants can be anything other than playacting on the part of a rapt and overenthusiastic audience.

Certainly it is true that this playacting, if it occurred during the studies themselves, would have been a factor in both candidates primed with oxytocin and those given a placebo (even noting, of course, that the experiments were conducted double-blind). The sheer power of the experimenter effect, how loaded the public media were with wild claims about oxytocin, the accentuated quality of the party atmosphere in which the studies were conducted, and the high spirits of the various participants, which may already have set a psychological context for increased generosity, the attitude of "high jinks" with "Doctor Love" must surely have had some not inconsiderable effects on the outcome of the studies, against all efforts to the contrary.

Precisely what such effects might have been is impossible to say without repeating the experiments in a more sober context. Even then, it can never be excluded that the media furor surrounding oxytocin might not be playing mischief with participants' expectations in these studies. As the various "placebo effect" studies seem to show, expectations are incredibly powerful shaping forces in the outcomes of such studies. Zak's incredible success publicizing his research on oxytocin may well have caused great difficulties for enacting the blindness requirements of further empirical study where oxytocin's morally related effects are concerned.[3]

Straw Man Virtues

Excessive Simplifications

In fact, one could write an entire book on the flaws in the methods used in the various studies of the moral brain. I would suggest that the points expressed here are sufficient to give the reader an indication of how decisively flawed, unrealistic, and unrepresentative such studies are. This will have no small impact on the sorts of claims made by moral enhancement enthusiasts such as Savulescu, DeGrazia, Hughes, and so on, who have simply referenced these various findings without subjecting them to the least bit of critical thought. However, there is another considerable problem with such studies: the very concepts and definitions of the moral traits these studies apply are themselves too simplistic to be representative of the moral goods in question. Therefore, we have another level of vagueness to consider—vague constructs of traits like "generosity," "empathy," and "trustworthiness" are being used as markers for the real phenomena, and statements with respect to the former are being used as indicative of properties of the latter. To put things in their simplest terms: *moral concepts are incredibly hard to define comprehensively.*

In itself, this is not an insurmountable or decisive problem. Many things are hard to define. We have to do our best to make a workable construct, and then we explore that construct as best we can. However, significant problems arise when the definition used is just so vague that it cannot be used to adequately distinguish what is being inspected from other related phenomena that might get caught in the investigative net. As we saw with aggression in the previous chapter, even seemingly simple, everyday moral or immoral traits admit of such tremendous inner complexity that they defy easy definition. This creates problems for the aspiration toward any fine-grained moral enhancement. And when the empirical neurological studies over-reduce these complex moral phenomena to crude straw men with no inner complexity, the implication is that they no longer represent what it is they are supposed to be investigating, or the variety of ways in which the words can be taken to have meaning. Two examples will suffice to illustrate this point. (While we will be picking on Zak, it should be clear that the very same definitional problems are a family problem for this neurobiological investigation into specific moral traits.)

Objectifying the Subjective: Altruism and Generosity Let us consider Zak and colleagues' studies into altruism and generosity. We are told that altruism can be defined as "helping another at a cost to oneself" (Zak, Stanton, and Ahmadi 2007, 1), and generosity is defined as "'liberality in giving' or offering more to another than he or she expects or needs" (ibid., 1). Defined in this way, it becomes possible for Zak et al. to present generosity as "a subset of altruism." Zak illustrates this by saying "one may give a homeless person 25 cents (altruism) or ten dollars (altruism and generosity)" (ibid., 1). With this, one has to hand a rather neat set of seemingly sound definitions.

Problems arise when one looks deeper. For example, the two definitions here are not of the same order. Altruism is defined in terms of "cost," a seemingly helpful economic term which can be used to measure things in objective terms; a "cost," for example, can be measured in terms of money (as Zak does), or time, and so forth. We have here a helpful, objective way of simplifying the idea of altruism into readily measurable terms. But what of generosity? "Liberality in giving" is not an even slightly objective term— what is generous or "liberal" is rather relative to a number of phenomena, and the manner in which persons measure costs to themselves differs substantially. Generosity, qua "liberality in giving," being so relative and subjective, does not present itself as a readily measurable entity at all. Indeed, it is hard to imagine a nonarbitrary objective standard by which this definition can be measured or unilaterally applied.[4] So what we have here in essence, then, is an utterly subjective term being presented as a subset of an objective term: two radically different measures are being presented as objects of the same fundamental order.

Indeed, look how vague this operationalization of "generosity" is. For one thing, it is measured as if money were its primary or only measure. This is a significant family problem with the research into generosity and the economic games we have encountered throughout. Money is used almost exclusively as the primary measure of the moral good involved. This ignores certain social attitudes that, in fact, giving money is actually a rather cheap way of being altruistic or generous, one which saves oneself from having to get one's hands dirty (it is easier, for example, to give money to a hospice or care home than to actually physically look after the elderly). Could it be that there are other things some persons value more than money? And is it right to assume, even if the study were a valid indicator of willingness to part with money for a good cause, that this measure of generosity would transfer onto other goods not so easily quantified, such as generosity with one's emotions?

As an illustration, suicide phone-line operators and listening services (e.g., the Samaritans) offer a peculiar kind of generosity not readily measured in the sorts of terms with which the studies into generosity tend to define their variables. Samaritan phone-line operators have to give of both time and of their selves in a particularly intimate way. One of the conditions of being a Samaritan, for example, is that one must perform an "all-nighter" approximately once a month. This can be an extremely grueling experience, during which one is called upon to be attentive and compassionate throughout. Yet, the universal testimony of the volunteers is that these are the most rewarding of all their shifts. While staying up all night can be a brutal experience, the calls received at 4 a.m. can be the most powerful. Samaritans are called to bear the weight of having to take calls from an abused child, knowing that their next call might be from a child abuser. Both must be attended to with compassion. Samaritans are called to listen with absolute respect for the autonomy of the caller while he decides whether he wishes to live or die, and must sit by, sometimes, to accompany a person who has swallowed a fatal dose of pills, unable to call the emergency services (they have no phone number tracking), keeping that person company as he or she dies, so that person does not have to die alone. This unfathomable burden is often referred to by such operators as "a privilege." So, precisely how are altruism and generosity to be measured in such circumstances? Precisely what are the "costs" here? How are they to be quantified? A sleepless night, sharing the pain of another? And how are these costs perceived? Are they perceived precisely as costs? Or are they experienced, sometimes, in a way in which rewards and costs are indistinguishable? In this instance, neither the idea of a "cost" nor that of "liberality" can so easily be objectified in the way Zak et al. have. What does "giving more than expected" or "more than is needed" mean here? What is being given might seem "liberal" to one person, but to the Samaritan him or herself, it might be "just a thing" they do, and after a while not felt to be liberal or generous at all. Indeed, they might come to see their work as a gift that has been given to them, something rewarding in a way that cannot be adequately captured using abstract concepts *at all*—let alone the simplified generic constructs unilaterally applied in the empirical studies. What happens to the idea of altruism and generosity when, in reality, it is differentially blended with feelings of reward? How does this contrast with an altruistic deed done out a sense of duty? Looking at generosity and altruism in a particular context helps to illustrate how vague and unrepresentative the constructs used by Zak and others really are.

It is to be kept in mind that such comments do not apply only to Zak and these particular constructs, which are but a cipher for the larger problem—these moral goods are contextual, and incredibly hard to define in meaningful generic terms that have representative, real life sense. The mismatch between what can be investigated empirically given these neurocentric methods, and the tangled nature of the reality of the goods in question is simply too great for these studies to be taken seriously. But they are taken seriously. And this is why such studies worsen our understanding rather than improving it. We have seen much the same complexity in creating meaningful definitions applies to terms like "aggression," and the comments regarding trust, fairness, and motivation above only amplify the significance of these points. Moreover, this observation points toward an important factor which has been almost unanimously neglected in the studies considered and those related to them: the abandonment of phenomenology; that is, the subjectivity of moral functioning, what it feels like from the inside to be moral and the manner in which the subjective experience interacts with objective factors in shaping a moral response or judgment. These subjective elements to moral functioning are, of course, much harder to get a grip on in empirical terms. Certainly they are not so readily accessible to neuroeconomic techniques—and this is precisely why they are bracketed out. Simplifying generosity and altruism in this way makes them readily measurable, but by doing so, they become unrecognizable to persons' own experiences. As such, all Zak et al.'s study shows is that, within the very idiosyncratic and unreal circumstances within which the experiment was conducted, people sniffing oxytocin gave more money to charity than persons not so sniffing. Generalizing these results is not defensible. This suggests that it might be more appropriate to examine altruism and generosity through more subjectively amenable, qualitative methods.

Will the Real Empathy Please Stand Up?   In addition to the intrinsically subjective nature of many of the phenomena in question, there is also the tremendous conceptual complexity of the moral ideals in question to be considered. We have already seen how vague Zak et al.'s constructs of "altruism" and "generosity" are. Let us look at the manner in which Zak et al. deal with the idea of "empathy." Empathy here, in effect, gets used in the sense of meaning nothing other than "I feel what you feel, so I'm going to be nice to you." This comes from Zak's reading of Adam Smith. In Zak's words:

Social creatures share the emotions of others. If I hurt you I feel that pain so I tend to avoid that. If I do something that makes you happy I share that joy and so I tend to do those things. ... Smith was right on why we are moral, I just found the molecule behind it ... this shows us how to turn up this behavior and how to turn it off. (Zak 2011b)

But like all moral traits, ideas like empathy conceal tremendous complexity and suffer definitional problems which have plagued philosophers from antiquity. Indeed, when it comes to defining empathy, there are many competing schools of thought. Amy Coplan (2011), for example, in her paper "Will the Real Empathy Please Stand Up?" illustrates the differences of opinion here by providing a very complex and nuanced exposition of a number of ways in which the idea of empathy might be constructed and their various strengths and weaknesses. Coplan writes:

A longstanding problem with the study of empathy is the lack of a clear and agreed upon definition. A trend in the recent literature is to respond to this problem by advancing a broad and all-encompassing view of empathy that applies to myriad processes ranging from mimicry and imitation to high-level perspective taking. I argue that this response takes us in the wrong direction. ... I propose that empathy be conceptualized as a complex, imaginative process through which an observer simulates another person's situated psychological states while maintaining clear self-other differentiation. (Coplan 2011, 40)

Coplan argues that it is important to keep the various kinds of empathy "theoretically and conceptually distinct" (ibid., 40). I would suggest that Zak et al.'s studies exemplify what can go wrong when one is not clear and specific about the kinds of construct one is dealing with and when one does not carefully separate the various possible ways it can be understood. For Zak, empathy is implicitly presented, as Coplan puts it, as "a broad and all-encompassing view ... that applies to myriad processes"; yet, at other times, Zak's use of the term seems to indicate something as simple as "emotional contagion" (ibid., 40; consider the study which shows an emotionally charged video to the participants before testing, and Zak's reading of Smith above). Even Persson and Savulescu have managed to differentiate notions of empathic imaginative capacities from merely sympathetic responses (Savulescu and Persson 2012, 13). Thus, before one even gets to the issue of whether empathy can unambiguously be associated with morality at all—Zak et al. treats empathy as if it were almost constitutive of the whole of morality, yet it is not even fair to say that empathy, regardless of the definition one takes, is always to be associated with moral goods at all; for example, some people high in empathy *avoid* morally demanding situations precisely because their empathy causes them inner pain they wish to escape from—one has to beware that the construct being put in place by Zak et al. is very vague indeed, and can sometimes move between very different uses of the term without realizing the important differences that such uses might entail.[5]

As throughout, there are three levels of lesson to be drawn here: (a) Zak et al.'s construction and definition of moral traits is so cavernously vague as to render their work highly questionable and limited in the extreme. (b) This reminds us of a key feature of moral traits that has to be taken into account: their tremendous inner complexity, the variety of ways in which such terms can be taken to have meaning and can be expressed and how this changes the nature of what one is investigating, and the variety of ways they have meaning in different cultures and traditions of thought. Finally, (c) the more general worry to be raised with respect to the scientific study of moral functioning is this: such empirical study can only ever be as sophisticated as the moral concepts being used.

Before investigation into the neurology of moral functioning is to proceed, it is crucial that more dialogue with philosophers of the caliber of Coplan be conducted in order to avoid the sorts of very shallow definitions that the likes of Zak et al. have applied. Fortunately, these concerns, while they certainly need to be addressed by investigators in the future, are not decisive. Increased capacity to provide more fine-grained concepts of the moral goods in question for empirical analysis is something that researchers in this field, which is yet in its infancy, can work upon in order to deepen and sharpen future study.

Conclusion

Having explored at length the sorts of methodological problems which seem to be rife with the neuroscientific investigation of moral functioning, a case has been made for the claim that the current investigatory paradigm and techniques are not only not fit for purpose, but actually the present brain-reductive approach to studying moral functioning might be intrinsically incapable of handling such a complex set of phenomena as moral functioning encompasses. Which is to say, our understanding of moral functioning may well be worse off for all this study than better. At the very least, for now it is safe to say that the studies we have explored do not really indicate anything beyond the bounds of their own context, and thus they make for a very poor foundation if moral enhancement enthusiasts want to present a credible scientific basis for the neuromodulation of moral behavior and judgment through pharmaceutical or other means.

Striking the right balance between simplifying environmental conditions such that experimental control is possible, without unduly ignoring the sorts of key mediating factors which interweave when playing their various roles in shaping moral functioning in real life—mediating factors such as personal history, culture, and the whole range of individual and social conditions that make up the world in which the embodied moral actor is doing his judging or acting—one realizes that these are not the sorts of things which can be simply bracketed out.

Little wonder such studies are so ambiguous and conflicting. They purposefully ignore virtually every pertinent psychosocial detail that might be influencing the decision or action being investigated. Rather than revealing neural correlates to moral functioning, these studies serve only to show us how hard it is to tie down any single aspect of moral functioning for empirical investigation on the neural level. As such, sadly, most extant neuroscience regarding "the moral brain" forms part of a self-enclosed game which only makes sense within its own bounds, a chimera which misdirects empirical investigators from looking at and taking seriously the primarily contextual nature of the moral functioning in question.

This does not mean that moral functioning is completely beyond scientific, empirical investigation. On the contrary, if there is a solution—and I think there is—the answer could be as follows: scientific investigation into the parts of moral functioning is best approached through a *qualitative* mode of enquiry. What is required is a person-centered approach, one able to take context on board and observe the living, beating heart of moral functioning in the real-life kinds of scenarios in which it occurs. By following ordinary persons, exemplars, mentors, and role models, qualitative studies are going to be the most appropriate tool for subjecting something so contextual as moral functioning to scientific, empirical inquiry. Until the biological and neurological investigation into "the moral brain" embraces a distinct shift in its assumptions and dominant investigative paradigm toward a more "systems biological" or integrative approach, it is the social scientific realm alone that can offer a keen insight into the contextual reality of moral functioning.[6] The brain-reductive approach, taking blood samples, using fMRI machines, playing games, and tightly controlled neuroscientific methods which rip moral functioning out of its meaningful contexts, strangle it through excessive control, and thus, by the nature of the investigative methods themselves, distort beyond all recognition and meaning the moral phenomena being investigated: these are simply not up to the task of investigating their object of study. As far as studying moral phenomena go, such methods are simply not fit for purpose.

III   Faith

6 Moral Education and Faith

In a recent conference focusing on "the moral brain," one member of the audience stepped forward during the time given over for questions and, a little bewildered, hazarded a significant question. The essence of it was this: given that moral formation has been associated with the various faith traditions for thousands of years, and indeed, that so many still look to these traditions for answers regarding virtue, formation, and self-control, why is it that no one ever wants to mention faith in the context of moral enhancement? Why has religion been completely left out of the picture?[1] Her bewilderment was entirely appropriate, considering that in most other areas of enhancement discourse there are a range of nuanced theological and religious responses, some more generally in favor of enhancement and some more skeptical about the entire business (quite like secular debate, though with considerably different rationales being presented on either side). It is curious, then, that perspectives on moral enhancement from within the faith traditions have not been at all forthcoming. It seems that neither secular nor theological thinkers wish to engage with the potential implications for moral enhancement created by the nature of our multifaith world.

The lack of a deep and focused analysis of religious or theological perspectives, for better or worse, is something that needs to be addressed. It should be obvious that the idiosyncratic elements of certain faith traditions are going to shape how one is likely to frame to oneself the nature of the moral questions at hand. The various faith traditions cater to persons who identify themselves as members of those faiths. These persons number in the billions. Whether it be seriously, or nominally, the faith traditions apply to the overwhelming majority of the human race and shape the lives, to differing extents, of all of us, whether that be directly through practice and self-identification, or indirectly through the vast and largely anonymous cultural and historical influences that the various religions have had on the present world.

As such, it is indefensible to attempt to bracket out the faith traditions from discussion of moral enhancement. Rather, the simple truth is that religious traditions and institutions cast a massive shadow over the entire debate. Moral enhancement, in so far as it manifests, will do so in a world where religious institutions and worldviews have power and tremendous reach. This alone should make things such that all commentators, secular and religious alike, those opposed to religious discourse as well as those in favor of it, should be interested in how moral enhancement is likely to be worked through in religious contexts. How will moral enhancement manifest in a world where the various faith traditions and their institutions have such a powerful voice?

Christian Focus, Secular Eyes

Before we begin to think about how various religious traditions might respond to the prospects of moral enhancement, or where moral enhancement might have a genuinely useful place in a religious context, it is essential to get at least the bare bones of an understanding of how, ideally and in practice, moral formation is understood and interwoven into life and faith. To this end, the present chapter provides an exposition of the manner in which Christian moral formative concepts are inextricably bound up within the tenets of the faith and the manner in which they are inextricably linked to the salvatory nature of Christian practices.

It is important that some defense be presented as to why the Christian faith will be taken as the predominant voice here. The answer is pragmatic. Simply put, there is a familiarity within the Western audience with matters of Christian faith, much of which is absorbed by osmosis, and often unconsciously and anonymously. So whether the reader is for, or against, or indifferent to Christian faith, its terms should be quite familiar and readily comprehensible. While other faith traditions will be drawn upon throughout this chapter and the next, a sustained and single-pronged inspection of but one faith tradition is necessary for articulating sufficient depths required to make the requisite points (a "comparative religion" text on moral enhancement is a topic deserving a book unto itself).

The main concern is that the primary focus on Christian faith might be read as a tacit, if not explicit, presentation of Christianity as an inherently superior viewpoint from which to survey the moral domain. If that were my intention it would, rightly, cause antagonism in at least two significant ways. First, such a view might antagonize those coming from other faith traditions, who might interpret the focus on Christianity as suggesting something of the order that Christianity is "the best religion," or that it has privileged access to the truths necessary for best creating moral people. Second, it might be read by those who are not religious at all as participating in the current "science versus religion" debacle, as putting a dog in that fight, saying something of the order that religious approaches to moral concerns are inherently better than those which scientific modes of explanation can provide (where "science" is so often taken as a euphemism for "atheism" by participants on both sides of that debate).

The hope is that such worries will be dispelled by making clear the manner in which the overarching expository voice to be employed here will be secular in nature. That is, the voice will be of one who is determined to provide a nuanced and sensitive appreciation of Christianity's various idiosyncrasies, someone interested in Christianity, while remaining agnostic with regards questions of superiority. The sorts of insights the discussion will explore do not necessitate readers to take upon themselves, nor shed, any particular religious commitments they may or may not have. It is hoped that the expository voice employed is both sensitive, though secular, enough to accommodate both those who do have strong commitments to a Christian faith, as well as those who do not, by giving a strong voice to the former without asking the non-Christian reader to make any commitments to Christian faith.

For those with a stronger antipathy toward religion more generally, the following chapters might be envisaged as a necessary if undesirable attempt at social archeology. The religious angle on moral enhancement simply cannot be ignored. So, whether it offends thinkers with an antipathy to religious subject matters or not (as Nietzsche put it, when reading the New Testament he felt like he needed to be wearing protective gloves—but read it he did), it is simply unacceptable to skirt the issue of how religious faiths shape—in very complex ways—the visions of moral living of those that inhabit these faiths when such persons make up such a huge proportion of the human population.[2]

Above all, the point that needs to be conveyed is the profoundly idiosyncratic nature of religious faith and how moral enhancement discourse is absolutely required to confront such idiosyncrasies. Such a confrontation can only be delayed for so long, and it is better for an engagement to occur at the conceptual stage, rather than later, in implementation, discovering that religious communities will make quite different use, and provide very different rationales, for moral enhancement than may have been proffered in the philosophical literature. Such faith communities are not going to be neutral on moral questions, nor upon questions regarding moral formation.

The idiosyncratic, nonornamental nature of religious thinking about moral living, moral interpretation, and the involvement of religious scaffolding has to be considered a part of the very foundation of practical moral enhancement discourse. As has been suggested throughout, moral enhancement cannot exist as a free-floating entity (as if apolitical, or here as a-religious), but rather needs to recognize the nature of the ground upon which it is to stand and build. One of the great strengths and weaknesses of liberalism is its tendency to assume a level ground for all. Yet for those who cleave strongly to a particular faith, culture, heritage, and set of moral ideals, this ground is not at all flat. Such ground is already shaped by concepts, cultures, and the moral outlooks of those who cleave strongly to them.

The biggest misunderstanding I wish to dispel in advance is the idea that moral enhancement can or will be done in a way that simply brackets out faith. This is another practical reality the discourse must face. Rather, a strong vision of moral enhancement will and must be understood in a way that can cater to the billions of persons who self-identify with one faith tradition or another, a great proportion of whose faith is not merely cosmetic, but rather part of their identities, and who will not be satisfied by a generic account of moral enhancement which attempts to simply ignore crucial tenets of their faith. The value of inspecting the religious groundwork upon which moral enhancement will come to be built will therefore help in providing a more realistic picture of the landscape, as well as a more realistic picture of the sort of moral enhancement that is likely to come about as it emerges from such a preshaped and contested ground.

Discipleship: A Distinctly Christian Account of Moral Formation

Then he said to them all: "Whoever wants to be my disciple must deny themselves and take up their cross daily and follow me."
—Luke (9:23, NIV)

The Shape of One's Life

One of the simplest ways of grasping what makes a Christian account of moral formation specifically Christian is to understand the manner in which it takes Christ and Scripture as its central reference points. In the broadest sense, Christian moral formation is concerned with the shape of one's entire life and is captured in the quintessentially Christian notion of discipleship—responding to Jesus' exhortation of "follow me!" Indeed, for many moral theologians, the creation of disciples is the very end and goal of moral theology itself.[3] What then is discipleship? One answer is provided by O'Connell (1998): discipleship involves an ongoing relationship with God and Jesus wherein one identifies oneself precisely as a follower. Discipleship is an ongoing engagement involving ongoing commitment, enacted fidelity, and affiliation with others similarly committed (O'Connell 1998, 142).

In its most ideal form, discipleship involves a total shaping of one's entire life, top to bottom, and it involves processes aimed at realigning one's attitudes along with efforts to break and reform one's dispositions and habits such that they are more appropriate to the example of Jesus Christ (albeit Christ as hermeneutically interpreted with respect to modern living). There must be a shaping of intentions, purposes, and ends such that they become commensurate with Christian faith. No less than all of this is required to make the process of discipleship one of specifically Christian character. Indeed, this list of requirements, partial as it is, still scarcely captures how total the notion of discipleship is, how fully it is to be interwoven into all parts of life and living—and, ideally, nothing is to be excluded from this process.

The Kingdom of Heaven

The next point that gives Christian moral formation its specifically Christian shape is its eschatological focus—its focus on our role in bringing about the Kingdom that God is said to have been glimpsed in the figure of Christ. It is important to note that the Christian narrative locates contemporary persons as part of a past, a present, and a desired future. In this way our values and understanding of moral functioning are not to be limited to the present times, to present culture (which, currently, is quite despairing with respect to the idea of some ultimate or final meaning, value, or truth), but rather Christians are taught to constantly understand themselves in relation to a history which shaped them, and toward a future whose realization they are meant to be working cooperatively to bring about.

A Christian moral enhancement must keep in mind the history and places from which its various traditions have come and have at least some sense of the larger good, the Kingdom, the bigger ideal toward which such enhancement is to be directed. That is, a hard moral enhancement in Christian context must absolutely make reference to the Kingdom, which must be its ultimate terminus and the gold standard to which the ideals of moral enhancement are to be held accountable.[4] This Kingdom is the larger eschatological context in which discipleship itself is embedded. Christ is understood as having worked to inaugurate the Kingdom and as having given us a glimpse of the ideals of peace, compassion, and reverence that we are to emulate in our own temporally appropriate ways, and we are to be his disciples in attempting to further that vision in some authentically Christian manner. Christian moral formation is thus, at least in part, about learning to locate oneself "within the overall ongoing narrative in which Christians themselves are called to live and play their part" (Wright 2010, 229).

Within this narrative are embedded certain specifically Christian values and specifically Christian formulations of how those values are to be expressed (these are "the new clothes" spoken of by Saint Paul: compassion, kindness, humility, meekness, patience, forbearance, forgiveness, and above all love; Wright 2010, 125). The key point to be noted here is not that these values are themselves Christian-specific, for one finds much the same values expressed and justified in a range of different creeds, both religious and secular. Rather, what makes these values specifically Christian is the manner in which they are modeled by and referred to Christ, the ways in which they are true to Christian scripture and the present world, and the way in which Christian history and its present relation to contemporary culture shapes the range of legitimate expressions they can take. Christian communities stand embedded within and in relation to "host" cultures, to which they are both accountable to and the judges of (Kelsey 2009, 6), and in relation to which they are to balance the poles of accommodation and sustained social critique. Christian moral formation then focuses on specific values, conceptualized in a specific range of ways, and stands against certain other value systems which might be more attractive, seductive, or popular in contemporary societies.[5]

The quintessentially Christian character of the virtues presented as Christian ideals, the manner in which they are not generic ideals but located inextricably within Christian narratives and within background Christian sense-giving frameworks is important—and we will go into considerable detail as to how the specific shaping that the various Christian discourses give to their value-ideals make them not only specifically Christian, but present difficulties with respect to how fine-grained a Christian-appropriate mode of moral enhancement intervention must be. It will later be argued that a great deal of moral enhancement discourse treats virtues and ideals as if they are "from nowhere"—that is, as if what makes for the appropriate expression of a particular virtue does not need some sense-giving background to shape it into the particular expression that it is. In fact, to do justice to Christian virtues a concerted effort must be made to engage with the very specific conceptual frameworks which make these Christian virtues the specific virtues that they are.

Christianity as Eccentric Existence

Another core feature of Christian faith and practice can be articulated in terms of its distinctly outward-facing, or "eccentric," nature. Kelsey writes of Christian faith that it presents

a picture of human existence as eccentric, centered outside itself in the triune God in regard to its being, value, destiny, identity, and proper existential orientations to its ultimate and proximate contexts. (Kelsey 2009, 893)

What this amounts to, in essence, is a way of "getting over yourself." It is a way of saying that your true well-being, what makes for the best kind of life you can have, is to think much less about yourself and focus more on God and the others around you. This is not so much about denigrating yourself; rather, it relates in part to the simple observation that self-absorption breeds an endless series of anxieties that persons embodying a more giving, generous disposition are less victim to. Indeed, emphasizing the outward-facing nature of Christian faith is a very contemporary theological antidote to self-obsession and the tremendous anxiety that comes along with the constant absorption in one's self and one's own life.[6] This outward-facing focus involves a very different way of thinking compared to the dominant consumerist model in which we are currently embedded. As Tom Wright suggests, Christian salvation is not about you, not about your happiness, not about your fulfilment, or your self-realization; it is about the kingdom, God, Christ, and giving yourself away generously. Consequently, the process of growing in virtue

is precisely to grow in looking away from oneself and toward God on the one hand and one's neighbour on the other. The more you cultivate these virtues, the less you will be thinking about yourself at all. (Wright 2010, 176)

As such, this outward-facing quality of Christian faith stands in direct opposition to the distinctly egoistic moral outlooks which implicitly pervade much of contemporary culture and thinking.

Religion and Salvation

These points about how the Christian idea of discipleship is intended to condition the very shape of one's life, a shaping which is to be drawn toward the idea of the Kingdom, make up a considerable part of the salvatory structure of Christian faith. While each faith tradition has its own idiosyncratic account of what salvation is and how to obtain to it, the idea of a salvatory structure to religious faith is one of the few universals in the established faith traditions more generally. Thus, the manner in which moral education and moral living are to be understood as necessarily intertwined with the salvatory impulses is idiosyncratic not so much to Christianity but, one might suggest, is one of the fundamental idiosyncrasies of religious outlooks per se. Indeed, some have argued (e.g., Hick 1990) that while the various religions may well bear a variety of family resemblances among themselves, the soteriological, salvatory structure of the faith traditions may be the one and only universal that can be drawn upon for defining a "developed" religion as a religion. Hick writes of the "developed world faiths":

They offer a transition from a radically unsatisfactory state to a limitlessly better one. They each speak in their different ways of the wrong or distorted or deluded character of our present human existence in its ordinary, unchanged condition. It is a "fallen" life, lived in alienation from God; or it is caught in the world-illusion of maya; or it is pervaded throughout by dukkha, radical unsatisfactoriness. They also proclaim, as the basis for their gospel, that the Ultimate, the Real, the Divine, with which our present existence is out of joint, is good, or gracious, or otherwise to be sought and responded to; the ultimate real is also the ultimately valuable. ... In each case, salvation or liberation consists of a new and limitlessly better quality of existence which comes about in the transition from self-centredness to Reality-centredness. (Hick 1990, 3)

So, while Christianity is certainly eccentric, the diminution of self-involvement seems to be a core feature that traverses all of the faith traditions in one form or another. Moreover, this eccentric feature of religious practice and philosophy is tied in directly and explicitly to the salvific or liberation-producing telos of the religion at hand. If we are to think through a religiously aligned moral enhancement, the very core of it will have to be enhancements designed to enshrine this idea of diminution of self, though in a way which is fine-grained enough to be amenable to the important differences in the conceptual framework idiosyncratic to the faith tradition in question.

For example, the diminution of self in Christianity interweaves with a very different set of ethical obligations such as one finds in, say, Taoism. While Taoism does indeed have a strict ethical structure, it does not arrange its precepts in the explicitly interventionist terms that one finds in the more missionary-based or evangelical traditions of Christian faith. Similarly, the diminution of self in Christian faith takes a completely different rubric for that diminution than is found in, say, Buddhism, where the diminution of self arises in part from the position that there is no enduring self to speak of anyway, only the circular, restless pursuit of gratification. As such, even though some movement toward self-diminution is universal to the faith traditions, no generic biomedical intervention is going to be capable of shaping that diminution in a way that can universally traverse those same faith traditions. The diminution of self is expressed, justified, understood, and brings ethical obligations in ways unique to each particular faith and the various denominations therein. These are precisely the sorts of complexities that a religiously sensitive, hard moral enhancement would have to contend with.

Again, it is in relation to such a salvatory background that such moral enhancement must understand itself. This larger background, this idea of religion as salvation, is completely lacking from the empirical work conducted on "the moral brain," for example, which makes no reference at all to the manner in which a person's religious faith may or may not be modulating their responses to the various tests that are applied. There is no trace at all of this idea that the moral life, for religious persons and communities, has a fundamental conceptual background. There is no accommodation at all of this ultimate motivation, which in no small part gives shape to the foreground moral valuations and visions of the moral life that are pursued as a part of that overarching salvatory narrative. These are not the sorts of concerns that are appropriately bracketed out from either philosophical or empirical study of moral judgment and action. Enthusiasts proposing a strong vision of moral enhancement must by implication propose some rationale for how moral enhancement might be understood as contributing to the salvatory structures idiosyncratic to the faith traditions of those upon whom such strong visions of moral enhancement are to be impressed.

"All the Way Through": A Virtue-Based Account of Christian Moral Formation

What Is a Virtue?

In his discourse on Christian virtue, Tom Wright uses the image of a stick of Brighton Rock, a candy which, wherever one cuts it, still has the same writing running right through the very center. While perhaps more than a little idealized,[7] this is a good image of how virtue and character are supposed to be—one has practiced and internalized the virtue in question to the point that it becomes part of one's very center: it runs "all the way through," and regardless of the circumstance, the virtue is manifest simply because one is so well trained, in such thorough congruence with the virtue emotionally, intellectually, bodily, in every way in fact, that it truly runs through to the very core of one's being.

To use more technical language, a virtue is to be understood as an "appropriate internal style," a "characteristic way of behaving," or "a developed skill" (O'Connell 1998, 40), whereby persons are able to perform acts with both authenticity and fluidity. A virtue is an integration of a variety of activities which the person is able to perform predictably, almost automatically, and with a high degree of success (ibid., 40). Herein, formation of virtues involves a honing of the will, a process of habituation, a focusing of intention, so as to make certain kinds of choices characteristic of the person in question. O'Connell continues:

> Virtues are integrated human skills whereby a person is able to, and tends to, perform genuinely good acts. Indeed, the person is able to, and tends to, perform these acts in a predictable and integrated fashion to such a degree that the good acts can truly be described as characteristic of the person in question. (O'Connell 1998, 42)

A virtue manifests then as both ability and inclination to behave in a certain way—and this is where the idea of "happiness through virtue" comes from (though "happiness" is a very poor translation of the original Greek word *eudaimonia*). The sense is that one has a conception of what is good for the human being to do, something quite unnatural compared to our usual conflicted animal impulses, and we then train ourselves to the point where we genuinely desire to accord with that conception of what is good. Once so trained, in principle, we are able to find fulfilment through doing the good.

Christian Virtue

Returning to the idea of discipleship, the project of inviting discipleship can be understood then as a project of cultivating virtue, by attempting to stimulate sincere inclination to live in the way of Jesus Christ. If the good for the human being is eccentric, located in God and God's desired future, the way to live a good life is about learning to align oneself with a higher way of being, which, once habituated, is profoundly more fulfilling than one's basic, immediate, aesthetic search for constant gratification. The idea, then, is to place oneself in God, through Jesus Christ and led by the Spirit. Thus, we can see clearly that Christian virtue is not simply a version of virtue ethics which has Christianity "tacked on," but rather the Christocentric and Spirit-led nature of the telos, the goal, the good to which one is aiming decisively to alter the nature of the task. Christian virtue is not simply about making habitual a certain value or good; it is about the entire reshaping of character in a specifically Scripture-oriented and likely Trinitarian way. Encouraging moral formation, then, is a project of providing opportunities for encounter and exploration, imitation and practice, which will nurture the ability to shape one's life and identity as a disciple.

The Habitus: Bodily Powers, Intellectual Capacities, and Affective Virtues

Continuing our own Christian "moral anatomy," we can mention that the development of Christian virtue and excellence in enacting various Christian practices require the acquisition of a variety of competencies and dispositions. Such competencies must be formed by various disciplines (Kelsey 2009, 355). Kelsey writes on this matter:

The lives of personal bodies can be formed in the most amazing variety of ways by various types of discipline of their rich and complex array of interrelated capacities. Such disciplines shape and nurture human capacities into definite competencies, both dispositions and powers to interact in determinate ways in [one's] proximate contexts. I use "act" broadly here to cover acts of intellect, perception, and the affections as well as intentional bodily acts.[8] (Kelsey 2009, 355)

In short, one begins with individual acts; over time, these acts form settled dispositions; and these settled dispositions knit themselves into a larger, more integral whole—character—when unified together under the heading of the project of discipleship. When formed, these virtues become interwoven into the very personhood of the individual so forming. Virtue, then, is not about isolated moral deeds, but rather about the creation of a moral identity, character, which involves the entire range of faculties and powers that we associate with the human person—and which results in the transformation of head and heart. As Cardinal Henry Manning put it: "It is the formation of the whole man."

This virtue-forming work is not something that can be done overnight, then; to the contrary, it is a long-term project, the work of a lifetime, and it involves honing the entire range of human powers, including those relating to bodily action, discernment, and intellectual capacities, as well as the affective capacities; that is, how one is to learn to be emotionally engaged by the objects to which one aspires. Arguably, this is never brought to completion; for the engaged Christian the process is one being constantly "in via" (McDonagh 1998, 15). Understanding moral formation as a long-term project, one which involves development, transformation, and above all time, is crucial to the moral formative question.

Renewing the Mind

If the mind is downgraded or lost then we are less than fully human.
—T. Wright (2010, 149)

One of the most famous exhortations one finds in the epistles of Saint Paul is to do with "the renewing of the mind," learning how to think properly, how to discern appropriately the nature of the good, and how to comport oneself appropriately before God and before others. The exhortation goes as follows:

Be transformed by the renewing of your mind ... so you can work out and approve what God's will is, what is good, acceptable and complete. (Wright 2010, 134–5, his translation; Romans 12:1–2)

One of the key and essential components to Christian virtue and Christian living more generally (no doubt some will be surprised by this, given the vocal prevalence of the more explicitly anti-intellectual strands of Christian thought) is the development of one's intellectual and cognitive capacities. Intellectual rigor, powers of discernment, capacities to think rationally and with sharp analytic skill are all central to a process of Christian wisdom (Wiseman 2013), which involves negotiating many countervailing cultural imperatives and problematic tendencies both outside the Church as well as from within it, as well as from within one's own community and one's own self. A refusal to learn how to think properly and in a disciplined fashion can manifest as a refusal to question authority or to question one's own problematic tendencies, and rather use poor rationalizations to defend what a more rigorously critical outlook should bring into question (even then, it is not as if intellectual rigor is any firm guarantee against such deceptions).

Apart from the arduousness of disciplined rational thought, it is unsurprising that anti-intellectual tendencies are highly prevalent in the more conservative branches of religious institutions (i.e., do not question authority), as well as in the more hedonistic excuses for spiritual practices (i.e., do not question your motives in seeking this gratification, just call it "spiritual"). Indeed, it has been seen on many occasions that certain practices (e.g., mindfulness and practices involving cutting off discursive thought) have been misappropriated and used abusively as means for making sure devotees do not question their "spiritual" masters, and many a scandal[9] has occurred based decisively on these two anti-intellectual tendencies—refusal to question authority and spirituality as a search for gratification.

Yet the emergence of such explicitly anti-intellectual strands of the faith traditions is likely to be a highly complex story. Christianity, for example, has a long history of pietistic, charismatic, and emotion-driven movements, yet none of these traditions which emphasize "the heart" are necessarily averse to intellectual knowing per se; rather, they seek to reinstate sincere affection and simple faith above excessive focus on doctrinal disputes. This reaction against the potential drowning of spiritual sense in heavy theological speculation is very different from the ultraconservative fundamentalism which absolutely refuses to engage with the sciences or philosophers, and which is the real target of the fundamentalist atheism that it shares so much in common with.

Indeed, this internal dispute between anti-intellectual tendencies and the all-out love of wisdom, knowledge, and truth is embodied in virtually every world religion. For example, with Buddhism, which has many traditions and forms, it is similarly difficult to pinpoint where exactly some of the more contemporary forms derive their excessive emphasis on the removal of "discursive thought." In fact, in authentic Buddhist practices, mindfulness is always to be supplemented with "the sword of wisdom." That is, the process of analytically discerning the nature of one's habitual and inward processes in light of Buddhist scripture is the necessary counterpart of silent observational meditation ("vipashyana, or awareness, in how to hear the teachings"; Trungpa 2003, 2; more specifically, how to hear oneself in light of the teachings). By point of contrast to those more contemporary Buddhists who overemphasize the "nondiscursive" dimensions of practice at the expense of all else, the celebrated and mischievous Tibetan Buddhist teacher Chögyam Trungpa (2003), in his aptly titled book *Training the Mind*, observed:

Sometimes talking to yourself is highly recommended, but it obviously depends on what you talk to yourself about. In this case, you are encouraged to say to your ego: "You have created tremendous trouble for me, and I don't like you. You have caused me so much trouble by making me wander in the lower realms of samsara. I have no desire at all to hang around with you. I'm going to destroy you. This 'you'—who are you, anyway? Go away! I don't like you." ... Talking to your ego, reproaching yourself that way, is very helpful. It is worth taking a shower and talking to yourself in that way. It is worth sitting on the toilet seat and talking to yourself in that way. ... Instead of turning on the punk-rock, just turn on your reproach to ego instead and talk to yourself. ... That is the best way to become an eccentric bodhisattva. (Trungpa 2003, 74)

As such, disciplined appropriation of the discursive faculties, study of scripture, and appropriate use of verbal contemplative techniques are as fundamental to an authentic and deep Buddhist practice as direct, nondiscursive, mindful awareness. The two are necessary complements, in fact.

Even more than with these traditions, Islam and Judaism are renowned, historically, for their love of knowledge and wisdom. Islam, of course, is known for preserving and redistributing the Hellenic classics unto the West. "Oh my Lord! Advance me in knowledge" (Ta-Ha 20:114): this Koranic verse is taken as the embodiment of the Islamic thirst for wisdom. What is most pertinent here is that Islam has made the search for knowledge a distinctly moral obligation. The pursuit of knowledge and wisdom is thus not an incidental part of Islam, according to the Hadith, which states that "seeking knowledge is obligatory upon every Muslim." Thus, it should be impossible to grasp a strong project of moral enhancement that is appropriate for devout Muslims without making reference to the pursuit of knowledge and the refinement of the mind.[10]

Judaism similarly has an entire genre of scripture known as "Wisdom"—literature which contains everything from the simplest proverbs, rules of thumb for life's day to day dealings; to the more existential musings of Job, with his insistence that though there be no ultimate justice in the human world, this world and its creator are to be loved in themselves and for their own sake anyway; and then onto the refined, quasi-Epicurean pessimism and view of life to be found in Ecclesiastes, that big dreams are all vanity and that the only things that really matter in this world are family, good friends, reverence, and enjoying the earthly fruits of one's labors. The practical wisdom and intellectual wisdom traditions of Judaism which span matters of both heaven and earth remain strong. Indeed, a love of hearty intellectual debate is a beloved characteristic of Jewish culture and family life to this day.

In sum, virtually all of the faith traditions, despite their very noisy anti-intellectual outcroppings, place the refinement of the intellect and practical wisdom at the very core of their religious practice. That is, the refinement of the mind and the gaining of knowledge are religious activity, a form of worship, a distinctly moral good, and so connected to the fundamental drive toward the salvific state that each of the faith traditions points toward it in their own ways.

We saw in part I that there is a debate in moral enhancement discourse regarding the overemphasis on emotional rather than cognitive dimensions of moral functioning. This is a danger for these anti-intellectual brands of the faith traditions. The appeal of a technology which applies only to the affective powers is liable to be particularly strong for religious groups who embody distinct anti-intellectual tendencies, who not only do not recognize the cognitive components of moral formation, but who reject outright the value and need for disciplined intellectual competencies. As Wright observes, love is a thought-out habit of the heart, not "undisciplined squads of emotion" (Wright 2010, 136); it is rather something in which the heartfelt emotions and mental discernment—both head and heart—have been brought to bear. While, of course, our intuitions are very important, the rational faculties must be brought to bear upon these intuitions in order to shape and refine them, or in fact to reject them when they go astray. In this

Wright places himself within a much longer philosophical tradition—from Plato, to Descartes, to Wittgenstein—one which recognizes that both intuition and ratiocination are required if fully wise discernment is to manifest (Osbeck and Robinson 2005, 79–80).

It could be that this diminution of the cognitive emphases in certain strands of the various faith traditions is a response to the immense discipline it demands, and thus is marginalized by institutional religions which are now very much part of the marketplace and who see themselves as having to "sell themselves" if converts are to be won. The advantages for authoritarian forms of religion in encouraging their followers to not think, and the advantages for certain persons who need to have their answers to life's questions predigested and spoon-fed to them suggests that there are numerous interests who benefit, superficially, from encouraging anti-intellectual readings of the tenets of the various faiths.

In either case, this importance of a cognitive dimension to religious practice and belief suggests that some justification for cognitive enhancement might be brought to bear in the religious moral enhancement debate. In general, the association of cognitive enhancement with moral enhancement has, rightly, met with little favor. However, when it comes to the faith traditions, actually, the cognitive elements of moral functioning, which have a broader salvific or liberation-directed background, are much more closely aligned. As such, a more modest case for John Harris' overlaying of moral and cognitive enhancement may well find greater justification in religious interpretation of moral enhancement than with respect to the more secular visions of moral functioning.[11] Which is not to say that religious persons are in need of cognitive enhancement simply by virtue of the fact that they are religious, but rather that those religious persons who are already anti-intellectual ought to be especially wary around visions of moral enhancement which further neglect, or de-prioritize, the use and development of cognitive and reflective powers in favor of affective manipulation. The problem is that one requires one's reflective and analytic powers to be able to fully appreciate the value of such powers in the first place.

Transforming the Heart

While the reasoning mind is an essential part of virtue formation, so too is bringing the heart into congruence with what has been discerned to be "the good." This transformation of the heart, or in more technical language, gradual refinement of one's various affective responses, is—as Wright suggests—a crucial tonic to the sort of ideal which favors unfettered spontaneity, unchecked emotion, and childlike (or childish) faith that what is "in the heart" is naturally good and wise and must be followed above all. Indeed, these crasser sorts of intuitivism are the bane of philosophers everywhere because they are so deeply entrenched in so many culturally valued goods. Placing too much focus on "the heart" is a convenient way of abrogating oneself of the responsibility of having to think clearly about what "the heart" is instructing one to do.

There is a big problem when it comes to questions of the heart. On the one hand, intuitivism points to something important; one does rely on one's intuitions for gathering morally salient information about one's situation. Moreover, we should reject the tendency, common in many strands of recent psychology, to attempt to demean intuitively gleaned information wholesale as products of some kind of naturally inferior process to the higher, rational cognitive powers (Osbeck and Robinson 2005, 79), rather than understanding them as necessary complements. Kahneman (2012) suggests, for example, that the majority of one's decisions are made automatically and pre-reflectively, rather than as a process of detached purposive reflection, so we rely on intuitive processes much more than perhaps many of us realize. If we had to reflect upon everything that our "fast and frugal" automatic thinking processes for us before acting, then we would never get moving at all. Most of the time, these automatic processes serve us well, but in more complex situations they can be less reliable.

The fact is that both intuitive and reflective processes can go awry. They both have their difficulties. Perhaps the best course, then, is not so much to belittle intuition per se, but rather to be more modest and simply grasp that it is not infallible. It is even problematic to suggest, as Wright does, that "the heart" is reliable only once it has been reformed or reshaped—for without initial intuitive guidance, we cannot even grasp what is morally salient and in which direction we are to go. The fact is that intuitions are the essential starting point for the moral life. What is problematic is valorizing the heart above all things or thinking that it does not need refinement and interrogation. I think the truth is that intuitive processes are both (a) messy and sometimes unreliable, and (b) indispensable in getting started, in providing the initial material with which a vision of the good can be formed, the initial material with which to reflect upon with one's analytic powers, and with which to aim toward. Note how this proposed characterization articulates a "complement/partnership" narrative rather than a "conflict/domination" narrative which separates intuition or emotion from reason and sets them against each other, as in the "faith versus reason," or the "faith is the *enemy* of reason" rhetoric so popular today. Throughout one has to recognize that, yes, faith can and sometimes does sin against reason, but that our reason is absolutely lost without access to this deeper, more refined, intuitive sense.

Quite often this discourse over the value of the heart comes down to two main bases: either, in practice, a general laziness that does not wish to use the mind in a disciplined fashion and thus elevates "the heart" as a way of patting itself on the back for just following knee-jerk reactions and not thinking deeply about anything at all; or, more philosophically, the issue comes down to two different views of human nature. Talk of "the heart" often polarizes into one of two extremes, an "optimistic" vision which sees human creatures as "basically good," and "if they can only follow their heart" they will find all to be essentially well; and a contrary, "pessimistic" vision, which sees the human heart as fundamentally corrupt and disgusting in nature. Of course, there are a range of mediating positions, and, indeed, neither extreme is particularly satisfying. Yet this split between optimistic and pessimistic assumptions about human nature is extremely significant when it comes to the moral enhancement debate. The extent and ferocity with which one espouses the urgent imperative to morally enhance is going to have some background against a certain view of who and what the human creature is. These variously optimistic and pessimistic visions have powerful religious correlates which drive them, and which, thereby, will drive certain visions of moral enhancement. Hughes' optimistic, liberal democratic techno-Buddhism is going to express a very different vision of moral enhancement than a version of Christian anthropology which is pessimistic and soaked in original sin.

In any case, the upshot is that the heart must be educated. Unless one takes a radically optimistic perspective, the implication will always remain that the heart can benefit from being reshaped to some extent, from being appropriately trained to cut off certain natural responses (e.g., pleasure in the suffering and failure of others), and from reinforcing various other natural human responses. Buddhism for example, takes the approach that human compassion is a natural state of our being, and thus its practices are not to create compassion but rather to expand that part of ourselves which is naturally compassionate anyway. Habits to be formed obtain, then, to that end.

Yet, it is not just that new habits need to be formed and old habits broken. Rather, it is that the very nature of the way we respond to things at the deepest, most visceral level needs to be reformed so it corresponds to the good. This means a fundamental change not just in our habits, but in something far deeper: our fundamental human responses. Throughout, this education of desire and the formation of persons is to be construed as a fully embodied phenomenon. As David Ford (2007) put it, if

Christian wisdom is concerned to correspond thoughtfully ... to God and God's purposes, the desire for this needs to be aroused; the heart and imagination must be moved as well as the mind. ... heart and mind are educated together and are stretched to engage passionately in their own search. (Ford 2007, 12–13)

This education of desire is described, in pastoral terms, as "value-preference modification" (O'Connell 1998, 99–100): inspecting and realigning the particular values one has invested with emotional significance is a crucial aspect of the project of moral formation. Affective transformation is a vital element in creating authentically driven personal moral responses, and without a subjective feel for the values one wishes to embody, it is very difficult to motivate a realignment of a person's moral behavior or character (O'Connell 1998, 74–5).[12]

Doing the Truth

The formation of virtue has thus far been construed as requiring habituation, and a process in which the various portions of the human creature are brought into a gradual alignment with what is discerned as good and worthwhile, as congruent with God's desires for his creation, and for the future Kingdom to come. However, this focus on the reformation of heart and mind as a co-extensive process might give the impression that the formation of virtue is a purely contemplative process. In fact, virtue is first of all an inherently bodily process. It is embodied and discovered in action, developed in action (in technical language, it is less about *theoria* than it is about *phronesis*—that is, practical wisdom).

The process of habituation, of developing these powers of discernment and refined affective response, are done in situ, in context. While detached reflection has an important place in consolidating and refining the knowledge garnered in experience, it is in fact the practice of virtue, the embodied element, that is the primary and prior component. This practical element is what McDonagh refers to when she says that making disciples is a matter of "doing the truth" (McDonagh 1998, 23).[13] As she observes: "It is in the doing that full moral understanding is achieved" (ibid., 78). Indeed, it is with "practical living and growth of the Christian as a person" (O'Connell 1998, 60) that moral theology and the development of Christian virtue are concerned. We have in Christian virtue the idea of man created in the image of God for wise stewardship over the earth (though admittedly, we have been more than a little unsuccessful with respect to this ideal).

Needless to say, the idea of habituation as a part of moral education is a universal among the faith traditions. The contemporary theological approach associated with the Yale school (and David Kelsey above), called the cultural-linguistic approach, is particularly sensitive to the analysis of the various religions less in terms of doctrines or beliefs and more in terms of the various *practices* which any community, denomination, or tradition encourages. It is into precisely these practices that moral enhancement would be embedded. This shift toward a vision of faith as a set of practices is very helpful in disclosing the immersive dimensions of religion, and, more specifically, the immersive dimensions of the moral life embedded within those religious traditions.

In this sense, written theology tends to be a step behind actual religious practice. As Wittgenstein famously observed, those most familiar with a given practice or idea are those often least able to give a verbal account of it. An understanding of the faith traditions as embodying practices, of moral living and moral formation as primarily something embodied—that is, of the moral life as something that is practiced—is an important shift which should be of great use in engaging moral enhancement discourse with moral theology.

As such, with practical realities always at the fore, we really need to be asking how religious communities actually go about enacting the moral life and how they go about thinking about it among themselves. This is where moral enhancement, in whatever form it takes, is going to manifest. The notion of virtue, or, at least, any account of the moral life which recognizes the importance of context, habituation, practical intelligence, and how moral choices actually play out in religious practices, is going to be a very useful way of bridging the practicalities of moral enhancement and the realities of moral education in religious communities. Not everyone is taken by virtue ethical frameworks, and this is entirely understandable; yet as a tool for communicating between moral enhancement philosophy and the various faiths (understood as systems of practices, immersion, and habituation within which moral education is meant to occur), virtue theory is very germane for conversation between the domains.

Summing Up the Personal Dimensions of Christian Virtue

Christian virtue is centered around specifically Christian formulations of certain values, in specifically Christian contexts, and orients the person's life so completely that an entire moral identity, character, precisely as a Christian, is created. This moral identity integrates and unifies the virtues into a fabric, an entire shape for the person's life, which runs through the very core of his personhood, thoughts, feelings, and actions, which he has taken great efforts to habituate into himself. Yet the various matters we have considered thus far only describe the individual or personal dimensions of virtue formation. If we are to do justice to religion as a set of practices, there is another dimension that is at least as crucial to Christian moral formation, as well as to all the other faith traditions. That is the social, or communitarian, dimension of formation. To this we now turn.

Community: The Essential Social Dimensions of Christian Moral Formation

The Body of Christ
According to the second Vatican Council, the twenty-first ecumenical council of the Catholic Church (1962–65), the church is to be considered:

Most basically a community, a group of persons called by Christ to live the faith, to undertake discipleship, to mediate salvation to a needy world, and to serve the Lord by serving one another and all creation. (O'Connell 1998, 128)

We saw that the overarching framework within which Christian moral formation takes its sense is the idea of discipleship—this is its personal, individual formulation. The overarching rationale embracing moral formation for the Church as a whole is precisely that articulated in Vatican II. Individual Christian formation takes its larger sense as an exhortation to enjoin in a community of discipleship whose organizing principle is that of service to God and love of this world God has created. The Church is then envisaged as a body of organized disciples working together in service of the God of peace. This is the ideal. This conception of the life of the Church as involving three activities—worship, community, and service (O'Connell 1998, 180)—helps make clear not just that there is a larger salvatory framework within which to understand individual Christian moral formative projects, but also that this larger salvatory framework has inextricable social dimensions. Herein, the processes of reeducation that constitute the formation of virtue and character are to be understood as occurring necessarily within a relational, community context.

More specifically, the transformation of the heart is to be understood as developed in the context of relationships and communities of worship, rather than being understood as purely individual projects of self-help, self-growth, or self-development. Indeed, the practices of communion, embodying the work of service and love so central to the Christian message, being relational in nature, are simply not things that can be learned in isolation. By definition, goods like service and love need to be honed in a public and relational context, for they concern nothing other than the way persons stand in relation to one another, especially in the face of their radical disagreements.

This reveals another issue that must be considered with respect to moral enhancement intervention. Most, if not all, interventions proposed by way of potential moral enhancement are envisaged as if they can be applied in private, in a dark room, as it were, somewhere far from any public context. However, if many (if not all) significant moral goods are modeled and learned primarily in social contexts, if these relational settings are required for the very shaping of a moral good—for a person to learn how to appropriately enact the moral good in question—then moral enhancement interventions must be made amenable, in some way, for their public use, and the possibility of their integration within these larger relational settings is necessary for modeling and shaping the good in question.

The suggestion, then, is that community is the primary context through which these relational virtues are to be practiced and developed in everyday Christian living. To use Kelsey's terminology, the formation of "communities-in-communion" is the primary work for Christians as they struggle together to find appropriate responses to their Creator God's ongoing attempts to reconcile them back to himself (Kelsey 2009, 1029–30)—the difficult work of creating communities and practices which can appropriately embody the "dual love" command: to love God and the world that God has made.

As such, discussion of Christian moral formation, and moral enhancement in a Christian context, must expand to make reference to relationships, communities, and practices—persons who are meant to be forming together, rather than imagining moral formation as occurring during some strange process of isolation. Returning to this idea of "the body," wherein the Church is understood as "the body of Christ," a given community can be seen to work best not as a uniformity, then, but rather in the Platonic aesthetic ideal as a well-ordered abundance of its various diverse parts. These individuals are then supposed to form a larger unified, organic whole which is bigger than the sum of its individual parts.

Of course, communities will inevitably have their problems, too, and if not properly formed themselves, if not properly organized and maintained, various extremely troubling distortions can arise. Wright makes the following observation:

Yes, there are many low-grade parodies of unity, totalitarian, forced conformism, the forced jollity of the like-minded small group ... [unlike] the real challenge, which is to allow the central Christian virtues of faith, hope, and love, and the fruit of the Spirit which is love, ... self-control to have free course in our relationships with one another, and to discover, as they work their way into our lives, the corporate virtues of mutual submission and mutual recognition of God-given gifts of leadership, teaching, and so on (Wright 2010, 184).

Putting such dangers in yet stronger language, David Ford writes that there are possibilities for group distortion also embedded within the best parts of scripture. He writes:

"The corruption of the best is the worst," and the energized, united community of praise that the writer of the Letter to the Ephesians envisages sets up dynamics whose power can be disastrously perverted. It may even be that the very rhetoric of Ephesians is totalitarian. It loves to use the word "all." Does it not portray a group identity that is imperialist, overwhelming and intolerant of others? Does its repeated advocacy of unity not call to mind the many regimes of domination that thrive on the suppression of difference? What about those who do not share this particular vision of God and God's abundance? Is there anything to prevent this letter being the charter for a thoroughly triumphalistic Christianity? Can it inform a Christianity that relates respectfully to others in a pluralistic society? Might the singing of such a community rather resonate with those many uses of song which create confident, aggressive solidarity aimed at coercive assimilation of others or confrontation with them? (Ford 1999, 130)

This need to recognize that communities themselves need forming, and that they can become distorted in ways which can have terrible consequences for its members and those around them, raises a very important point: the group is to be understood as an entity in and of itself, with a moral status of its own, not just as a conglomerate of individual persons who happen to be standing together (McDonagh 1998, 35). In other words, entire communities themselves can be good or bad, praiseworthy or shameful, wise or foolish, and this is going to put a significant limitation on moral enhancement—which makes most sense, if at all, in relation to individual persons as moral agents rather than at the level of group agency. This is a point hinted at in chapter 3 when discussing the aspiration for a moral enhancement which can manage qualities relating to mass social action, which is precisely what is needed to deal with many of the most pressing problems facing the world today.[14]

   This group focus is then an extremely important consideration with respect to moral enhancement. For if we are looking to religious practice, then we are looking at something which is public first. That is, for something to be established as a practice requires a group of persons to have established such a practice precisely *as* a practice. In this sense, even a practice which involves a person in solitude is still public, since it is something that has been mutually agreed to be an appropriate practice in advance within the community of followers. If our focus here is on presenting the faith traditions in terms of the practices into which moral enhancement must be embedded (thus, that moral enhancement needs to fit, not just with the intellectual or doctrinal parts of religious faith, but more importantly, within the very ways in which those faiths are put into practice by its members), then we must look at this communal dimension of religious faith as fundamental. The public nature of practice thus points us to community by implication. Needless to say, this idea of community as the core of religious practice, more importantly for our purposes, of community as a formative moral core for the salvatory visions embedded within those faiths is not at all limited to Christian faith, but again is universal to any faith tradition, or community within that tradition, that has established practices embedded within it.

Communal Dimensions of Moral Formation: Where Can Moral Enhancement Fit In?

As soon as one begins to grasp the profoundly idiosyncratic nature of Christian moral formation, be that on the individual level, regarding the idiosyncratically shaped dimensions of moral formation qua the development of habitual, intellectual, affective, and bodily action along specifically Christian lines; or communal, with respect to the manner in which Christian moral formation is envisaged as participatory and nonseverable from central Christian practices such as worship, prayer, and the revitalizing of common Christian narratives, one can begin to get the sense that a deep and thorough integration of moral enhancement technology and religious moral formation is going to be more than a little awkward.

In moral enhancement discourse there is a lot of lip service paid to the idea of moral enhancement *complementing* traditional means of moral formation. The word "alongside" is used often. The more circumspect of the enthusiasts comment that they do respect traditional means of formation but see no reason why moral enhancement might not be explored as a possible avenue which might work "alongside" such traditional means. But what does this "alongside" really mean? Until one gains a sharp grasp of the intricacies of traditional moral formation—in even but one Christian guise—it is all too cheap and easy for commentators to say that these two modes, traditional and technological, might complement each other. If one is going to take a complementary perspective such as this, one must at least provide some sort of rationale for how moral enhancement and traditional moral formation are to work side by side—else the claim is merely vacuous. Before moving onto our attempt to provide a rationale for a possible complementarity between technological and traditional Christian modes of formation, let us look at how recent pastoral theology has dealt with the actual concrete practice of motivating moral formation in the Christian context.

Contemporary Pastoral Reflections on Moral Formation

Value-Preference Modification

We have looked at the more thoroughly theological aspects of Christian moral formation, but what of the actual practice of motivating such formation? In responding to the importance of this question, moral theology has already enjoyed a long and rich encounter with the social sciences.[15] This conversation, in particular with psychology and sociology, has been of invaluable assistance in developing more refined pastoral interventions for motivating moral formation. Before moving on to the next chapter, where we will try to understand how moral formation might suitably work alongside Christian moral formation, it is worth getting a view on the nature of such empirically informed pastoral means for encouraging Christian moral formation. If nothing else, this will help articulate many of the crucial social-environmental dimensions of moral formation and help us to see why making the biological aspects of moral functioning primary leads us away from the most salient, meaningful, and potent motivators for such formation. It was suggested earlier that part of the process of Christian moral formation can be understood as a project of "value-preference modification." O'Connell emphasizes the importance of communities in providing and helping inculcate values. Indeed, as O'Connell insists:

Values reside more in groups than individuals. Thus groups, like individuals, can be characterized as good or bad depending on the value preferences that they embrace. (O'Connell 1998, 78)

If this is so, then pastoral strategies for moral formation should be more effective if their focus is more on communities than individuals. Again, this is not to diminish the individual's contributions to moral formation and action—quite to the contrary. It simply recognizes that there is a very complex dialogical back and forth between individuals and the groups they are a part of. In fact, group norms often make a powerful formative impression on individuals growing up in or moving into such groups.

If it is true that moral development arises, in part, out of the emotional interaction and experience of moral-social conflict between members of groups ("social dissonance" or "disequilibriation," to use Norma Haan's terminology; Haan 1985, in O'Connell 1998, 146), then it is those group values that need to be the primary focus, rather than individual moral development. If Christian moral formation is about making disciples, and the group is the primary home of values, then the question of Christian moral formation really becomes a process of "creating communities of discipleship, homes where the value priorities of the disciple flourish" (O'Connell 1998, 85–6).[16] The pastoral approach is one which sees the process of motivating formation as involving nothing less than providing an individual or group with an entire range of new experiences. O'Connell observes that it is one's experiences which for the most part form the values and judgments one tends to make. Without offering the means for individuals and groups to be challenged by new experiences, inviting them to reflect upon and grow from these experiences, it is very difficult to understand how a credible and lasting change to a person or group's value preferences can be sustained.[17]

Identity Formation
The next important task in motivating the life of discipleship is that of helping fashion in the person and group an identity wherein being a disciple is understood to be important. Unless there is some background scaffolding wherein persons want to identify themselves as disciples, want to form themselves in a specifically Christian way of living, then it is going to be very difficult to motivate them to take on the identity of disciple. Indeed, the problem of motivating persons to take up the hard path of discipleship, to imitate the behaviors into which they are being apprenticed, requires a closeness of relationship such that the desired identity can be given the sort of salience, or emotional power, required for such motivation to arise. As with so many of the considerations presented here, it is this background scaffolding, the need for a "circumambience" of the values and identities that are meant to be passed along, that is so crucial.

To create disciples, then, one must facilitate the creation of persons' identities as disciples. O'Connell draws on the idea of "role identity" (Stryker and Serpe 1982) to describe the manner in which emotional investment in certain identities is so important in motivating persons to enact them. While persons carry a variety of identities with them throughout the various parts of their lives (e.g., teacher, parent, friend), it is the salience of any given identity which largely determines the amount of commitment that a person or group will invest in that identity (O'Connell 1998, 98). According to identity theory:

The greater the commitment involved in a particular identity, the more salient that identity will be. And the more salient an identity is, the more likely that role-related behaviors consistent with the identity will take place. ... The more that my relationships depend on my having a particular role, the more that role will be central to me [and] ... the greater the likelihood that in role-related settings I will behave in accord with that role (O'Connell 1998, 98).

So, where ministers intend to call people to a life of discipleship—to live out roles in all the myriad behaviors that might express authentic forms of discipleship—identity theory indicates that people will tend to behave in these roles to the extent that the role of disciple is salient to them. By extension, the relationships of persons in these communities of discipleship are utterly critical to the formation of their personal identities. Relationships then constitute a critical part of moral formation. O'Connell writes that people are influenced by relationships whose continuance depends on their embracing the role that the behavior enacts (ibid., 99). He continues:

People live out abiding identities. ... So, ministers have the job of inculcating an identity; this means that ministers are in "the important-relationship-cultivation business"—to the extent that relationships are truly valuable to a person and to the extent that the continuation of those relationships depends on my embracing a particular identity, to that extent will I tend to make that identity my own. ... Religious education is an enterprise in the shaping of personal identity. It is about formation, not information. Indeed, whatever information is shared in religious education ... is shared precisely because it can serve the process of identity formation. (O'Connell 1998, 143)

As such, this relational dimension, the importance of moral formation being found precisely in the relating of persons, has implications for moral enhancement discourse, which has proceeded for the most part as if moral formation can be understood as a purely private activity and can be perfected in isolation or in "solitary self-discipline." Indeed, this is one of the problems of thinking of morality in primarily biological terms: it has concealed the manner in which moral learning is enacted in relationship. While this is not a decisive impediment, it is certainly a bias that stands in need of correction. Moral enhancement discourse needs to be realigned such that it can at least make sense of the idea of formation as having a primarily social background, a prior social background in fact, and enthusiasts need to think about how their proposed interventions are to be enacted in a relational manner. If moral enhancement cannot be satisfyingly realigned, such discourse will remain inherently myopic in focus, and, arguably, quite misleading with respect to where it places its focus.

Exemplars and Apprenticeship

We saw O'Connell mention the idea of "important-relationship-cultivation." This points to the final social dynamic that we will be considering in relation to practical moral formation: the roles of exemplars—that is, exemplary individuals who have the power to teach and inspire through their own practice. It is perhaps an obvious point, but it is important to note that people are influenced by significant others in their lives; indeed, people are involved in a never-ending process of observation and imitation of those around them (O'Connell 1998, 88). The powerful role played by exemplars in aiding Christian formation is perhaps best captured in Hauerwas' (2007, 2013) concept of "apprenticeship." For Hauerwas, learning to be a Christian is, in a sense, like learning a trade. Before one can learn to think or act for oneself, one must become knowledgeable and proficient in performing core elements of the faith. This idea of apprenticeship has been taken up and developed by David Ford. For Ford, Christian worship and living is not simply a matter of theory or believing the truth of certain propositions; rather, it is a matter, first of all, of practical mastery. He writes:

Apprenticeship is following particular paths and learning how to negotiate them in all weathers. It is a matter of the quality of experience needed to understand and make judgments. ... A common problem is that of trying to short-circuit apprenticeship by taking an overall view. That makes a basic error, assuming that the social sphere is somehow transparent, accessible without the specific disciplines of "fieldwork" which require that one play a role in the group or practice studied (Ford 1999, 141–2).

In apprenticeship one finds many of the core elements we have been considering—habituation, practice, identity, creative responsiveness, community and worship—all knitted together in participation. However, participation in complex practices requires that there be some person, or persons, to show one how to participate properly. One needs to be shown the various "rules of the game," taught the key "steps of the dance": shown what is required, what is desirable, and what is not permitted in participating in the practices of the faith and appropriate moral action.

Thus, apprenticeship is a practical, interpersonal activity that relies on both repetition and spontaneity. It involves learning the various rules for participation, but the goal is not to turn individuals into robots; there must be a constant creative responsiveness which refreshes, within limits, the nature of the practices in relation to the changing world the members of the community inhabit. But there again, knowing the limits of creative responsiveness, knowing how far boundaries of appropriate expression of one's faith can and should be pushed, is part of the process of apprenticeship too.

Apprenticeship to exemplary individuals—following others as they attempt to respond to Christ's call to follow him—is thus an absolutely critical part of the various social dynamics making up the external scaffolding by which Christian moral formation is mediated. The influence of powerful and charismatic exemplars, for good or bad, simply cannot be overstated in the morally formative domain.

How experienced exemplars manifest their faith and their own life of discipleship, insofar as the relationship had with them is emotionally significant, has immense power in establishing the identities and challenging the value preferences of those forming in relationship to them. Whatever one's views on the Church, it is impossible to deny that the sorts of comments drawn upon here are enormously potent factors in moral formation, in the Church and more generally. Without even claiming that such accounts are sufficient to describe fully what is going on with moral formation, they at least point toward the extent to which the social-environmental background, the scaffolding of moral formation, involves an incredibly complex and mercurial web of factors.

Concluding Remarks

O'Connell summarizes the goal of pastoral moral theology thus: to understand how people embrace the values by which they live, so as to help them better embrace the values of the Gospels. The central transformative process is one of value-preference modification. The challenge is to get persons "to habitually and skillfully prefer certain values and behaviors to others" (O'Connell 1998, 99–100). Given that these value preferences are often profoundly affective in nature, the manner in which persons subjectively engage with their values is a primary subject of concern on this point.

Moreover, in presenting a specifically virtue-based account of Christian moral formation, we have made clear that the totality of the human creature's bodily powers must be brought to bear. A moral identity or character is required, which can only be formed through engaged practice, the shaping of the reasoning mind and congruence of the emotions formed through habituation, and done, necessarily, with the support of a larger social scaffolding, the worshipping community, which, in ideal circumstances, has the power to motivate the taking on of specifically Christian values and is capable of sustaining the identities of its members precisely as Christian.

This social dimension of moral formation is deeply significant to moral theology, which suggests that groups, and not just persons, are moral subjects to be judged and formed accordingly. If values reside more in groups than in persons, and these persons and the communities they belong to are in complex dialogical relations, then a significant part of moral formation must surround the creation of communities, the education of their desires and values, and the formation of competencies and attitudes which encourage in that community certain ways of living. Yet Christian moral formation is also part of a larger, all-embracing shaping of one's life in terms of discipleship. There is a profoundly practical and immersive element in moral formation, a "doing the truth" which is interwoven with all aspects of moral formation. This all serves to make moral formation a total project which draws together every part of a person's life and activity. Transformation and reeducation of the heart and mind together is required for this project, in which subjectively held values are to be reshaped.

All this articulates for us a far deeper social-environmental backdrop to the idea of moral enhancement. Up till now we have treated moral enhancement largely on its own terms; we have met it in its own arena and worked from within its own dominant assumptions and conceptual frameworks. Taking a look at theological and pastoral approaches to moral formation changes the game completely. Whether one believes in God, whether one is a Christian, whether one buys into the idiosyncratically Christian elements that are part of the picture that has just been presented is not of particular importance for the present task. What is important, in the end, is that one comes to understand precisely how varied and numerous the conceptual and social-environmental influences on moral formation can be, and how complex, manifold, and interwoven such influences are in their embodied human reality. Moral formation as described here respects the tangled, embodied, socially embedded nature of such development. Moral enhancement must do so too. *But can it?* Is it even capable of doing so, taking the neurobiological and individualized, atomized, compartmentalized approach that it does?

Yet we have only covered a highly truncated version of the theological, religious, and spiritual dimensions of the social-environmental context of moral formation. We have not even looked at political and economic realities, nor even at the complex psychological dimensions of moral formation, nor yet any of the other subtle, hidden, and nuanced ways in which the various manifold social-environmental and psychosocial factors play into the enactment of moral functioning and moral formation. All of these varied and manifold influences and the manner in which they should affect our reflections on moral enhancement have gone unmentioned.

Before this chapter, we were thinking predominantly from within the dominant assumptions of the philosophical and empirical work done on moral enhancement and moral functioning, respectively. Yet look at the comparatively simplistic and superficial manner in which these biologically oriented approaches deal with the matter, offering perspectives which have done precious little more than relating moral functioning to things like "the levels of serotonin in the brain," "the role of the amygdala in emotional processing," or "the role of dopamine in causing certain behaviors." That sort of analysis does seem rather paltry and superficial when stood against the integrated and conceptually rich diversity of possible influences that are at play in moral living. It makes one wonder whether the current terms of moral enhancement discourse are even up to the task of doing justice to the complexity of the matter at hand.

This diversity of influences needs to be given voice. Such neuroreductive foci—the amygdalae, neurotransmitters, and hormones—treated of in isolation tell us precious little. As soon as one even gets a sense of the interactive nature of brain, person and context, the complexity of the various external factors which influence moral functioning and with which such neurological aspects are interwoven, it becomes harder to take too seriously these neurobiological studies which just bracket out this immense wealth of supervening social-environmental influences, as if it is all just completely extraneous to the subject matter at hand. What is required now is to deal with the question of whether these biologically based reflections that have made up the body of parts I and II of this book can be integrated with the social-environmental factors considered. And if they can be so integrated, particularly with reference to an idiosyncratically Christian vision of moral formation, what might this integration look like? In short, can Christian moral formation and biomedical moral enhancement work together side by side? If so, how might this cooperative picture be envisaged?

7 Moral Enhancement in a World of Faith

We have just glimpsed how the idiosyncratic nature of Christian faith can complexify the task of putting moral enhancement to work. Moral formation involves a considerable amount of scaffolding which cannot be ignored when thinking about a workable moral enhancement that is to be applied in a world where billions profess their accountability to one faith or other. In this chapter, we will inspect some of the challenges that integrating biological moral enhancement within such larger scaffolding poses. Recall that, given the prevalence of faith in the present world, any attempt to provide a strong vision of moral enhancement without referencing how its interventions might be integrated with matters of faith has bracketed out a practical concern fundamental to the inquiry. One needs to confront the following question: what is realistically possible with respect to moral enhancement in a religious context? There are no easy answers here, neither for critic nor enthusiast.

The suggestion to be forwarded in this chapter is that the possibilities for integrating moral enhancement with matters of faith will likely be quite superficial in nature, but not insignificant for that—indeed, likely all the better for it.[1] More significantly, this need for a certain brute quality to the biological angle of such integrative, multipronged interventions may mean that peak sophistication regarding moral enhancement design may already have been reached. This could be a good thing. Generalized, brute forms of intervention, insofar as they might be able to leave room for the supervening idiosyncratic elements of the faith tradition in question to do the work of shaping such goods, might be the best option for a hard moral enhancement. As such, what needs to be explored in this chapter are the ways in which moral goods are shaped into the moral goods they are by supervening scaffolding forces. Room must always be left for this shaping if the goods are to be enacted as authentic to the faith tradition they are a part of. If this is so, it is only superficially that room can be made for the supervening input of genuine personal moral agency, as well as social scaffolding. That being so, the superficiality of the intervention must be matched by a profound grasp of the moral good one is attempting to enhance. In this way, it may be possible to construe some version of moral enhancement as a nonreductive, positive, and potentially desirable support to moral formation or moral living.

In this vision, working "alongside" traditional moral formation, any potential biomedical intervention would need to be envisaged only as a supplementary support mechanism for those already inclined to pursue a project of moral formation. This caveat of the predisposition of will is fundamental. In this more modest but realistic light, moral enhancement (we are referring in this chapter to hard, explicit, and strong forms specifically) would be best understood as broad-scope, biologically based interventions to support moral formation in persons already forming themselves through more traditional moral educative means. Moral enhancement would bolster existing activity rather than being able to motivate or compensate for that predisposition of will.

Dealing with Christian Virtue Ethics: Obsolescence and the Replacement Thesis

Practical Knowledge and Neuroplasticity

Can biotechnology ever come to make religious moral formative methods obsolete? James Hughes hints that it may be possible:

> Eventually we will have the capacity to change genes that affect the brain permanently, and install neurodevices that constantly monitor and direct our thoughts and behavior. At that point the distinction between traditional methods of self-change and neurotechnology may become moot. In fact, eventually it is quite likely that the changes of thought and behavior that neurotechnology will enable will be much quicker, far surer, longer-lasting and more targeted than traditional religious practices. (Hughes 2013, 31)

Let us take a final look at the notion of Christian virtue—and the idea that such a phenomenon might be produced by biological means alone. As we saw, the idea of "virtue," in the more philosophical use of the term, has a very specific sense. It will not do to talk of, say, honesty as a virtue unless one has in mind a person who has trained him or herself in the habit of truthfulness. Virtue is a particularly strong way of describing moral functioning, since the standards by which a virtue is measured largely have to do with how much of a disposition toward that behavior there is, and the extent to which that disposition has become, and explicitly so, a part of the very identity of the agent in question. This stands in contrast to a "weak" view of moral functioning which is only interested in looking at isolated acts such as, say, stopping to help a person who has fallen over in the street. So when we talk about Christian virtue and come to say that it is impossible on principle to engineer a Christian virtue by technological or pharmacological means alone, we have to keep in mind that it is the strong sense we are referring to.

The key point religious practice raises with respect to virtue is that it involves practical forms of knowledge. Indeed, it is this reliance on "how to" knowledge that creates so many problems for the idea of "virtue engineering." This dimension of practical knowledge is a crucial point with respect to moral functioning, and it deserves more attention than it has been given in the moral enhancement literature. We have mentioned throughout that one of the major points of contention in the philosophical debate has been over the "cognitive content" of moral enhancement. Thinkers like John Harris find the idea of moral enhancement through "emotional modulation" to be deeply unsatisfying. For Harris (2011, 2012), moral functioning mostly, if not always, relies on matters of discernment, reflection, and judgment—figuring stuff out. His point is that if we just change persons' emotions, making them more likely to do a good thing, then they have lost out on something very important: understanding why what they are doing is right or wrong. Worse still, we would then lose the ability to reflect on right and wrong for ourselves, which is (like a virtue) something which needs to be practiced.

These are very important points. However, the practical, enacted dimension of virtue complicates things even further. In fact, there are numerous practical elements to virtue, of which powers of discernment are but one. As we saw, virtue draws upon the entire range of a person's bodily powers: intellectual, affective, physically active, and social-interpersonal. All these are embodied in the human person's moral functioning as one integrated unit. What of this range of bodily powers that virtue draws upon? Are not these other powers entirely missing from the cognitive/emotive debate? Is moral functioning not being treated of, in this particular debate, as if it were a mere summation of these two terms "emotion" and "cognitive content," such that one or other, or, at best, some combination of both were sufficient to encompass the entire business? The embodied nature of moral living must be taken into account.

Consider any skill; being able to paint, for instance. An aspiring artist may have all the motivation in the world; he may even have the knowledge and intellectual power to discern what makes for a good piece of art (for example, a good eye and a degree in fine art). Would this make the person in question an artist? There is one way, and only one way, in which practical knowledge and practical mastery can be gained, and that is through practice. Practice, then, is the embodying vehicle through which cognitive and affective elements are integrated with the rest of a person's powers. Talking of moral enhancement in either/or terms of cognitive enhancement or emotional modulation, or even as a combination of both, entirely neglects that larger embodiment which supervenes on both cognitive and affective dimensions, and which unifies them, for better or worse, in their enactment.

How is biomedical moral enhancement to cope with the practical dimensions of moral functioning, with its embodiment and the practical mastery it demands? Is there any realistic, foreseeable scenario in which a biological intervention can produce in a person this kind of "how to" knowledge that virtue relies upon? This gaining of "how to" knowledge, the immersive dimensions of moral learning and moral functioning that are requisite for a skill to be raised to the level of a virtue, is precisely what biomedical moral enhancement, considered on its own, can never produce.[2]

The development of practical knowledge might find a neurobiological correlate in the phenomenon known as neuroplasticity (Pascual-Leone 2001)—the idea that changes in the brain's neural pathways and synapses can be discerned in response to changes in behavior, environment, and neural processes. In other words, the very anatomy and physiology of the human brain alters in response to the way persons behave and in response to the nature of their experiences. This idea of the brain's self-rewiring through practice—a change in the very substance of the brain that requires, above all things, time—makes it such that no immediate virtue-producing shortcuts on the biological level are possible.

One can introduce into the brain any pharmacological agent one desires, apply any technology, but there is simply nothing which can manifest this rewiring from nowhere, for we are talking about physiological changes in organic matter that are unfathomably intricate in nature. Not only does this process of rewiring and habituation, require time, but it is a wiring which occurs in relation to the specific set of situations in which one has practiced, in relation to the brain connections that have been made over the course of that person's entire life (and thus are both personal and unique), invoking, as it does, the particular emotions and affective responses that person had when immersed in those scenarios, as well as the various reflections and insights gained throughout the practice. All of these factors, never the same for any two persons, are wired into the brain in its own peculiar and immeasurably complex ways. Obviously, the idea that it is "quite likely" "neurodevices" will be devised which can emulate this process is not very well thought through.

Moreover, once rewired for the desired virtue in this way, it is not as if the brain will keep its state forever, for the brain is constantly rewiring itself, and if one does not continue to practice the virtue in question, then the physiological bases of the habitual actions will simply wear away. Like a castle in the sand, strong as it may be in its own terms, left unmaintained, each progressive tide, each wave, will gradually reform the structure accordingly. What habituation and its physiological bases offer the practice of virtue is for the most part the assistance of inertia once formed—the possibility that in a given circumstance the impulse to perform the appropriate moral action will be generated spontaneously and automatically, whether one is in the mood or not. What is required, above all, is some supervening way of motivating the ongoing practice of the virtue, and this, again, biological interventions alone cannot proffer—what one requires is a worldview, or at least some strong deeply meaningful reason(s) to carry on being virtuous, to carry on sustaining the habit.

Indeed, this need for ongoing practice is precisely why having something like the larger salvatory background involved in the various faith traditions can be so helpful (of course, in principle any strong-minded person, secular or religious, with a thorough life philosophy, one with deep motivational salience for them, would do just as well). In specifically Christian terms, the manner in which moral formation is a part of an entire "shape of life," an element of the larger project of ongoing discipleship, provides (at least in principle anyway) a rationale for motivating ongoing efforts toward virtue formation, and the community of believers in which that person is embedded would ideally help sustain continued practice.

In other words, one does not merely learn a virtue and then leave it at that. Christian virtue ethics represents a set of skills which one is to make use of on a regular basis as part of the life of discipleship. It is this background impetus to carry on using the virtues, virtue integrated with a background salvatory scaffolding, lived in a social-immersive context, that helps provide the durable, temporally extended foundation for moral growth that transhumanists are expecting pills and neurodevices to make manifest.

Whatever one thinks of religious narratives, it is hard not to appreciate their power to sustain ongoing efforts.[3] The integrated and fully embodied nature of virtue, the requirement of practical mastery, means that a replacement thesis—the prospect of moral enhancement as something which might one day make traditional formative means obsolete—is simply a nonsense to be put aside. Let us ask then: if traditional means of moral formation as the vehicles for embodied practical mastery cannot be got rid of, how might the integration of moral scaffolding and biomedical interventions be realistically envisaged?

Integrating Moral Enhancement and Christian Moral Formation

Necessary Superficiality: Integrating Biomedical Interventions with Idiosyncratically Shaped Moral Goods

We have seen that moral enhancement discourse has tried, quite successfully, to remove any and all reference to specifically religious concerns that might modulate how such interventions will need to be construed in a multifaith world. Moral enhancement is generally talked about in a "one size fits all" manner, in completely universal terms, as if any potential moral enhancement intervention would be like type O blood, readily assimilated by all persons regardless of any and all differences that might separate them. The empirical study of the moral brain and the philosophical discourse surrounding it has managed, much to its detriment, to put itself in a position where it need make no commitment at all to recognizing differences in the deep conceptual shaping of moral goods across the various faith traditions, to recognizing the various cultural elements particular to the communities who endorse such moral goods, nor yet to the various idiosyncratic practices which make up such traditions. Yet these are the basic ground in which such proposed moral enhancement will be enacted.

As far as moral enhancement goes, what needs to be kept in mind is that the more complex the moral functions one intends to enhance, the less likely it will be that a technology might realistically be devised to deal with the problem in any nonsuperficial, fine-grained way. Above all, we need to constantly remind ourselves of the way in which scaffolding is required to shape moral goods into the goods that they are. If this is so, we might hazard the following proposition:

It can only be the surrounding psychosocial context—personal intent and scaffolding—which can shape any given enhancement intervention into an appropriate expression of the moral good being aimed at.

As such, it is likelier that the more generic and universal kinds of moral enhancement intervention be better molded to fit within the more idiosyncratic elements of the faith traditions. It is only as such that room can be made for the shaping influence of the supervening environmental and personal dimensions of moral functioning. What has been presented as a flaw for biomedical moral interventions—their brute nature—is a strength when it comes to hard, strong visions of moral enhancement, since nothing other than a brute intervention can, in principle, leave the appropriate space that only scaffolding can be expected to fill.

This being so, one still needs to be aware of what one is attempting to enhance. While a broad-scope, nonspecific approach may be the ideal here, *not any old intervention will do.* One still needs to be aware of the landscape one is attempting cater to, and we will see shortly how a broad-scope approach which does not have any grasp at all of the complicated nature of the moral good in its sights cannot be presented as a worthy candidate for enhancing such a good. We need to take a look at the rather complex ways in which the moral goods to be enhanced are interwoven into their background sense-giving frameworks, and what this means when it comes to envisaging the possible relationship between moral formation as it occurs in a faith context and biomedical moral enhancement possibilities. We can illustrate this need for a sophisticated awareness of the moral territory by looking at phenomena like empathy or compassion, which have a distinct and complex conceptual shaping and which fit into background frameworks necessary for giving them their sense.

Now, certainly, it is true, as we suggested in chapter 3, that there is a commonality and overlap between the various moral goods that the human race aspires to (it is too implausible to think that cooperation between persons would have been possible across the centuries if there were no semblance of commonality among our moral sentiments and moral goods). At the same time, one cannot neglect the fact that manifestations of such moral goods also stand in relation to traditions which have very particular views about their nature and appropriate use. To some extent, then, one can only explicitly recognize a particular act as being compassionate, for example, if one already has in mind some sort of template through which one can recognize given acts as compassionate. We cannot be talking about compassion "in the abstract," or empathy "in general" (whatever that might mean), but rather compassion understood in a particular sort of way. For example, when we were discussing Paul Zak's work on empathy, we were forced to ask, "What particular concept of empathy did Zak have in mind?" No clear answer was forthcoming. Radically simplistic ways of thinking about moral goodness will not suffice for moral enhancement purposes. Take the following assertion from Zak:

Oxytocin orchestrates the kind of generous and caring behavior that *every* culture, *everywhere* in the world, endorses as the right way to live, the cooperative, benign, pro-social way of living that *every* culture *everywhere* on the planet describes as "moral" (Zak 2013, x; emphases added).

Now, clearly, there is more than a little simplification going on here. Certainly, if one wishes to stimulate moral formation in a specifically Christian fashion, the sort of very generic approach used by Zak is not going to do. Such generic constructs are not going to be able to function as a one-stop-shop for moral development. Such interventions need to be broad, yes, but one still needs to understand what it is one is enhancing. One needs to be responsive to the background frameworks and conceptual shaping that the community in question is working within, and to find some way in which these background shaping elements can be integrated into the enhancement intervention being proposed.

Let us give a clear example of why a sufficiently deep understanding of the moral good in question is necessary even with broad-scope approaches. Let us take the case of charitable giving—a moral good investigated by Zak and his various collaborators. We saw that charitable giving was one recurrent means through which Zak and his team attempted to locate oxytocin's role in modulating generosity and altruism.[4] Now, oxytocin might arguably make persons more generous, but Christian charity has a complex inner dynamic which involves more than just saying, "I feel like giving more money now." Unfortunately, this is all that Zak's various studies seem to boil down to. Apparently the true measure of moral goodness in almost every case is appropriately and adequately measured with no other question than "How much money do I feel like parting with here?" The answer always seems to be that of dosing oneself with oxytocin, with the expectation that this will increase one's empathy and make one want to be more charitable, generous, trustworthy, and so forth.

But what of explicitly motivational and conceptually shaped concerns? What of the idea of charitable giving as an expression of humility? What of charitable giving as an expression of affiliation? Or what of charity as a means of developing an attitude of self-sacrifice? What of charitable giving as a way of expressing one's belief that there are more important things in this world than material gain? Or what about charitable giving as an expression of gratitude (another very complex theological concept)? Or, less palatably, what of the more charismatic religious notions of charity qua giving monies in transactional terms (e.g., "giving money = receiving healing," or "giving money = I love God")? None of these highly diverse motivations are captured within Zak's highly superficial equation:

More oxytocin = more empathy = more money given to charity.

As soon as one understands the inner complexity of something even as apparently simple as charitable giving, a proposition as superficial as Zak et al.'s "oxytocin makes you more generous" ceases to make sense. At the very least, it is hard to imagine how such propositions can be taken as providing any kind of fine-grained basis for enhancing the moral good being investigated. But in this way, Zak is the perfect cipher for even the most circumspect of philosophical enthusiasts discussed in part I. Very little thought has gone into wrangling with the inner complexity of the moral goods set forward, nor the idiosyncratic backgrounds which shape them into the goods they are.

Moreover, just as the simplicity of Zak's conceptualization of his moral goods helps highlight the hidden crudeness of the various enthusiasts' contributions, so too the theological complexity of the moral goods in question highlights the hidden depths of what appear to be perfectly simple, universal moral goods. The background to the concept of the moral good in question, the motivation of the giver and the framework within which the charitable giving is being enacted, is crucial in determining the nature of charitable giving as the specific moral good that it is, which can have a number of possible motivations, many of which are almost entirely unrelated to empathy as cause. Nor do they necessarily have anything to do with the idea of "wanting" to be charitable or "feeling like it."

Broad-scope and generalized as such an intervention would be, oxytocin will most likely be counterproductive with respect to Christian charitable giving. Part of the discipline of charitable giving in a Christian context is that one learns to give generously especially where one does not want to. In other words, oxytocin, insofar as it makes one want to give generously, robs one of the opportunity to participate in the *discipline* of giving generously. The conceptual scaffolding of charitable giving involves aspects of duty and of discipline, which represent independent moral goods that have nothing at all to do with "wanting" to be generous. This is not a matter of Christian self-punishment, but rather it is about awareness of the simple fact that living a moral life involves doing things that one does not want to do, and this is something that needs to be practiced.

This is why it is so fundamental to grasp the territory of what one is attempting to enhance. Oxytocin, and by extension Douglas' entire noncognitive approach to moral enhancement—which relies entirely on the idea of making one want to do the moral good in question—falls foul of moral goods that have to do with "doing the right thing" where one does not want to. Such theses neglect the reality that many, if not most, moral systems attribute at least some sort of moral value to doing what one does not want to do. Self-discipline is one of the highest moral goods, and in a way parent to the rest. For even in the most idealistic Aristotelian proposal that one has learned to genuinely want to do the good, getting to that stage involves a somewhat torturous process of habituation. Many moral goods, like charitable giving, are performed not just because they are themselves good, but because they contribute to one's development of self-discipline. They are moral goods integrated with other moral goods. Self-discipline is a tremendous moral good, the value of which gets suppressed when we begin to think of moral enhancement solely in terms of making one want to do the good.

Returning to oxytocin and generosity as a case in point, saying that oxytocin may "work alongside" traditional moral formation methods in relation to generosity or charitable giving fails to appreciate the very diverse ways in which such generosity can be construed and presented as a moral obligation or moral good. While we should definitely contend that superficial moral enhancement interventions are the best one can hope for here, again, this does not mean that any old superficial intervention will do. We still need a very fine-grained understanding of the nature of the moral good in question and the framework in which it is embedded to make sure we are not selling short the moral good being enhanced, or, worse, offering biomedical means which are subtly antagonistic (and thus self-defeating) with respect to the moral good when it is more sharply understood.

Faithfulness to Creation

We can take another example of how the conceptual shaping of certain moral goods can make things very difficult for those aspiring to integrate biomedical moral enhancement interventions within more fine-grained, traditional moral formative practices. A particularly clear example of the idiosyncratic nature of moral goods can be found in the idea of "faithfulness." Faithfulness is a complex theological idea which, in the Christian faith, is based on the idea of human existence being situated in a world created and sustained by a Creator God. Yet it has parallels in virtually every contemporary moral system, its focus being upon looking after the well-being of the natural world and the various creatures in it (including ourselves and the institutions we create). Theologian David Kelsey (2009) describes faithfulness as involving a dual involvement with respect to God and the world God has created. Faithfulness involves a range of competencies and skills (relating to intellect, affect, and bodily action), and demands involvement in a range of practices which help foster "existential orientations" (that is, very profoundly formed dispositions which orient the way one explicitly decides to live one's life and understand one's place in the world). Kelsey writes:

When human creatures are understood in ways guided by canonical Wisdom litera-ture's background creation theology, they are understood to flourish in wise action that seeks the well-being, for its own sake, of the quotidian, which includes them-selves. (Kelsey 2009, 324)

As such, "faithfulness to God's creation" represents a "standard of excel-lence" for judging the appropriateness of one's practices as responses to God (ibid., 194, 332). In this picture, human creatures' glory[5] resides in their "being dedicatedly active" for the well-being of the lived world (ibid., 315). The reference to scripture and the triune Creator God make the theo-logical virtue of faithfulness specifically Christian and, indeed, extremely manifold. Kelsey continues:

The distinguishing feature of the intrinsic standard of excellence, the normative edge, of human practices as understood in canonical Wisdom literature lies in the practices' intentionality as response. These practices are responses to God's call to human creatures. ... The relative excellence of enactments of these practices is mea-sured by their appropriateness as, precisely, responses to God's call. (ibid., 194, 197, 310, 402)

This intentionality of response is what defines Christian faithfulness as pre-cisely Christian. Let us now contrast this vision of faithfulness with similar overlapping ideals to be discerned in other traditions of thought. For one can readily find certain parallels to this idea of faithfulness in various value systems, cultures, or ideologies—perhaps some sort of deep ecological com-mitment to the biosphere, or in some sort of pagan heroic sense of "faith-fulness to the world." Unless the world that one has in mind as the object of one's faithfulness is envisaged as part of a created order sustained and related to by the Christian God and intended for wise human stewardship, there remain fundamental noncosmetic differences between the accounts which should not be overlooked. These are differences which shape how faithfulness as a moral good is appropriately made manifest.

Now, if one stands far enough back, one can see that there are overlap-ping ways in which such faithfulness might manifest in the different ideol-ogies, such as consideration for the creatures of the earth and their habitats, respect for the natural resources found within the earth, and so on. But as soon as one takes a fine-grained perspective, important differences will emerge. A pagan faithfulness is liable to invoke animistic or pantheistic forms of reverence, in combination with an often profoundly realistic and balanced grasp of both the cooperative and destructive powers of nature.

There is a vast array of concepts of what it is that makes up "nature" or "the natural order" contained within the panoply of pagan and neopagan forms of spirituality. And since faithfulness to such a natural order is the aspiration, how one goes about conceptualizing the natural order (e.g., in terms of Gaia worship, nature as divine, nature as living, nature as home, nature as mother, to mention but a few pagan nature tropes), will then shape how one understands the moral good of faithfulness. Indeed, one might say more generally that one's construct and understanding of the natural order will be an important conceptually shaping piece of scaffolding that will undergird how one articulates to oneself one's responsibilities toward the environment and the world in which we live.

Thus, a Buddhist or Jain vision of faithfulness might manifest as an extreme refusal to do any kind of harm to the earth's creatures. Such an idea of faithfulness might be expressed in terms of a strict vegetarianism (this might contrast with a more pagan view of hunting for food as a sacred and necessary part of the natural order[6]). A radical and extreme deep ecological faithfulness might manifest in a range of ways, from veganism to acts of ecoterrorism, as forms of suprapolitical action attempting to bring awareness back to issues facing the natural world. We can see, then, what makes faithfulness a moral good is defined in no small part by the scaffolding in which the action is embedded.

In chapters 2 and 3, we saw that an ecological concern is one often mentioned by moral enhancement enthusiasts. This brief inspection of the complexity through which a concern for nature is understood by various cultural and faith groups should go some way toward demonstrating how very difficult, if not impossible, it is going to be to motivate such concern in any unilateral manner (or, at least, in a manner that might speak directly to the underlying motivations and conceptual frameworks that would inspire the person in question to action). Therein is the issue: persons are treated of as if from the outside with moral enhancement, as generic "brain carriers." None of the enthusiasts in the philosophical literature want to get inside the heads of those who are to be enhanced. Yet people are motivated by different things, they understand their moral goods in different ways, and they need to be spoken to in different ways. This needs to be taken into account when thinking about moral bioenhancement.

So we cannot overstate the importance of having a close understanding of the nature of the moral good in question before attempting to generate in oneself an appropriate biological context. For example, some moral enhancement intervention related to empathy might be useful as a means to encourage a vegan approach to faithfulness, but this might not be particularly helpful if alternative kinds of expression are required. Someone who felt ecoterrorism to be the only appropriate moral solution to the neglect of the natural world might not self-select an intervention to make themselves more compassionate. And too great a specificity would make any given intervention incapable of dealing with the various legitimate conceptual backgrounds through which the ideal of faithfulness might be taken to have sense.[7]

We have mentioned charitable giving and faithfulness as things which, on the surface, seem like fairly simple, universal moral goods. What could be more simple than the idea of giving money to charity? What could be less exotic than the idea of faithfulness to the world we live in? One would be hard-pressed to find any moral outlook that cannot find language for expressing the importance of looking after the natural world and the various creatures within it. Both charitable giving and faithfulness fit easily within the range of common, universal goods mentioned in part I (i.e., are amenable to a "minimal" vision of moral enhancement). Yet look how quickly they yield a tremendous inner complexity and admit of so much potential variation given how many background constructs of the concepts are possible. The frameworks within which these goods are embedded and the obligations that they impose upon person and society are clearly very different, and sometimes antagonistic, when going between ideological scaffolding. The various kinds of responses that are characteristic and approved of within the respective background frameworks come from being immersed within that framework, and thus a scaffolding is the fundamental, definitive feature when it comes to defining moral goods as the moral goods they are.

Recall Savulescu and Persson's list of moral goods that are worthy objects for moral enhancement:

moral enhancements which (can) increase altruism, including empathetic imagination of the suffering and interests of others, coupled with sympathetic response to this, together with greater preparedness to sacrifice one's own interests, greater willingness to co-operate, and better impulse control. (Savulescu and Persson 2012, 13)

How are we to make sense of such acontextual generalities in light of all this? So long as the philosophy and science of moral functioning continues to treat moral goods as if they are "ideals from nowhere" (which is precisely how they are treated), then it has done no justice at all to the particular good in question, nor the group, society, faith, or culture which holds that good as something to be so highly esteemed. Rather, if moral enhancement is to work with Christian values or any idiosyncratic set of values (are there any other kind?), it must make a concerted effort to engage specifically with conceptual background or practice and shape itself from there.

The Promise of an Integral Technological and Traditional Project of Moral Formation

While it is possible for a number of rationales integrating biomedical and traditional approaches to moral formation to be forwarded, we can propose the following in light of the comments made throughout part III: that one integral approach might be envisaged as the utilization of (a) broad-scope biomedical means to set a general biological context, which would (b) allow space for a particular value or moral good to be sharpened into the fine-grained, enacted, conceptually formed, idiosyncratically specific value or moral good that a culture or group recognizes it to be, through (c) primarily relational means, that is, occurring within the context of a relationship, be that a mentoring relationship, family, or potentially even a therapeutic or group support relationship, which would then (d) shape and guide the development of the relevant salient identities, the gaining of new experiences, and the overall process of formation and reformation embodied in (ideally) the best expressions of that culture or group's vision of the moral good in question. All this scaffolding, combined with an existing predisposition of will on the part of the agent so enhancing, would help us envisage one potential form of integrative, hard, strong moral enhancement—one in which moral enhancement would be an explicit, self-chosen, and temporally extended project of ongoing formation.

What we are not saying is "let us take some SSRIs and this will make us less aggressive," or "let us take some oxytocin and this will encourage us to give more money to charity." Rather, what is attractive about our proposition is the way it invokes both personal responsibility and the need for external scaffolding as the dominating structure through which any given moral enhancement intervention would be operationalized. And then, again, not any old intervention will do, but rather one that is grounded upon a profound and subtle understanding of the moral good in question, and the background scaffolding in which the intervention would be applied. This proposal is a long way from simply saying "moral enhancement needs to work alongside moral formation." Such a vague proposition tells us precisely nothing. In wrangling with the question "how are such methods to work alongside each other?" it becomes clear that we have quite a complex task ahead of us in figuring out, for any given intervention, how it is to be applied in real life situations. Such a proposal demands integrative moral enhancement be understood not as a solitary project of self-discipline to be learned as if in isolation from the world. Instead, it is primarily understood through a structure which involves relationships, face to face practice, and active, personal engagement. Nor does this approach proffer any oversimplistic notions of the values or moral goods in view, for the idiosyncratic nature of such goods is to be refined precisely within that context, which opens up room for the exemplar or mentor in question to indicate, through practice, what is expected of the person being apprenticed.

By extension, this means that the entire range of bodily powers available to a person have potential to be used, given that such practice is embodied and enacted in real-life scenarios (rather than simply limiting our grasp of moral functioning to, say, "emotional modulation"). Nor would such an approach be biologically reductive, for the biological element of the intervention would be limited to a support function which only exists to augment already existing efforts to engage in the process of moral formation that the relationship entails. We have proposed, then, a provisional but relatively rich rationale for envisaging how hard, strong moral enhancement and traditional means of formation can be brought together in a potentially desirable manner. In chapter 9, we will show how such a proposal might play out with respect to the treatment of alcoholism and drug addiction.

A Final Word on the Problems of Social-Environmental Dominance in Moral Formation: Appreciate but Do Not Valorize

An overemphasis on the social-environmental dimensions of moral functioning is also a form of reductionism. Yet there is a difference between saying that the social-environmental dimensions of moral formation are the dominant factors with respect to moral formation, on the one hand, as we have been contending throughout, and, on other hand, valorizing them beyond their worth. One must encourage the former without necessarily falling victim to the latter. Up to this point, there has been no interrogation of the quality of given social-environmental situations, only a weighing of their significance to counter the neuroprimacy so prevalent in the moral enhancement domain. Yes, social-environmental contexts are weightier, but this says nothing at all about the actual goodness or badness of particular social setups. One must be attentive to the manner in which social-environmental dimensions of moral formation can foster evil as much as good. It is the power and range of the social-environmental dimensions of moral functioning that justifies saying that they have a supervening role over biological factors, but it is also what makes them, at present, so damaging in fostering evil.

One must be aware of how social-environmental factors can lead to the distortion of the persons inhabiting the various environs which scaffold them, as well as distort their moral faculties and the way they articulate their faith or culture's moral goods. There are plenty of well-established psychological studies into such phenomena, the Zimbardo prison-experiments (1971) being the most famous. There is little need to reiterate these findings. Suffice it to say that a person's social context can legitimize and encourage truly evil behavior, regardless of the moral character of persons involved (Zimbardo gives the example of the Abu Ghraib torture as a recent example of this phenomenon; see Zimbardo 2007). As a result, it is because the social-environmental surroundings of the individual in question are so influential that it is so important not to valorize social-environmental scaffolding dimensions in and of themselves without critical reflection.

An environment which facilitates evil (as in the case of, say, "the rape of Nanking," where the systematic rape and murder of women and children at the rate of 1,000 a night over a six-week period was perpetrated by Japanese soldiers in 1937) is only one dimension of the problem. A potentially more disturbing prospect, with respect to the present concern of moral enhancement anyway, is the power of the mind (individually or in groups), to delude itself into thinking its evil deeds are actually moral goods. Social scaffolding can facilitate evil under the delusion that it is actually sustaining moral goodness.

This is a big problem for moral enhancement. What happens when groups' and cultures' very expression of moral goods are despicable? Or what happens when groups and cultures appeal to value systems which by their nature denigrate the use of faculties necessary for nuanced moral functioning (e.g., a religious fundamentalist group that fosters anti-intellectual tendencies, or a Nazi-esque society cleaving to virtù, respecting only pitilessness and the elimination of empathy)? There are ways, then, in which a group's background scaffolding factors can pervert or even destroy the moral faculties of the group in question.

The same morally formative methodologies we have seen in the pastoral reflections of the previous chapter—the formation of identity, providing new experiences, realigning perceptions, cultivating new significant relationships, and value-preference modification—are only as worthy in any particular instance as the particular identities and values being impressed upon the individual and group. Again, these means are morally ambiguous. Creating a moral identity is only a worthwhile task if the identity in question is not abominable. Precisely the same approaches discussed there can be appropriated by cults to be used in processes of brain-washing and conditioning. The exemplars drawn upon to lead the development of the persons being apprenticed to them are only as good as the exemplars in question.

We mentioned in the previous chapter the existence of many extremely conservative groups, many of which are religious, that demonize independent thought. There are explicitly anti-intellectual groups which either overemphasize fuzzy "intuitive" thinking (i.e., not thinking at all), or who overtly valorize conformity to strict group norms. What good will a moral enhancement be in a social-environmental scaffolding which does not prize discernment and rigorous rational thought? How are a person's powers of moral reasoning, self-criticism, and independent thought to develop in a society which does not prize such goods? This is a very important point. Members of ultra-conservative religious groups are all too often not educated to value such thinking but rather to despise it in favor of conformity. Indeed, members of such groups have been explicitly educated not to think.[8]

Moral enhancement in such contexts could in fact do immeasurable harm by galvanizing problematic tendencies precisely under the misguided belief that such tendencies are in fact moral goods to be idealized. This is another expression of "the bootstrapping problem" we encountered earlier. What good is moral enhancement in a social context where the guiding shape of moral behavior is abominable? And what do we make of the fact that so many areas where morality is a strong and explicit force in the public space often have monstrous visions of what moral behavior consists in (e.g., Taliban-dominated areas)?

The problem here is a powerful one: though moral enhancement's functioning would depend on its scaffolding, no moral enhancement can create a moral scaffolding. This is a meta-problem beyond moral enhancement's power and scope, but upon which the value of moral enhancement entirely relies. Moral enhancement might help augment a given vision of the good, but it cannot itself create a vision of the good, and relies on there already being a worthy vision of the good in place to scaffold its use. One would need an already morally laudable scaffolding if the prospects for moral enhancement are to be appropriated in a morally laudable way.

Let us look at the ways in which certain groups' scaffolding generate despicable ways of defining moral concepts and moral goods. Take the notion of "honor," in ideal circumstances a moral good which is profoundly noble (by definition), yet also a concept that has been used throughout the millennia to give moral justification to the outright murder of anyone who causes the slightest offense. The ancient Buddhist notion of "compassionate killings" (the doctrine of poa) exemplifies in a darkly comical way the manner in which homicide, even genocide, can be justified as a laudable moral good if conceptualized in the right sort of way.

The notion of compassionate killings, employed by invading Buddhist kings to rouse their armies to brutal murder, and used more generally to justify the assassination of basically anyone that one does not like, hinges on the idea that it is morally permissible, laudable even, to murder a person if that murder prevents them from doing something that would accrue bad karma. Since accruing bad karma results in a lower rebirth, something requiring many lifetimes as a lower creature to atone for, one does a person a great favor by murdering them in order to prevent them making such a terrible a moral mistake (Jones 2008a, 53, 82–3).[9] To us, with our sense of distance, such a practice can be seen to be what it is, a convenient self-deception to justify murdering inconvenient persons, invading their countries, and taking their resources. Yet do we not all have, as societies and individuals, the same self-deceptive tendencies to dress up as morally laudable highly questionable deeds that we wish to get away with?

Returning to the idea of honor, even today this concept is in many countries taken to brutal extremes. Honor is taken as a rationale for maiming or murdering members of one's own family—for, say, not marrying whom the family patriarch decrees. So-called honor killing (defined as "acts of murder in which 'a woman is killed for her actual or perceived immoral behaviour'"; Hassan 1999, 25), is deemed a socially appropriate and morally praiseworthy way of resolving many perceived slights performed by non-docile and "disobedient" women in various countries across the globe. Such slights include refusing to submit to an arranged marriage, but also women seeking a divorce, receiving phone calls from men, failing to serve meals on time, or a woman "allowing herself to be raped." Adam Jones of Gendercide Watch writes:

> In the Turkish province of Sanliurfa, one young woman's "throat was slit in the town square because a love ballad was dedicated to her over the radio." (Turgut 1998, in Jones 2008b)

The thing to be noted is the distinctly moral grounds which are appealed to here to justify such actions. It is not uncommon for recalcitrant young ladies (of predominantly Islamic situation), be that of Pakistani, Jordanian, Palestinian, Israeli, Balkan, Serbian, or Albanian origin, to be so turned upon by their families. Mothers will turn on their own children with homicidal fury, inciting and demanding their sons, often already willing, to torture and slay their own blood.[10] We have, then, murder characterized as a moral good.

The social and institutional scaffolding (gender biases in law, in the police forces, in the judiciary, and government indifference to honor killings; Amnesty International 1999) which embraces these notions of honor are manifold and significant, and we need not go on at any great length about the psychological tools used for twisting evil acts into moral obligations, objectifying and then demonizing out-groups or marginalized members of society. Despite an almost universal moral outrage, Western powers are loathe to get involved with such killings, and there are many cynical reasons for this. So let us ask, then, what good moral enhancement technology will do in a world where the international political status quo is quite openly one of realpolitik, where the philosophy of "might is right" and self-preservation are the primary driving forces?

Again this is very sticky territory, for one is involved in judging other cultures, when one's own culture is hardly above reproach. We are not isolated in this world, and expressions of moral goods are in constant conflict. In Israel, for example, a largely secular state, there is currently a resurgence of an ultra-orthodox form of Haredi Judaism (Sharon 2013). Such groups are highly insular, sectarian, and utterly unafraid to impose their will on what they see as a diabolic encroaching secularization of Israel. They are to be found on buses harassing the general public, physically enforcing a segregation of men and women on public transport (such segregation will likely touch on the emotions of certain groups in the West, and shows how quickly progressive tolerance can give way to evils that we associate with a shameful past we naively assume "can never happen again in today's world"), and violently haranguing women who have fled their oppressive, misogynistic family structures and the enforced marriages they were subject to.

So here is the real challenge for moral enhancement: if the social-environmental dimensions of moral formation are the dominant power, as has been asserted throughout, what good will moral enhancement be in an environment that scaffolds highly distorted visions of the good? What use is an intervention to generate empathy in a society which rewards and valorizes cruel, self-serving, aggressively competitive behavior? In a society where "looking after number one" is a primary supervening good, where can empathic interventions fit, and what good can they do against the weight of the larger social pressures in which they would be embedded?

The more distorted the scaffolding within which moral enhancement interventions are to be applied, the less effective we can hope them to be. Will an intervention to increase empathy do any good in an honor killing scenario? If a mother is such that she is willing to torture and murder her own children, do we really believe that a quick dose of intranasal oxytocin and a steady course of SSRIs will do the trick? Moral enhancement cannot be our salvation here—social distance and critique, coordinated deliberation, discussion, and mass social action on every level possible, at the very least, are what is required. We have to be absolutely realistic about the extent to which we think moral enhancement can even contribute to the resolution of the great evils of our time. Indeed, we have to be aware of the possibility that it may well end up galvanizing such evils and making them infinitely worse by framing evils as the very moral goods to be enhanced, then attempting to "lock them in" through biomedical means. If moral enhancement is the weaker partner in the interaction of biology and scaffolding, and the scaffolding in question is despicable, then are not the greater evils of the world better dealt with on the level that the problem truly resides in—the scaffolding and attitudes which sustain such evils? If the quality of moral enhancement relies on the prior question of the quality of its scaffolding, then the scaffolding must absolutely be remedied before any strong moral enhancement be proffered.

Conclusion

Throughout this chapter, we have been developing the suggestion that, as far as an integrative, hard, strong vision of moral enhancement goes— one wherein persons are purposefully using biomedical means to augment their moral performance—a superficial, brute intervention may be the best option. This superficiality is required to allow room for the scaffolding, which is the stronger part of moral enhancement, to give shape to the moral good being aimed at, to direct that enhancement toward its appropriate range of idiosyncratic expressions. Though a superficial biomedical intervention may be preferable, we must be adamant that we need a very sharp understanding of the moral good in question in order that such brute approaches, generalized though they are, do not end up going counter to the moral good being augmented.

The hope for the enactment of moral enhancement would then be that it is embedded within more person-centered kinds of relationships which can, in principle, give idiosyncratic shape to the general biological context produced by whatever biomedical intervention would be applied. The implication is that, if sufficiently fine-grained interventions cannot likely be devised, then the aspiration should rise no higher than that of creating interventions of a more purposively superficial order.[11]

This is not as pessimistic an appraisal as it sounds. To the contrary, actually—for as soon as one recognizes the facts that no virtue or moral good can simply be engineered from scratch, out of nowhere, and that the social-environmental dimensions of moral formation are supervening, shaping, inextricable, and noncosmetic elements of moral formation, one then has the basis for proposing a realistic rubric for moral enhancement.

This suggestion is quite a radical turnaround. The usual skeptical perspective is that moral functioning is too complex, and therefore any current possibilities for such intervention which are inherently superficial and overly broad (e.g., lobotomization, chemical castration, psychiatric-grade sedation) cannot reasonably qualify as genuine moral enhancement. The turnaround is this: the superficiality of moral enhancement interventions is not what makes them weak proposals; to the contrary, it is only as superficial that a given intervention can hope to contribute to a realistic project of hard moral enhancement. By the standard of hard moral enhancement, it may very well be the case that there is no standalone intervention that could possibly count as genuine moral enhancement. Genuine biomedical moral enhancement, by this standard, can only ever manifest if it is integrated as one prong among other nonbiomedical approaches (and then, again, not any old biomedical intervention will do).

Ironically, then, contrary to the popular skeptical view that a superficial moral enhancement technique might lead to the possibility of biological reductionism in moral enhancement, in fact the very opposite may be true. In fact, it would be a fine-grained approach to moral enhancement that would reduce moral functioning to its biological and neural components, since it would be attempting to do everything on biological-neural terms. Yet the attempt to tackle moral problems on the primarily biological level will always have a limiting factor: actual reality. Since moral goods cannot be understood in biological terms only, but always require shaping, any attempts to treat them as such will result in failure. We need not rely on speculation here. Numerous illustrations exist: electroconvulsive therapy or transorbital lobotomization can be appealed to as tried-and-failed attempts to deal with extreme aggressive behavior on the purely physical level. Not only are such means barbaric, but ultimately, they did not even work. At least, not in any way that leaves over some semblance of the personality of the subject in question.

Contrary to the assertions made by many commentators that moral enhancement is a futuristic idea whose realization is far from the present moment, in fact the very opposite is likely true. Not only is moral enhancement happening now, but if we have in mind explicit, self-chosen projects of moral bioaugmentation, then we are likely already near to the very peak of what any credible, desirable moral enhancement could hope to achieve. It is not so much that we have already obtained all the potential interventions that might be envisaged (assuredly this cannot be so), but rather that it is *the degree of penetration* of the interventions that has likely reached its peak. One would suspect that it is both possible and likely for "more of the same" sorts of interventions to be produced as we have encountered (pharmaceutical, neuro-manipulative, and so on). Moreover, it may very well be that through some great ingenuity, some entirely new intervention be devised (if optogenetics shows anything, it is that remarkable ingenuity is indeed possible).

Yet, as we have seen, regardless of how sophisticated or impressive the technology that might be devised, in anything but the most fantastical prospects for the future, there will always be certain walls that biomedical approaches simply cannot breach. The nonbiological dimensions of morality that we have been considering throughout—context, ambiguity, moral scaffolding, predisposition of will, and the dynamically interwoven quality of all these various influences—are all fundamental to the shaping of moral goods as they manifest in moral living.

These elements are fundamental to the moral life, and so, too, their interweaving. Yet these features are biologically incommensurable. Moral scaffolding, the need for conceptual shaping for our moral goods, is not a factor which can be penetrated by biomedical intervention. Thus, moral enhancement cannot be a matter of "hoping for future technology." The level of sophistication of the technology in question is not the ultimate concern at this point.

Until we get a technology that is capable of turning us into remote-controlled moral robots by means of CPUs that contain contextually sensitive processors able to interpret conceptual goods and apply them on an ad hoc basis, I would say that we have already penetrated about as far as we can by means of technology alone. Certainly for now, the walls mentioned here are simply too steep to allow for any further penetration of moral enhancement technology. We may have more of the same—more drugs, more technologies of potentially incredible sophistication—but without integrating these biomedical means within a more interwoven framework appreciative of the complex multiplicity of influences, we have little room to progress. The implication of all this is that our focus needs to be not so much on developing new technologies, but rather upon thinking through some meaningful ways of assimilating what we already do have within a practicable, integrated framework. Chapters 8 and 9 will provide an attempt at thinking through how such an integration might be envisaged.

IV   Praxis

8 Enhancement or Remediation?

At the beginning of the book, we saw that there are a number of ways of construing moral enhancement, a set of distinctions which are helpful in getting clear exactly what we are talking about. That being so, the book has been a story about two dominant modes through which this idea of moral enhancement can be understood: as "hard," attempts to explicitly improve the moral functioning of persons along one or more dimensions of such functioning; or as "soft"—that is, less overt ways of manipulating a person's behavior and judgment (usually without any explicit mention of morality being made). What happens when the effects of a medical or mental health intervention have considerable overlap with morally related powers—though without taking "improving moral functioning" to be the rationale or purpose? An intervention to help treat, say, aggression, or an affective disorder, may well have considerable overlap with the sorts of concerns we have been dealing with throughout the present book. Yet there may have been no intent at all to enact such measures as distinctly "moral enhancement."

Such soft moral enhancement, it will be suggested, has numerous forms which are already presently being enacted. Some of these forms are more morally problematic than others. Having inspected the difficulties and moral issues relating to hard, explicit moral enhancement, it is time to change our focus. It was mentioned in chapter 1 that both narrow and broad definitions of moral enhancement are required to get a thorough grasp of the potentiality of the domain. Shifting our focus away from hard moral enhancement, we can begin to expand our scope to include more subtle means for manipulating morally related behavior. Indeed, part I concluded with the assertion that this softer kind of moral enhancement is not only more realistic in terms of its real world potential, but that such interventions are already in play, and it is likely inevitable that this continue to be so. The aim of this chapter is to explore specific ways in which soft moral enhancement is already a reality in the public space and to evaluate the sorts of reductive traps such soft moral enhancement might fall into when considered through biomedical lenses.

We shall begin this exploration by meditating upon another core distinction that was raised at the beginning of the book, that between "positive" and "remedial" kinds of moral enhancement (Agar 2013). It matters a great deal, for example, whether we are intending moral enhancement to actually improve a person's moral functioning (positive), or whether we are attempting to resolve certain morally problematic "undesirable" behaviors (remedial). This is a very important distinction, but, like the others, it is also quite tricky. Not only can it be rather hard to disentangle enhancement from treatment, but even when so distinguished, there are problematic ways in which both treatment and enhancement can be construed.

This distinction between enhancement and remediation, difficult as it may be, is a very important one for the moral enhancement discourse. This is so for at least three reasons. First, it is implicitly applied by commentators throughout the discourse. Second, the overwhelming majority of even semiplausible interventions proposed in the literature are of a remedial nature. And third, the semblance of feasibility that these remedial interventions display mean that separating treatment from enhancement might also help to separate more realistically workable interventions from those that are more fantastical.

Given the manner in which the positive/remedial distinction subtends the literature, the intricacies of how it might work in relation to moral enhancement deserve discussion. Treating moral problems in medical terms is, as we shall see, fraught with dangers. What is needed, above all, is some way of grasping a person's moral powers, without objectifying that person into a "biological machine" and without treating their moral agency as something to be managed in the way that one deals with a bacterial infection or some kind of unsightly growth. Given the extent to which many commentators in the field implicitly divide positive from remedial moral enhancement as if the distinction were the easiest, neatest, and most conceptually unproblematic to make, it behooves us to look more deeply into the implications that thinking in this way provokes.

*

We will begin by defending a version of the treatment/enhancement distinction, which, though fuzzy, serves as a helpful conceptual heuristic for thinking about how certain morally related issues might be managed using pharmacological/technological means. As should be clear by now, moral enhancement can be quite dangerous if thought of in medical terms only. Moreover, we will see that the medicalization of morality, of "undesirable behavior," shares much territory with the more dubious elements in behavioral control, social engineering, and paternalism. These are touchy points, for there is an extent to which paternalism, behavioral control, and social engineering can themselves be distinctly moral goods.[1] The question with these issues is always where the lines of permissibility are to be drawn.

A similar point is to be made with respect to treating moral problems in medical terms. Indeed, the suggestion will be made in the present chapter that much of what is currently considered moral enhancement in the literature is actually more appropriately considered under the rubric of mental health intervention. This is a pragmatic point. Many of what are being considered as moral problems in the enhancement discourse are already treated of as mental health issues (e.g., alcoholism, antisocial personality disorder). Given the serious pharmacological prescriptive dimensions involved in the idea of remedial moral enhancement (and who else would have the appropriate authority to prescribe "moral enhancement drugs"?), the existing mental health domain, with its already defined structure of accountability and its existing structure for managing such morally related issues, presents itself as the most natural institution for assimilating the many concerns we have been considering throughout this book. Yet such a prospect is not without dangers, and these dangers need to be explored.

Revisiting the Treatment/Enhancement Distinction for Purposes of Moral Augmentation

What are medicine's moral limits? At what point can we say that a particular medical intervention has transgressed the bounds of what it should be dealing with? When it comes to questions of enhancement (we are going beyond the idea of moral enhancement now and referring to other forms of human enhancement, e.g., genetic enhancement, cognitive enhancement, physical and performance enhancement), then perhaps one of the most bitterly fought moral lines is that which may or may not exist between therapy and enhancement.

There are many voices critical of enhancement projects generally—we need look only to the anti-Brave New Worlders, the Agrarians, and the bio-Luddite religious right (Lawler 2005) to find what are, in some cases, relatively mainstream groups who condemn the idea of enhancement outright. These groups espouse various rationales to justify their dislike of enhancement projects; for example, the anti-Brave New Worlders fear that technology will cause humans to surrender their humanity and languish like Nietzsche's "Last Man," a contemptible creature rotting in a state of subhuman contentment, persons of endless leisure, endlessly gratified and too satisfied to ever do anything worthwhile with their lives (ibid.). Alternatively, some groups have the sense that man's only hope is to return to an earlier stage of technology, to live simpler lives. In agrarianism, recovering the connection with nature and local community is articulated as the true source of human dignity and happiness (ibid.). Or, there is the more Christian fundamentalist sense among religious bio-Luddites that "trying to improve nature and valorizing human intelligence are abominable hubris certain to lead to apocalyptic consequences" (Hughes 2004, 107).[2]

The problem such critics face is rationally justifying the existence of a moral boundary between that which is completely acceptable (i.e., treating persons who are in a bad way), and taking persons beyond the limits of good health, making them "better than well." And while, on the surface at least, the treatment/enhancement distinction does seem to have some obvious, intuitive sense, there are a range of pro-enhancement advocates who take a very different view. Such enthusiasts (e.g., Harris 2012) present a number of strong arguments for suggesting that the distinction is completely incoherent, or simply not the appropriate way of dealing with the question of permissibility in an enhancement context to begin with.

It is very interesting that the treatment/enhancement distinction is so thoroughly embedded within the moral enhancement literature. Because a great deal of its use is implicit, the distinction is applied for the most part without the slightest concern for the problems such a distinction raises. One might suggest that the use of this distinction without subjecting it to any close inspection constitutes a significant gap in the conversation. For if the distinction does turn out to be incoherent, indefensible, then this is going to have rather strong implications for the discourse. Indeed, as we shall see, the distinction as used in a moral-medical context raises considerable issues idiosyncratic to itself. Yet I will suggest that it is worth keeping the distinction around if for no other reason than to raise and inspect the various dangers that the idea of a "moral sickness" in medical terms might imply. For in thinking of man's moral capacity as something which can be "abnormal," which can conceivably "malfunction" or "break down"— or, worse, "become sick"—a range of perturbing conceptual issues are raised which rely on a very questionable use of metaphors regarding the human person.

Problems with the Distinction

Defining the Treatment/Enhancement Distinction The first challenge that arises when attempting to use the treatment/enhancement distinction is the tremendous difficulties that arise in actually defining the key terms themselves. At first sight, the distinction between therapy and enhancement seems fairly uncomplicated. It can be broadly stated as follows:

> Therapy is generally defined as the prevention or cure of disease, or as the restoration or approximation of return to normal physiological function. Enhancement is defined as the alteration of individual (or group) characteristics, traits, and abilities (both health- and non-health-related) beyond a measurable baseline of normal function. (Lustig 2008, 41)

Yet this split between "curative" or "preventative" medical therapies and those intended to make individuals "better than well" is a little more complex than it first appears. Pinning down what exactly constitutes the appropriate domain of treatment is not easy, and terms like "health" and "normalcy" cannot be defined without controversy. Juengst, for example, gathers together three significantly different ways of understanding what might count as a treatment. The three main definitional paradigms are as follows (Juengst 1998, in Carson 2000, 641):

(a) Malfunction (disease-based accounts): a treatment in this view is aimed at restoring the function of an ailing organ or body (e.g., in the case of heart disease or kidney failure);

(b) Physical norms (statistical normalcy accounts): there is a given range within which human normalcy can be located, while treatment is aimed at anything outside of that range and is aimed toward bringing persons back within this range; and,

(c) Sociological norms (ideological accounts): in this view, something is a treatment if it is generally "regarded as such" within the social context in which it takes place.[3]

All three of these ways of defining terms like "health" have been used at various times and have implications for the way one defines the concept of treatment one is using. And these definitions are more or less value laden. For example, the sociological norms approach (e.g., giving hormone replacement therapies to menopausal women [and middle-aged men now, too] is in no small part the treatment of a socially constructed malady that exists as a malady in no small part because of contemporary Western attitudes about how middle-aged persons should be living their lives) proffers an extremely value-laden construct of health; we are talking here about "health" as a normative concept rather than a purely descriptive one. Health in this sense has prescriptive dimensions (i.e., middle-aged persons should be vigorous and active, thus replacing the hormones of their youth constitutes a medical treatment), and whether this be explicit or not, treatment can be mandated on value grounds, not to manage sickness but simply to remedy something that society considers to be "unnatural," undesirable, or some kind of aberration. Here the moral and the medical are blurred to a considerable degree. Depending on the context, such sociological definitions of treatment can be perfectly legitimate, or not.

In contrast, the disease model of health is a lot less complicated and more value-neutral. For example, a heart is meant to pump blood to the rest of the body. If it does not fulfill this function, then it becomes an appropriate target for treatment. This does not raise too many controversial concerns. The physical norm account is more ambiguous, and any reference to norms as definitive of health (since norms change) suggests a scale which is located, relative, and mutable. This also sneaks value judgments into the notion of health, because persons—usually those at the bottom end of the norm curve (e.g., very short people)—are considered in bad shape for no other reason than they are at the bottom end of a scale which is shifting and relative in nature.

At the very least, the ambiguity of these terms complicates the task of defining the treatment/enhancement distinction, shows how hard it can be to even get off the ground in using the distinction as a morally significant boundary, and demonstrates the need to get very clear what one is talking about so one does not unknowingly switch between the different visions of health and confuse the various implications the different definitions evoke.

Too Muddy to Work?   Even defenders of the treatment/enhancement distinction will admit that there are some serious boundary issues in applying the distinction. Indeed, as Sparrow (2010) notes, many commentators feel the distinction is simply too muddled to do any meaningful work. The normative power of the distinction has received sustained philosophical attack. Some commentators, such as John Harris, have gone so far as to argue that there is no distinction there at all. Harris (2012) writes:

Scientists and doctors like to pretend that there is a distinction between therapy (treating illness and curing people) and enhancement (improving upon normal functioning), but in reality that distinction collapses. Let me give you an example: people don't die of old age; they die of the diseases of old age. Treating diseases is therapy, but if you could systematically treat all the diseases of old age people wouldn't die.[4] ... That would be enhancement. So you can see that things that start out as therapies actually constitute enhancements because of their effects, so actually the distinction is illusory and not useful. (Harris 2012, 38)

This is a little too strong. The fuzziness of the distinction does not mean that it is illusory, nor does it mean that it might not be useful in other ways. Harris is certainly correct in saying that in many cases, if not most cases, no firm boundary between therapy and enhancement can be drawn. There are many cases (e.g., embryo selection on the basis of genes, "when the same technology is used as either a treatment or an enhancement"; Lewis 2003, 269), that cause the distinction to "blur beyond recognition" (ibid.).[5] Indeed, once we add the fuzziness of the distinction to the manner in which no clear and uncontroversial definition of the key terms "health," "disease," or "normalcy" can be settled upon, it would seem that the distinction only complexifies rather than simplifies the various boundary issues it is meant to resolve. It is for reasons such as these that various commentators have suggested the distinction be abandoned and alternative conceptual approaches be sought.

In fact, the entire distinction is surrounded with fuzziness on all sides. James Giordano (2011) describes the fluidity of the situation and the difficulty of getting fixed or sharp definitions as follows:

Medicine is not enacted in a social vacuum, and therefore it is important to examine the ways that current and proposed diagnostic schemas can affect, and are affected by social and legal meanings, values and attitudes. This becomes ever more meaningful if and when diagnostic architectonics shift, and we are forced to confront changing constructs of normality and abnormality, and the various clinical, cultural and ontological dimensions that are impacted by such distinctions. (Giordano 2001)

In other words, definitions are in constant flux, and the technologies that become available further affect the definitions one has. What occurs then is a "diagnostic creep" in which a condition is defined as a disease simply because an intervention exists to ameliorate it (Dees 2007, 377). This idea of diagnostic creep is particularly relevant for moral enhancement. This is part of the problem of medicalization, the process of expanding the area which medicine takes to be its purview, and the difficulties in drawing a clear line between enhancement and treatment can be illustrated by looking at the way this medicalization shifts the "constructs of normality and abnormality," as Giordano put it.

Take, for example, the partial deregulation of the pharmaceutical industry by the FDA with the "Modernization Act" of 1997, wherein the use of pharmaceuticals for maladies not part of the original intent of the drug in question was made legal. This means that drugs designed for one purpose, say treating depression, can now be marketed as treatments for other disorders, such as anxiety (Conrad 2005, 5–6). Conrad, a vocal critic of this process of medicalization, writes:

Regulations allowed for a wider usage and promotion of off-label uses of drugs and facilitated direct-to-customer advertising, especially on television. This has changed the game for the pharmaceutical industry; they can now advertise directly to the public and create markets for their products. ... Marketing diseases, and then selling drugs to treat those diseases, is now common in the "post-Prozac" era. Since the FDA approved the use of Paxil for SAD (Social Anxiety Disorder) in 1999 and GAD (Generalized Anxiety Disorder) in 2001, GlaxoSmithKline has spent millions to raise the public visibility of SAD and GAD through sophisticated marketing campaigns. (Conrad 2005, 5–6)

Therein, what was once considered an enhancement is redefined as a therapy for a newly characterized disease (Dees 2007, 377), and "anything that makes us feel better thereby becomes a therapy. At that point, the boundary between what is a therapy and what is an enhancement is completely blurred to the point of uselessness" (Dees 2007, 377). As this creep continues, terms such as "average" or "normal" functioning continue to narrow toward more and more perfectionist extremes. Andy Miah (2008) makes a similar point. Observing how enhancement slides into therapy as definitions of normality shift in light of medical possibilities, he writes:

If one sought enhancement to raise height or even intelligence to a level that corresponds with the upper sector of the normal human range, the consequence of this would be a shift in what is regarded as normal. ... A highly idealized or perfectionist view of superior or normal traits would mean that a trait we ordinarily take to be normal would count as "defective" (Brock et al. 2000, 105). ... Individuals who are below the normal point of this enhanced range could find themselves subject to the medical diagnosis of needing enhancement. (Miah 2008, 7)

Salvaging the Distinction With all this in mind, it is easy to see why many commentators wish to do away with the treatment/enhancement distinction altogether. Certainly, there are useful alternative conceptual approaches to draw upon. For example, there are a range of proposals which look at permissibility in terms of a related constellation of ideas such as "identity," "authenticity," and "essential nature." The core of such approaches is to generate a picture of what the human being is, what the human being *should* be, and what it is exactly about the human creature that needs to be preserved in the face of increasing possibilities for augmentation.

By way of example, some will suggest that enhancements which alter a person's height, or their memory capacity, or the number of fingers they have on each hand, while not insignificant as alterations, are completely acceptable forms of enhancement because they do not rob the human being of that which is fundamental to his dignity. In contrast, making a change to a person's demeanor—for example, using deep-brain stimulation to cure depression[6]—alters their identity in more significant ways. Such interventions make more fundamental alterations to a person's being, changes in the nature of their very personhood; these are considered more morally problematic. Here, it is a question of determining what is fundamental about our present being, what is worth keeping, and what is permissible to change.

However, while alternatives to the distinction are certainly welcome, and do provide important exploratory lenses through which to further examine the subject matter, none of the alternatives are free from problems either. Defining what is essential and worthwhile in human nature involves posing questions of personhood which are arguably even more tangled and as interminable than defining terms like health and normalcy. As such, these approaches are not only arguably as fuzzy,[7] but they are all dependent on alternative arguments beyond themselves for their moral force. Indeed, I would suggest that all of the approaches we have encountered here have their uses; they do important conceptual work, bringing to the fore important considerations that need to be brought to bear on the discussion. If this is so, the treatment/enhancement distinction is worth preserving, with certain qualifications. I think Richard Dees (2007) has put this best:

Although the boundary between therapy and enhancements is murky and fluid, the distinction is nonetheless real. We do not discard the distinction between black and white because there are so many shades of grey that we cannot always draw a clear line between them. Like the distinction between black and white, the therapy-enhancement distinction is useful, even if the boundaries are fuzzy. (Dees 2007, 377)

Where one has to agree with critics like Harris is that the treatment/enhancement distinction cannot be said to work as a morally significant boundary. Beyond that it is worth retaining as an exploratory lens, a conceptual heuristic, and a means of bringing into focus powerful issues otherwise concealed within the discourse. Using the distinction for merely descriptive purposes, to articulate some shades of difference between kinds of moral enhancement project, or as some pragmatic distinction, is perfectly defensible. As we shall see, the distinction helps clarify further issues in the moral enhancement debate, the fuzziness of the compulsory/elective distinction, the dangers of diagnostic creep, the dangers of medicalization, and the dangers that use of medical metaphors present when thinking about moral remediation. This elucidative power alone should be sufficient to justify using the treatment/enhancement distinction.

Implications for Moral Enhancement

From the Treatment/Enhancement Distinction to the Voluntary/Compulsory Distinction

The discussion of the treatment/enhancement distinction is germane, not just for the use of the same distinction with respect to moral enhancement, but also with respect to the application of the voluntary/compulsory distinction. In fact, the two distinctions share quite a lot in common, and many of the critiques of the treatment/enhancement distinction evoke similar problems for the voluntary/compulsory distinction too (see Wiseman 2014c). The question of the permissibility of moral enhancement usually turns on whether that enhancement is presented as voluntary or compulsory. The likes of Persson and Savulescu are condemned because their vision of moral enhancement is one in which no one has any choice over whether to take it or not. The offerings of Douglas, Hughes, and Rakić are more palatable because they defend moral enhancement as a choice. That is the perception, at least. One chooses to take a moral enhancer or chooses not to. What could possibly be wrong with that? We see quite clearly then that the voluntary/compulsory distinction is being used to draw very clear lines of moral permissibility (compulsory enhancement = morally bad, voluntary enhancement = morally permissible). In fact, the same conclusion we have just reached regarding treatment and enhancement, that the distinction is useful on a descriptive level, but much too fuzzy and vague to work as a morally significant boundary in and of itself, works just as well here.

While the elective quality of the offerings of Douglas, Hughes, and Rakić look more palatable on the surface, things come apart very quickly as soon as one observes that, actually, in a world where moral enhancement were a reality, it might actually be better for some interventions to be compulsory. A humane intervention to constrain predatory and sadistic child rapists would be (as Rakić himself observes) more responsibly set forward as compulsory. Something more like taking oxytocin to increase one's prosocial tendencies would be something better off kept as voluntary—it would be highly suspect for a government to mandate such non-necessary drug use for every last one of its citizens in this case. So, to put things in their simplest terms, if some interventions are better off being compulsory and some are better off being voluntary, the distinction does not have any decisive moral power in itself. The whole set-up of the moral enhancement debate, insofar as it premises the moral permissibility of such enhancement upon whether an intervention is voluntary or not, loses its integrity because one is constantly having to look at the particular factors on the ground (e.g., the intervention itself, the target audience, the danger posed by that segment of society, the side effects, the larger bioethical implications, and so on). In other words, what makes a given intervention morally valuable or not has little to do with whether it is voluntary or compulsory, but must rely in each case upon a range of prior factors beyond itself.

In other words, as far as the moral worth of the distinction goes, any attempt to use it as a way of sorting what is morally legitimate from what is morally problematic has got everything backward. It is not whether the intervention is voluntary or compulsory that gives it decisive moral significance, but rather it is the facts on the ground that determine whether the intervention should be voluntary or compulsory. Suggesting that just because an intervention is voluntary one has something more morally desirable than the same intervention would be if made compulsory is to abstract out from the inquiry virtually everything morally relevant about the particular intervention and target in question. We will see in chapter 9 that very idiosyncratic issues are raised with regard to whether the particular intervention in question is to be made voluntary or compulsory. As we will see, these idiosyncratic issues absolutely cannot be settled in the abstract by looking to the voluntary/compulsory distinction in isolation.

Again and again, an appeal to a case-by-case analysis must be made, and a constant eye must be kept out for the practical realities of the environment in which the proposed intervention is to be instantiated. Simply saying, as Persson and Savulescu have, that the world is in a precarious position, and thus enacting compulsory moral enhancement is a moral obligation, is to make a proclamation which has no literal sense. Such statements are just too generalized, too vague, and abstract. They are utterly detached from any ground-level realities of the world in which such a prospect is to be enacted. Remember, the distinction cannot tell us about whether "moral enhancement per se" is morally worthwhile, because *there is no such thing as "moral enhancement per se."* Something is required that is better differentiated. Rather than just asking the abstract question: "should moral enhancement be voluntary or compulsory?" we need to be asking which *particular* intervention is best understood in voluntary terms and why the particular facts on the ground make things so. As such, like the treatment/enhancement distinction, although talking about voluntary or compulsory moral enhancement does have important descriptive significance and an important part to play in the discourse, we have to conclude that it cannot be taken as a morally significant boundary in and of itself.

Diagnostic Creep and the Illusion of Choice

Like the more dominantly discussed voluntary/compulsory distinction, the treatment/enhancement debate raises significant questions for moral enhancement. In fact, the two distinctions are subtly intertwined. We have seen that the more easily a certain behavior can be presented as "defective" or as occasioning some extant pharmacological treatment, the more likely that perfectionist views are to proliferate, and the range of normal human behavior (or in the moral case, the range of acceptable human weakness) to narrow to an ever-increasing degree. Rob Sparrow (2014a) has made a case against the desirability of moral perfectionism and moral elitism, making the claim that any egalitarian society must be wary of the perfectionism that this sort of diagnostic creep can entail.[8] In light of the reflections just considered, such worries seem all the more justified.

The point with diagnostic creep is that it places strong but concealed social pressures on persons to enhance solely to keep up with the bell-curve, or to compete at its upper edge. This is a kind of compulsion that can be more insidious in nature than overt legislative compulsion. It is true that self-centered incentives to morally enhance might well be different than for other forms of enhancement (they would be less attractive, by definition, if we are talking about enhancements to do with going beyond self-centeredness). For example, the opportunity to enhance one's intelligence is much more immediately attractive in our competitive world than the opportunity to enhance our capacity for kindness. This would make it less likely that people will be so aggressive in seeking out moral enhancement interventions.

As such, it could be argued that diagnostic creep is less likely to pose a problem as far as moral enhancement goes. However, things are a little more complex than this. With remedial moral enhancement in mind, there are numerous subtle coercive forces that might be generated by the introduction of interventions to treat various morally related social ills. As Hughes (2006) has pointed out, it is not at all hard to imagine a circumstance in which a person might be deemed negligent if he or she did not take a intervention had one been available and harm was done as a consequence. If a person knows him or herself to be an addict, or violent, and interventions exist to manage these problems, not taking treatments in these cases might lead to more severe judgments being put upon them for transgressing the moral or legal boundaries they knew themselves to be prone to transgressing. As such, it would be very easy to get into a situation where the mere existence of a moral enhancement intervention could generate severe social pressure for certain groups of persons to engage in such enhancement—whether legislation is in place to compel such enhancement or not.

Again, the treatment/enhancement and voluntary/compulsory distinctions are subtly intertwined. Read in light of this notion of "diagnostic creep," and the manner in which such creeping can create powerful social pressures (which, though not legislated for, generate powerful compulsions nonetheless), the lines between voluntary and compulsory enhancement begin to look even less sharp than presented above. There are, after all, many ways in which social forces can generate subtle but powerful manipulative forces. For example, the means used to pressure parents to use Ritalin in schools as treatment for their children's ADHD (a more blatant example of behavioral control of children on a mass scale is hard to find) have been well documented—for example, making attendance for some "problem" pupils contingent on their taking the drug, in-house therapists using their authority to demonize parents who refuse to put their children on the drug, and claiming that they are "bad parents" not acting in their children's best interests (Breggin 2008, 220). This is not even to mention those parents who have gladly endorsed the application of what is, essentially, an adult anti-psychotic for use on their children. Thankfully, with the public profile of the drug and its various stunting, degenerative side effects on children's development having been raised to the national level and with several high-profile voices speaking out against its overprescription, the Ritalin bubble seems to be bursting. All the same, the process of manipulation and abuse of medical authority remains a persistent problem, with people who defy medical advice being labeled as "difficult."[9]

There are other ways in which the compulsory/elective distinction might be blurred. Semi-legal obligations may be produced, for example. Taking the case of drug addiction, judges may offer a choice to offenders during prosecution either to go jail or to take addiction remediation drugs. This would be voluntary in principle, but the other option, incarceration, might be so unpleasant as to make things such that there is very little real choice here at all. This is not morally problematic, necessarily, but it is yet another potential subtle process that challenges the neatness of the distinction. So, while any given remedial moral enhancement might not be legally obligatory, this does not mean that powerful forces of compulsion might not exist anyway, amounting to a compulsion in effect.

Covert Moral Enhancement? A Rose by Any Other Name

At the end of chapter 3, we concluded with the idea that "soft" moral enhancement can be understood as something that is already happening under the guise of paternalism as an overlapping set, though not an equivalent thereof. We can argue now that there is a sense in which such "soft" moral enhancement is also already taking place under the guise of medicalization. Throughout this chapter, I have made reference to the manner in which social norms end up being presented as objective, medical diagnoses—especially where treatments for these "disorders" have been devised. We have seen that there are a number of ways of defining health and disease; not all of these are so thoroughly ideological, but this cannot diminish the fact that ideological diagnosis does in fact occur. In other words, what we have is a process through which value judgments, which include moral judgments, are being presented as objective medical conditions to be remedied by medical means. The authority of medicine and medical science continue to be used as a means of objectifying deviance and "undesirable behavior" in culturally specific or politically convenient ways.

This is not a merely hypothetical concern. Soviet Russia was particularly famous for its use of "punitive psychiatry," wherein various "dissidents were labelled schizophrenics, thrown into psychiatric hospitals and drugged just for questioning the government" (Krainova 2011). Krainova has reported on the manner in which much the same process is occurring today, she writes:

Old habits die hard. ... Poet Yulia Privedyonnaya was hospitalised in a psychiatric facility in February 2010 for a month as part of a criminal investigation into charges that she formed a militant group under the guise of a poetry club ... two club members were sentenced to psychiatric treatment ... Sergei Kryukov, a reporter with the Chechen separatist website Ichkeria.info was hospitalised in a psychiatric facility by a court for a month in June 2009 at the request of the Federal Security Service ... Rifkhat Khakimov, a candidate in elections in the Urals town of Pervouralsk, was put into a psychiatric hospital for a month ... in April 2009 as part of a criminal case against him on defamation charges after he accused local law enforcement agencies of corruption. (Krainova 2011)

Several liberal activists, poets, and professors have had their sanity questioned by local authorities for their various political critiques, and it is still the case that "regional authorities use psychiatric examinations as part of intimidation campaigns against people who file lots of complaints in courts and other state bodies trying to instate justice'" (Krainova 2011). Apparently, "some people who pester the authorities ... suffer from 'querulent and litigious syndrome'" (Krainova 2011).

This kind of abuse of psychiatry is a widespread and present problem. In China, in 2002, it was estimated that "as many as 15% of people held in Chinese psychiatric institutions may be in custody for committing political offenses" (Munro 2002). By late 2000, there were documented more than 100 cases of Falun Gong practitioners being "detained by the authorities and forcibly sent to Ankang hospitals—special psychiatric hospitals run by the Ministry of State Security and the Ministry of Civil Affairs" (Bezlova 2002). According to the "Dangerous Minds" report published by Human Rights Watch:

China is holding and torturing thousands of political and religious dissidents in mental institutions. ... Among those held in mental asylums are ... independent labor organizers, whistle-blowers and individuals who complain about political persecution or official misconduct. (Bezlova 2002)

This systematic and institutional abuse of psychiatry is pervasive and has also been documented in Cuba, India, Japan, Norway, and, more disturbingly, very close to home. Overt violations are numerous, and in some cases not isolated but institutional.

For example, in 2009 New York Police Department veteran Adrian Schoolcraft was forcibly committed to a mental institution for pursuing claims that his department was involved in corrupt activities (a "Meritorious Police Duty Medal" winner, Schoolcraft used secret recordings, the now infamous "NYPD Tapes," to evidence his accusations that arrest quotas were leading to corrupt activities in his department). After being harassed and reassigned to a desk job, Schoolcraft was "taken in handcuffs to a psychiatric ward" after an Emergency Services Unit forced entry into his apartment and detained him. Then, "after what he describes as a frightening, involuntary hospital stay, Schoolcraft was suspended from the force" (Associated Press 2010; Dwyer 2012). The report into this affair, carried out by Queens District Attorney Richard Brown, concluded that the detention had been no crime, and Schoolcraft was offered not an apology but a hospital bill (Dwyer 2012). Whatever the truth in the Schoolcraft affair, one can clearly see the potential for medical authority to be abused in order to stifle troublesome behavior. That labeling someone as "mentally ill" destroys any credibility they may have in courts of law is a fact easily taken advantage of by powerful authorities, regardless of the life-long destructive consequences for the persons so labeled.

Given the way, as we have seen, medical diagnoses overlap in complex ways with our socially relative ideas of normal and desirable behavior, and the manner in which institutions have used their authority to treat deviance as a medical or psychiatric condition requiring segregation and treatment, it is very important for us to look at the more subtle ways in which soft moral bioenhancement is already being applied. The question is not what moral bioenhancement might look like in the future. Rather, we need to be asking about the extent to which we already have moral bioenhancement without even knowing it. That is, we need to examine the extent to which biologically based solutions, medicines, and technologies are already being applied, knowingly or not, toward "disorders" that are moral in nature—or, at least, that have strong morally related dimensions. How far has medicine already encroached into morality? And to what extent does it have the right to do so?

A very clear example of this process is the flagrant manner in which the DSM-V (the American Psychological Association's Diagnostic and Statistical Manual, the diagnostic "Bible" of US psychological practice) goes about turning behaviors traditionally understood as completely normal and natural, in some cases, into "disorders" requiring treatment and medication. We have "tobacco use disorder," "gambling disorder," "restless legs syndrome," "insomnia disorder." These are real "disorders," apparently, included in the most recent psychological diagnostic manual. This classification process is quite disturbing. One need only take any behavior that is perfectly natural, normal but presently undesired (e.g., insomnia) or even slightly outside the norm, and attach the word "disorder" to it[10]; suddenly, one has a diagnosis.

Insofar as some of these behaviors have moral dimensions (particularly the ones to do with addictions and aggressive behaviors) and pharmaceutical treatment protocols are available, we have a very clear case of morally related concerns being given objective classification as mental illness. Morally related concerns become mental health concerns, particularly when pharmaceutical treatments exist to treat them. Then we already have moral bioenhancement by proxy taking effect.

The DSM-V has come under wide and sustained criticism. There have been a range of financial conflict of interest claims raised regarding the panel members involved in the development of the DSM-V and their various ties to the pharmaceutical industry. It has been suggested that 69% of the panel members had such ties (Cosgrove and Drimsky 2012, 2), and one need only look at the institutional structures being put in place to facilitate such diagnoses and justify "treatment" to see the manner in which various financial interests, mental health, and moral issues have come to be fused. Speaking of the United States, Breggin (2008) writes:

The nation is rapidly rushing toward ... the systematic "mental health screening" of schoolchildren and even infants. The federal government's New Freedom

Commission supports both Early Mental Health Screening in the schools, and the Texas Medication Algorithm Project, a pharmaceutical company attempt to enforce guidelines necessitating the use of its products. Because most children will be referred for medical evaluation, virtually assuring the prescription of psychiatric drugs ... "TeenScreening" in our schools ... is moving toward "toddler screening" and even "infant screening." In Minnesota, legislation has been introduced calling for "socioemotional" screening of toddlers before admission to kindergarten ... to "assure that all children ages birth to five are screened early and continuously for the presence of health, socioemotional, or developmental needs" ... As "ToddlerScreening" and "TeenScreening" take hold, they will become a psychopharmaceutical steamroller. Even without these latest impositions on our children, millions are being pushed into becoming lifetime consumers of psychiatric drugs. (Breggin 2008, 220–1)

We have already seen in part II that much the same "early intervention" programs are being advocated at the legislative level in the United Kingdom, with brain scan propaganda and "voodoo neuroscience" being used to justify the screening of children and intervention in parenting practices along implicitly morally related lines (O'Connor and Joffe 2012).

Similarly, we might look at the federal "violence initiative" of 1992, an explicit attempt to link inner-city crime with genetic markers by identifying "potential violent children." The estimate by Dr. Goodwin (psychiatrist director of the National Institute of Mental Health) was that 100,000 children as young as five would be identified "for psychiatric interventions." Goodwin fell out of favor with the National Institute of Health when news of his comparison of the "high-impact inner city" to "a jungle and its youth to rhesus monkeys who only want to kill one another, have sex and reproduce" (Breggin 1992) was made public. The suggestion then was that "the violence initiative scapegoats black children and excuses society from facing racism, poverty, unemployment and other problems that cause crime and violence" (ibid.). Once again, we find the attempt to discover "easy answers," biological solutions to intractable social difficulties.

What else is all this but soft moral enhancement? We have overt attempts at the highest institutional levels to associate morally related issues, violence and crime, with genetic and biological markers. These associations are then used to identify persons for preemptive psychopharmacological intervention. If we are only thinking of moral enhancement as something like how to make people nicer, or more considerate to one another, or if we are only thinking of moral enhancement as targeting what are traditionally considered to be virtues (e.g., altruism, moderation, restraint), then we overlook how profoundly morally laden are the targets of such initiatives. We do not mention "morality," but rather crime and violence. We do not talk about making people "good," but rather we talk about ways of reducing to the taxpayer costs accrued by those who drink alcohol excessively. Insofar as biologically based interventions to such issues are made manifest, we have nothing other than soft moral bioenhancement implicit therein. How do those who think moral bioenhancement is a merely speculative, futuristic, and nonsensical area of enhancement discourse respond to the reality of political initiatives such as these, which are occurring at the very present moment?

Throughout the book we have hinted at the manner in which our narratives about human agency have been guiding our neuroscientific inquiry from the outset, and the manner in which neuroscientific study is being drawn upon by legislators and public policy makers. In light of this process of the medicalization of moral concerns, we have substantial reason to think that a subterranean form of moral bioenhancement is already in force. Medical metaphors are being fused with moral language, and value-laden terms do become intermixed with medical terms which give the appearance of purely objective language. As Wolpe (2002) writes:

> Scientific inquiry is guided by culturally determined standards of what traits we think are valuable to explore and what behaviors we think are desirable to control or eradicate. [For example,] the attempt to localize criminality and explain it as the function of a specific pathologized section of the brain is itself an agenda of a particular cultural and historical moment, and one with significant moral implications. ... Neuroscience today is also built on a series of fundamental assumptions about human nature and worth. It is not possible, and perhaps not desirable, to purge neuroscience of moral presuppositions, dealing as it does with fundamental aspects of identity, personality, free will, and other value-wrought concepts. (Wolpe 2002, 387)

Quite so. Yet the key point for us is not just that value-laden assumptions are unavoidable in neuroscience, as with any scientific inquiry and mode of interpretation. It is that one is already getting a subtle and covert moral enhancement by proxy simply by virtue of the projection of moral values and social mores into the very definitions of what counts as health and normalcy. It is certainly important to note that there may very well be cases where such covert, soft enhancement is justifiable, desirable, or even necessary (soft moral enhancement is not necessarily problematic in principle). All the same, we must absolutely make ourselves aware of the various dangers and vested interests that make up the reality on the ground. The consequences of not recognizing the moral and value-laden edges of the terms used in present scientific discourses can be great. Savulescu (2012) describes the dangers of projecting values in medical terminology:

People can be "treated" for social reasons. Homosexuality was defined as disease by the American Psychiatric Association until 1973. It was even retained by the DSM until 1986 as "ego-dystonic homosexuality." Homosexuals were subjected to the painful aversion therapy depicted in Stanley Kubrick's *A Clockwork Orange*, in the 1950s and '60s to "cure" them. (Savulescu 2012)

This encroachment of the medical into the moral is certainly something to be borne in mind. The problem is that so many of our values are taken for granted. They are part of the air we breathe and thus not even visible for the most part, let alone subject to critique. Until one takes a critical stance toward one's social values, it can be very difficult to appreciate the lines that exist between one's own socially relative values and that which is more objectively "defective," and thus in need of medical or psychiatric care.

Remediation: Moral Sickness and Medicalization

Is There Such a Thing as a Moral Sickness?
At the start of the chapter, it was suggested that, in thinking of man's moral capacity as something which can be "abnormal" or can conceivably "malfunction," "break down," or "become sick," a range of significant conceptual issues are raised. Let us hone in on this idea of "moral sickness" and the use of medical metaphors when talking about morally related issues. All of these concerns are evoked as soon as one begins to think in terms of remedial moral interventions, and the sometimes unpalatable implications that remedial moral enhancement engenders have not been particularly well thought through by advocates of such enhancement.

How might we make some sense of this quite shocking term "moral disease"? Is there any literal sense in which the term can be understood? This is debatable. Let us examine two ways in which a moral sickness might be said to exist:

(a) cases of organic damage to a person's brain, or some neurological deficit which is organic in origin; and
(b) some deeply disturbed psychological basis, for example, extreme antisocial behavior.

Starting with the former case, we might point to something like psychopathy. Inasmuch as some persons subject to this diagnosis can be seen to have distinct neurobiological deficits in the organic matter of their brains, might such a condition count as worthy candidate for the appellation "moral sickness"? It is hard to say. On the one hand, as we saw Sinnott-Armstrong (2012) suggest in chapter 1, there is no part of the brain which deals with morality as its sole or primary function. There is no "morality spot" in the brain. Moral functioning, depending on the function in question, requires a complex integration of various neurological elements. As such, appeals to organic deficits or brain damage, such as one sees in the famous case of Phineas Gage (a pleasant man who was impaled through the brain with a railroad spike, causing severe changes in his personality), or cases of lobotomization (fusing together parts of the brain considered to be instrumental in regulating certain behaviors), are problematic as bases for defending the idea of moral sickness. If there is no part of the brain which deals exclusively or predominantly with moral functioning, then the appeal to organic damage or deficits as ways of justifying talk of moral sickness is less persuasive.

But the argument is not clear-cut. Such persons do have a serious medical issue. They do have genuine deficits in brain areas essential to certain forms of moral functioning, so this sickness does impact on their moral functioning. In this sense, the label "moral sickness" does have some intuitive sense. The point, though, is really about getting one's labels very clear. When dealing with issues of psychopathy, medicine, and treatment (particularly in legal contexts, such as determining legal responsibility, and whether offenders should be imprisoned or incarcerated in mental facilities), it is absolutely essential that one get straight what one is talking about.

Psychopaths (insofar as the psychopathy in question is in fact caused by organic deficits) have deficits in areas of their brain relevant to moral functioning, but because those same brain parts are relevant to many other functions, it is debatable whether one can really call it a distinctly *moral* disease. It would be like saying a person who had his arms amputated has an eating disorder, because arms are used in eating food. But arms have all sorts of functions, and isolating just one of those functions is not really appropriate, particularly when the diagnosis one presents shapes one's ideas of the nature of the patient so treated. Calling someone "morally sick" can be used inappropriately in at least two ways. First, it involves applying a powerfully stigmatizing label, and second, it can be used to absolve from criminal responsibility those who are potentially worthy candidates for criminal incarceration. Making the situation appropriately clear requires, at the very least, a balancing of the following:

(a) respect for the fact that the brain areas involved in functions requisite for certain aspects of moral living have been damaged, and that the person does in fact have genuine medical problems; with

(b) recognition that this disability that affects his or her moral functioning is just one aspect of a larger set of brain deficits which extends beyond moral functioning.

One can say that such persons are indeed sick, and we can say that their sickness has an impact on their moral functioning (among other things)—but whether this amounts to a "moral sickness" is a matter for debate. Given the powerful implications of such a label, one ought to be very careful about how it is used.[11]

Dangers of Medicalization in the Moral Enhancement Context

In the best places, where straightjackets are abolished, doors are unlocked, leucotomies largely forgone, these can be replaced by more subtle lobotomies and tranquilisers that place the bars of Bedlam and the locked doors inside the patient. Thus I would wish to emphasize that our "normal" "adjusted" state is too often the abdication of ecstasy, the betrayal of our true potentialities, that many of us are only too successful in acquiring a false self to adapt to false realities.
—R. Laing (1990, 12)

In dealing with moral problems in medical terms, there are a range of mapping issues that arise. Such terms as "treatment," "prevention," "inoculation," and "resilience" will have to mean different things in relation to morally related problems than they would in relation to something like a viral infection. In talking about moral remediation, specifically in the context of biomedical interventions, it is easy to forget that moral problems, or behaviors with morally related dimensions, are extraordinarily complex. Such issues are complex in ways that usual medical issues simply are not. Organisms such as bacteria, or genetic disorders such as Huntington's chorea, have a profound complexity of their own—but that is a very different kind of complexity than moral functioning.

Medical science has a powerful authority. This is, of course, an authority that it deserves. By nature of this authority, though, when moral issues come to be put in medical terms, the language of medicine can very easily come to dominate the discourse. Even when disorders are not appropriately reduced to medical terms only, the tendency is for the biomedical aspects of the disorders to be considered primary. This invites the application of medical language in ways which are not necessarily appropriate, and which, with regard to moral functioning anyway, may well be counterproductive. If a given issue (say, addiction) is a result of the interaction of a plethora of causes, then simply thinking of such an issue in purely biomedical terms may well cause more harm than good.

Part of the problem here is that there are many interests who have serious stakes in wanting this issue presented in such reductive terms. Szasz (2010) notes the manner in which psychiatry and the state have self-serving interests in expanding their power; they do so by reducing behavioral issues to the level of physiological sickness, thus meriting medical treatment protocols and involuntary incarceration (Szasz 2010, 281). More recent commentators such as Conrad (2005), Healey (2004), and Breggin (2008) place the focus on the bottomless greed and Machiavellian tactics employed by the pharmaceutical industry, which has a massive cash incentive to promote and advertise their products (as indeed they have done, investing many millions each year), and to promote various reductive ideas about persons' ailments.

For example, the multipurpose notion of the "chemical imbalance" (a term whose proliferation is so widespread that it has become part of common language, such that few even think to question it) has become a very convenient means for reducing massively complex psychological difficulties to the level of "that which can be treated with a tablet." Given the proliferation of this term, it can be quite shocking to find out that it has only the most dubious basis in scientific fact (see Kirsch 2009). The idea of the "chemical imbalance" is one of the "cherished false truths" that proliferate modern society, a commonly accepted idea which serves too many interests to be abandoned or brought into question. Indeed, as soon as one gets beyond this idea that complex psychological problems can be adequately thought about in monocausal, biomedical, or pharmaceutical terms, the entire notion of a "chemical imbalance" as the cause of such difficulties becomes prima facie dubious.

I think it would be naïve to underestimate the extent to which powerful financial interests invested in a physiological model of psychological disorder can influence the way in which medical practice can be shaped. Lucke et al. (2011), for example, observe that:

Big pharmaceutical companies have been reasonably criticized for unethical conduct and dubious marketing strategies such as ghostwriting scientific articles, suppressing negative or inconclusive findings, and engaging medical professionals involved in regulatory processes as consultants, thus producing less than objective regulatory decisions. (Lucke et al. 2011, 40)

Indeed, psychiatrist David Healy (2004), mentioned by Lucke et al., has much to say in lamenting "the input from academics at the interface between Big Business and Big Science" (Healy 2004, xv), such as the manner in which a disturbing proportion of scientific studies which healthcare professionals (i.e., the very doctors that you and I seek advice from) use to educate themselves are ghost-written by the very pharmaceutical companies who will benefit financially from such science being publicized.[12] Within academia, there are those at the most prestigious institutions whose entire careers have been premised on encouraging the connection between pharmaceutical company investment and academic research. Having academia and medical science funded by Big Pharma cannot but introduce bias into the results, at the very least because only "desirable" projects will obtain funding, and the various respective faculties will then seek to promote certain "desirable" lines of research in order to make themselves more attractive to funding bodies (it happens all the time in academia, across the board). The desired results then have the names of prestigious institutions attached. Advocating for the greater association between Big Pharma money and academia yields tremendous financial benefits for those in favor of it, and such voices become powerful opinion-shapers within academia and the public sphere. It is not particularly surprising when such persons use this authority to quell debate and distort arguments over the appropriateness of the primacy of pharmaceutical approaches to treating complex human difficulties. Healy writes:

The training of academics is subsidized by the communities from which they come. … But academics who were once the substantive movers in healthcare now serve as ornamental additions to business. How else can we interpret the fact that probably at least 50 percent of academic publications in therapeutics are now ghostwritten … in particular those in the most prestigious medical journals? How else do we interpret a circumstance where some of the most eminent figures in the academic health-science research community appear party to the suppression of data on lethal side effects of treatments, and such behavior has become something of a norm? (Healy 2004, xv)

Again, it is a matter of the authority of science being abused. Studies conducted using scientific methods are right to stake their authority, but the dissemination of such studies and their arrangement into publicly visible pictures of scientific truth (i.e., what publicly comes to be called science) is mediated by a tremendous number of nonscientific influences which distort this public perception and diminish the authority such studies ought to have. Studies can be arranged selectively for ideological or financial reasons. Contrary voices struggle to be heard.[13] Ben Goldacre (2012) has written and lectured extensively on this phenomenon of "publication bias," something finally being accepted by the mainstream as a genuine and serious problem within medical science. Goldacre writes:

We only hear about the flukes ... we don't hear about all the times that persons got stuff wrong ... this phenomenon of publication bias has been very well studied ... we can see what a staggering difference there was between reality and what doctors, patients, commissioners of health services, and academics were able to see in the peer-reviewed literature. *We were misled.* And this is a systematic flaw in the core of medicine. ... Publication bias affects every field of medicine. About half of all trials on average go missing in action ... If I flipped a coin a hundred times but then withheld the result from you from half those tosses I could make it look like I had a coin that always came up heads ... this is exactly what we blindly tolerate in the whole of evidence-based medicine. ... this is research misconduct ... research fraud. (Goldacre 2012; emphasis added)

When the accepted scientific standard is plagued by ghost-written articles with blatant conflict of interest issues and a tendency to simply exclude studies which do not plot positive correlations between pharmaceuticals and desired physiological effects, it is very clear that neither medical professionals, nor the media, nor the general public are getting an accurate or full picture of what is going on with such disorders.

It is to be noted that there is a significant backlash against such tendencies coming from within the respective sciences themselves. There are plenty of appropriately placed individuals who are perfectly well aware of these problems and are attempting to provide a fair counterbalance (James Giordano, Paul Root Wolpe, William Casebeer, to name but a few). Yet, against this counterforce, the pharmaceutical companies have billions of dollars to spend annually on furthering their interests, and while there is certainly a backlash in certain circles against such reductionism, the resources and legislature are not at all in their favor.

What rarely gets mentioned, however, is arguably the most damning factor in this process of reductionism—the general public. While Big Pharma and the various institutional forces are financially involved in creating markets for their products, it could well be argued that the general public is often entirely complicit in such deception (Wiseman 2015, 31). For in addition to the various financial interests at play, there are also various psychological reasons for the lay public to wish to see complex human problems, including moral functioning, reduced to such biomedical terms. Reductionism in this context leads persons to believe that their problems can be dealt with by taking a pill, after which they can simply go on their merry way. The reality is that human problems, especially moral problems, are massively complex and require extensive analysis as well as arduous efforts at self-change. The idea that a pill might exist to resolve our problems is one readily seized upon by a massively busy and overworked general public. As Breggin (2003a) puts it:

We are spared the painful search for the personal and psychological causes of our suffering in our lives as children and adults. We are relieved of the necessity of finding more valid and meaningful principles of living. We do not have to face our conflicts with our husbands or wives, fathers or mothers, children, friends, coworkers, or bosses. We do not have to seek more meaningful work and more satisfying relationships. Heroism and determination in the face of our suffering becomes irrelevant. *We are only responsible for taking our medications as directed.* (Breggin 2003a, 44; emphasis added)

Again and again, what one sees is a reduction of mental issues to physiological causes, the elimination of any other mediating factor as cause—the medicalization of mental issues as treatable in literal medical terms. The biopsychiatric model which is currently so in vogue involves nothing else but this. Every problem is forced into the physiological or chemical order of explanation, even if no such explanation exists. It is far easier to reduce things to this level and to generate a tablet to "treat" it. As soon as one begins to talk of moral functioning in precisely the same terms, then one invites precisely the same interests to co-opt the debate, to invest in and dominate the discourse. And at the bottom of the entire situation, invisible to the real persons who seek medical help for their various maladies, is the implicit ideological conflict that is embedded within the very practice of medicine itself. Breggin continues:

When people choose to become patients of a psychiatrist who prescribes drugs, they are doing a great deal more than merely "seeing the doctor." They are subjecting themselves to a very specific and limited model of thinking about human suffering and failure. ... It demands that we think of ourselves as broken machines or flawed mechanical devices. It requires blind faith in doctors and scientists, combined with a materialistic faith in molecular causes and manipulations. In the biopsychiatric model, we are mechanical devices similar to computers or other machines. Our suffering is caused by genetic and biological factors beyond our control. When we cannot seem to find a solution on our own, we place our fate in the hands of technicians who know how to tinker with our machinery. In this mechanical model, we have very little personal responsibility for our condition. (Breggin 2003a, 43–4)

This view, the basic norm, conflicts with an entirely different way of construing the patient-doctor relationship, particularly in mental health contexts:

The prevailing professional tendency is to conceptualize the conflict between psychotherapy and drug treatment as a scientific one; but it is at root a conflict between two different views of human nature. ... In the humanistic, existential approach, human beings are seen as endowed with unique capacities, yearnings, and aspirations. They seek to overcome and transcend suffering through self-understanding, ethics, community, and enriched lives. In this model, people must take personal responsibility for their lives, including the quality of their mental condition and relationships with others, including children. The corresponding psychotherapeutic model ... is more focused and even practical: First, the human suffering dealt with by psychiatrists and other mental health professionals is almost always psychological, existential and social in nature, rather than biological; and second, psychotherapeutic rather than biological interventions are safer and more effective for these problems. (Breggin 2003a, 34, 40)

It is not hard to see why the general public are so readily complicit in the biomedically reductive way of conceptualizing our problems. In a fast-paced society such as ours, who has time to properly deal with their problems? While many individuals find the constant stream of pharmaceuticals prescribed for the least malady to be unattractive and turn instead to natural remedies, unfortunately this is to have missed the point altogether. The problem is not so much in taking pharmaceutical versus "natural" drugs. The problem is in attempting to reduce such problems to the level of that which can be solved by simply "taking something." It matters not a great deal, in this context, whether one is taking an herbal remedy or a pharmacological remedy—the presumption is the same in either case, namely that there is some "substance = X" which if ingested will solve the problem. In neither case are the root problems being dealt with.

While this meditation on the issues of medicalization, the biomedically reductive ideology subtending it, and the plethora of powerful influences—institutional, corporate, social, and psychological—which sustain such tendencies has been lengthy (though arguably this is just to touch the tip of the iceberg), it is absolutely essential to understand that the idea of remedial moral enhancement cannot and will not exist in a vacuum. To the contrary, any remedial moral enhancement that will come to exist, or which arguably already does exist, has its force in a climate which is almost inevitably reductionistic. And while there are many voices against such reductionism, most helpfully coming from within the domain of the biological sciences themselves, the countervailing forces in reality have a dominance that means remedial moral enhancement will be confronted by such forces and shaped by such forces—not least of which is the general public's desire for easy solutions to what are essentially complex problems requiring time, analysis, care, and action.

A realistic grasp of the actual territory in which moral enhancement would have to be enacted is an essential precondition for grasping how it might realistically be enacted. The sorts of biases and powerful interests that will be seeking to spin the respective moral enhancement situation to their own benefit need to be elaborated. While, of course, there will be benign influences in attempting to manifest moral enhancement, be that in "hard" or the arguably already existing "soft" form, it is to be noted that, realistically, such prospects will be thrown into a complex ocean which will involve swimming not just with dolphins, but also with sharks.[14]

Conclusion: Moral Enhancement in Mental Health Contexts

Despite all of this, if remedial moral enhancement is to go forward, I would still suggest that embedding it in the context of mental health intervention is the only appropriate way forward. This is by no means an exhortation that moral enhancement should be brought to bear in this way, but if we are already getting moral enhancement by proxy, and this is to some extent inevitable, the best solution may be to drag the whole thing out into the open and critically inspect the process in the full light of day. If some forms of medical and mental health treatments will always have morally related aspects or societal moral judgments embedded within them, let us make these judgments explicit and attempt to find some way of integrating them within an acceptable code of practice—something which ensures that the therapeutic context is appropriately person-centered in nature and nonreductive, and that the healthcare professionals involved are appropriately directed and sufficiently well-armed against the dangers raised above.

If a moral enhancement intervention is applied within the suitable context, with a trained professional who views that patient as an actual human being, who engages with the larger features of their environment and takes note of the various psychological and environmental issues that might be influencing their behavior, then I think that moral enhancement is on safer ground. But these are very specific criteria. The difference between moral enhancement biomedically reduced, and a biological intervention applied in a mid- to long-term person-centered context that focuses on the entire range of bio-psycho-social elements relevant to the problem makes all the difference. The former is dangerous beyond words; the latter is something that could be more desirable.

This is the crux of the matter with moral enhancement—contrary to those purely skeptical of the idea, I do think that, done properly, moral enhancement holds some interesting potential and is something that in certain circumstances might be worth exploring. Yet, contrary to the enthusiasts, one must maintain that there is so much that can go wrong, so much room for distortion of this process, so many extant obstacles in the present social setup, that one must be very, very cautious in articulating the conditions through which a worthy moral enhancement might be actioned and recommended. One must have powerful institutions in force to secure the relevant industries against the very real dangers that might arise.

Given all this, so far as "remedial" moral enhancement goes, the proposition being made here is as follows: that if such intervention takes place in a mental health context, in a person-centered and fully bio-psycho-social fashion, one which respects the value and influence of personal agency, cultural scaffolding, and quality relationships, then we have begun to outline a context in which moral enhancement might be put to work in a positive and desirable way.

The worst thing of all would be to have "moral doctors" whose sole purpose is to arbitrate moral decisions for persons and prescribe medications to encourage what they deem to be moral. To some extent, doctors will always be involved in behavioral issues with moral dimensions (for example, the refusal by medical practitioners to tell families from certain ethnic backgrounds the gender of their unborn child due to the preponderance of abortions carried out on female fetuses within that demographic). Indeed, the state can and should in certain cases prescribe to persons what they can and cannot do.[15] All the same, the idea of setting up "moral doctors" seems intuitively suspect. For who would train them? And to what value system should they adhere? The ethical issues here would be interminable.

So the need for a mental health context in applying remedial moral enhancement is actually a more pragmatic point than anything. When it comes to behavioral issues, problems with strong morally related dimensions, such issues are already treated in coordination with mental health interventions. Antisocial behavior, for example, has its mental health label—"antisocial personality disorder," which is defined in the ICD-10 (the International Statistical Classification of Diseases and Related Health Problems, tenth revision) as involving at least three of the following criteria:

1. Callous unconcern for the feelings of others;
2. Gross and persistent attitude of irresponsibility and disregard for social norms, rules, and obligations;
3. Incapacity to maintain enduring relationships, though having no difficulty in establishing them;
4. Very low tolerance to frustration and a low threshold for discharge of aggression, including violence;
5. Incapacity to experience guilt or to profit from experience, particularly punishment;
6. Marked readiness to blame others or to offer plausible rationalizations for the behavior that has brought the person into conflict with society.

In short, the morally related problem of inappropriately expressed violence is already met and dealt with in the mental health context: the institutions are already in place, the guidelines for appropriate practice and treatment are already in place, the healthcare professionals are already there, ready-trained for dealing with such issues.

Much the same can be said of addiction. While there is much debate as to whether addiction is actually a genuine mental health or medical problem, it is treated as such anyway. Again, the institutions, healthcare professionals, and practice guidelines are all already in place. This is a potentially attractive position for remedial moral enhancement. On the other hand, such a framework for describing remedial moral enhancement is not without problems. For example, one needs to ask whether these issues really are mental health problems or whether they are in fact moral problems. To what extent are they both? And if they are moral problems after all, is it not entirely inappropriate to be thinking of them in mental health terms at all?

Psychiatrist R. D. Laing is particularly vocal about this idea: that what we call mental illness is actually just behavior we don't like, that mental illness is simply a label we give to inconvenient persons in order to do away with them. Laing (1990) observes:

In the context of our present pervasive madness that we call normality, sanity, free-dom, all our frames of reference are ambiguous and equivocal. ... A man who prefers to be dead rather than Red is normal. A man who says that he has lost his soul is mad. A man who says that men are machines may be a great scientist. A man who says he is a machine is "depersonalized." ... Psychiatry could be, and some psychia-trists are, on the side of transcendence, of genuine freedom, and of true human growth. But psychiatry can so easily be a technique of brainwashing, of inducing behavior that is adjusted, by (preferably) non-injurious torture. (Laing 1990, 12)

For Szasz, as we have seen, psychiatry has turned into a process of using physiological language to justify inappropriate treatment protocols regard-ing matters better construed as concerns of personal responsibility. Given this overlap between "mad" and "bad," if—as Szasz and Laing insist—there are good reasons for worrying about whether the concerns that are pres-ently being treated as mental health issues are in fact moral issues after all—then one response to my proposal is that I have got everything entirely the wrong way round. So far from making moral enhancement a mental health issue, I should be doing exactly the reverse, and continue to push for the re-moralization of what has been falsely called mental illness, thus restoring the discourse of personal responsibility and control. Indeed, given all the manifold dangers elaborated above, the suggestion that certain cases of remedial moral enhancement be put under a mental health umbrella might then seem to be the worst possible suggestion of all.

In response to this, it is important to note that Szasz and the various critics drawn upon here are all still therapists, psychiatrists, and scientists. The main issues at stake for them are not to do with therapy per se, but rather the labeling, the reductionism, the ideologies covertly smuggled in under the guise and authority of objective scientific research. So the problem is not therapy; nor, crucially, is it therapy as a vehicle for dealing with issues that are moral in nature. Indeed, quite the contrary in fact, humanistic psychiatrists like Breggin and Szasz are keen to emphasize the moral dimension of life during treatment as a means of counteracting the crude materialistic and mechanical worldview at the basis of the reductive biopsychiatric treatment model currently in force.

In fact, the therapeutic environment is one extremely valuable context through which the moral dimensions of larger psychological issues can be addressed in one way or another. The therapeutic context, if properly applied, proffers an opportunity to deal with clients' morally related issues humanely, personally, and in a relational, integrated (and thus nonreductive) way, as part of the larger fabric of the many dimensions that make up people's lives. Treating persons' problems in this humanistic way is a matter of looking at the broader framework of one's life, something unique to each and every person seeking help. This is a framework which necessarily involves moral dimensions and attempting to change thought, feeling, and behavior such that the patient begins to identify what it is in their lives that is causing their problems. Then biomedical interventions might be applied, judiciously and with constant monitoring, to generate a fully integral bio-psycho-social package. It is in this context that moral bioenhancement makes most sense, certainly more sense than the idea of simply giving people drugs to make them "want to be good."

Yet the key point is to provide a nonreductive approach which deals with moral issues as one dimension of a larger person-centered approach, one which recognizes moral functioning as complexly and thoroughly interwoven with all aspects of a person's life—thus, as something to be dealt with on an ad hoc basis during the course of the requisite therapy. Such an approach may, if safe and efficacious, and not counterproductive to the treatment goals, use pharmaceuticals as an adjunct to such treatment as part of a more sensitive, humanistic, and nonmechanistic framework for remedial action. In this way, moral bioenhancement would blend in seamlessly with the way our most humanistic therapists already do manage mental health issues in their patients—though whether, in this proposal, such forms of moral enhancement simply disappear into mental health practice is an important question, and we shall return to this in chapter 9.

It is true that thinking of moral issues in remedial terms invites medicalization. However, unfortunately, nearly everything in a clinical context invites medicalization. The key lies, as Hopkins (2012) and Miller (2012) suggest, in having responsible and informed care professionals who are aware of the dangers of medicalization and reductive pharmaceutical strategies and are not swayed by the many billions of dollars in incentives waved around yearly by the pharmaceutical companies. Above all, we need responsible institutions in place, along with healthcare professionals who are not swayed by the general public's desire for inappropriate shortcuts and easy remedies to complex problems.

9 Treating Addiction: Moral Enhancement in Practice?

We have encountered many suggestions for moral enhancement throughout this book. Many of them have been dismissed as impractical and not well thought through. We have seen time and again that there are significant practical impediments, practical realities, which have been simply neglected by many of the various philosophers involved. The lack of practical focus is perhaps the biggest failing of moral enhancement discourse. In this chapter, we shall be producing an extended meditation on a single possibility for moral enhancement, one which seems entirely realistic and feasible, not least of all because the intervention in question actually exists, has been tested, and is under some circumstances an accepted part of medical practice. We will be looking at pharmacological interventions for helping in the treatment of alcoholism (and to some extent substance addiction more generally, since it raises many parallel issues). We will explore pharmacological interventions into treating alcoholics as a vehicle for drawing together many of the threads that have been explored throughout the book. We will come to see pharmacological interventions for treating alcoholism as exemplifying the following:

(a) The need for a moral enhancement that is but one dimension of a larger treatment protocol, one which treats of the issue in its full bio-psycho-social context, without reducing the matter at hand to the purely neuro-biological level only;

(b) The overlap between mental health and moral discourses—how there can be on occasions a great difficulty in separating the one from the other, and the extent to which this causes difficulties for defining moral enhancement;

(c) The manner in which spiritual and religious dimensions of living might be integrated within a broad project of treatment which is moral enhancement "by another name."

Above all, it is the practical realities involved in the application of pharmacology for such ends that are to be the guiding force here. We have seen that many of the so-called possible interventions for moral enhancement have been simply bandied about without any real sense for how they might be applied in any real-world context. It was mentioned in part I that the likes of Jotterand (2012) and Kabasenche (2012a), though skeptical of the power of technology to actually make persons more moral per se, have yet agreed such enhancement might be able to set some kind of "biological context" (Kabasenche 2012a) through which reflective growth may be facilitated. Moral enhancement technology in this view would then be like fertilizer: it would facilitate and amplify pre-existing efforts toward moral growth and the development of a moral identity. As such, it is moral enhancement's critics, and not its enthusiasts, that forward the most realistic, down-to-earth possibilities for its enactment.

What will not do, I would suggest, is the sort of "list" approach we discussed in part I—simply saying "let us use oxytocin to make persons more empathic and generous" or "let us use SSRIs to make people less aggressive," and just leaving it at that. These utterances are simply too question-begging. *How* are we to use these substances in this way? Through which means and institutions, and by what persons are these substances to be dispensed as moral enhancement agents? What would be the contexts in which they would be used, and how would their use be monitored? What would be the standards of excellence for assessing whether such uses were successful or not? How might they be integrated within a broader social context? And, what do we really expect such interventions to achieve? There are idiosyncratic considerations for any particular intervention, or target population, that require sustained reflection on their own terms.

One thing that is important here is to notice how *pedestrian* such approaches would turn out to be. This may well be very disappointing for those who are expecting recourse to some fantastical gadget, a neurobiological "magic wand" to be waved such that a terribly naughty person be transformed in a moment into an avuncular and generous sage. We have said that moral functioning is simply too complex and interwoven for such a thing to ever come about.[1] We sacrifice fantasy for something that might actually be of use, here and now. So, it is within this very limited remit that we might have something workable: moral enhancement, as broad-scope, a partial intervention, to be understood as part of a larger person-centered approach, which does not neglect psycho-social dimensions of the person's life, but rather sees moral enhancement as needing to be subtly integrated within that complex web.

We will focus here on a remedial situation, such as treating alcoholism is, for it is in the domain of treatment that the more realistic and more readily assimilable aspect of moral enhancement can be found (simply because we already have institutions, trained healthcare professionals and relatively nonreductive standards of practice, and many of the various morally related problems that moral enhancement concerns itself with are already treated of in such mental health contexts). Moral enhancement in this case would then be constituted as an adjunct to therapy—some process involving pharmacological or technological means for augmenting the larger therapeutic processes involved in these broader mental health interventions.

Alcoholism as a Moral Problem?

Is alcoholism even a moral problem? While the act of drinking alcohol is arguably not itself evil (some disagree; see below), and many people can drink without doing anything terribly destructive, yet the consumption of alcohol is obviously one factor related to a range of other behaviors which certainly are immoral—like domestic violence, general belligerence at large, unintentional violence (as in driving one's car over a woman and her child while intoxicated). It is also to be noted that intoxication diminishes one's faculties for moral reflection and self-control; in this sense, one might say that alcohol by its very nature lends itself toward *de*-enhancement of one's moral capacities. There might be exceptions, perhaps a drunken person, being disinhibited, might be more likely to intervene in preventing a street robbery or rape where a sober person might be too afraid, but at the very least we can say that alcohol abuse, in general, is implicated in a number of profound evils.

Social attitudes to alcoholism also have to be considered. Despite various social imperatives to imbibe alcoholic beverages, getting lost in drinking to the extent that one becomes an alcoholic is frowned upon, and often in distinctly moral terms. Rightly or wrongly, an alcoholic is considered a "degenerate." It is true that attitudes toward alcohol and alcoholism are malleable—not even a century ago it was quite acceptable for a prominent politician to be a drunk. Winston Churchill, the most celebrated of British Prime Ministers, was an inveterate drinker. Today the idea that a publicly visible alcoholic could be elected Prime Minister is unthinkable—it is simply not congruent with the current image of moral uprightness that prominent politicians are now expected to exude. Likewise, in certain cultures and faiths, Islam for example, the idea of drinking alcohol at all is considered taboo, an explicitly immoral deed, a moral prohibition from which one can gain no remittance. Thus, whether or not alcoholism is a moral problem in some more "objective" sense, certainly it has been and is treated as an explicitly moral problem within a variety of social contexts.

While there is a great deal of contemporary vacillation between describing alcoholism as a "disease," on the one hand, and as a moral problem that is to do with personal responsibility on the other (it is possible for it to be both a moral and medical problem), alcoholism falls squarely within the remit of moral enhancement. There are very few credible voices who suggest that alcoholism has no dimension of personal responsibility (i.e., inapt for being judged in moral terms), and the de facto treatment of alcoholism as a moral problem in various cultural contexts means that alcoholism poses for us a viable object for moral enhancement intervention, even if we only have a "soft" version in mind.

So, thinking now in terms of "soft" moral enhancement—that is, treating alcoholism as, say, a public health issue—we do have a very serious problem to hand. Consider the following statistics relating to alcohol abuse:

According to the World Health Organization's 2011 global status report, alcohol is the world's third largest risk factor for disease and disability, and the harmful use of the substance leads to 2.5 million deaths annually. ... In the U.S., the number of alcohol-induced deaths totalled 26,256 for 2011, slightly higher than 2010's count, according to a preliminary report by the Centers for Disease Control and Prevention. U.S. Congressional findings indicate that an estimated 10 million Americans are problem drinkers. (Gates 2013)

Treatments for alcoholism, viewed through the soft lens (as a matter having clear morally related dimensions, yet not necessarily couching its rationales in explicit moral terms), touches on various notions of physical and mental health (though in addition, given the costs to taxpayers, health services, damage to public property, to the general ambiance, and so on, alcoholism touches on more metaphorical notions of health qua civic, family, and economic health, too). As a practical point, though, alcoholism is treated under the mental health umbrella, and this is where our focus will be.

In chapter 7 we made very clear the point that, if we are to have a meaningful kind of moral enhancement, then we must understand what it is we are attempting to enhance. As such, we must attempt to penetrate through to the depths of the problem, to understand its nature as best we can in order to see where the pharmacological or technological means for enhancing or remediating such problems might fit in. This being so, the first step in dealing with the issue of addiction is having a working definition of what addiction is. Young, Armas, and Cunningham (2009) proffer the following:

The American Society of Addiction Medicine defines addiction as a primary, chronic, neurobiological disease that is influenced by genetic, psychosocial, and environmental factors and that is characterized by strong cravings, compulsive use, and a lack of control to stop despite the harm it may cause. (Young, Armas, and Cunningham 2009, 59)

This is a helpful start. While this definition does present a neuro-primary account of the matter, it is at least inclusive of broader scaffolding factors. Let us also remind ourselves that the complex manner in which such influences overlap and intermingle must also be a factor here. While each individual must be treated precisely as an individual, one whose problems are specific to him or herself, we must also recognize that medical definitions must be general enough to include an appropriate range of persons. The fine-tuning, ideally, comes in the personal or group therapies that are enacted (this is the ideal that was presented at the end of the previous chapter).

Yet, in addition to these main features, addictive behavior involves a range of secondary features. Young and colleagues continue:

Research has revealed that people who suffer from addictions often have higher rates of depression and anxiety ..., trauma histories or posttraumatic stress disorder ..., anger management problems ..., family and relationship problems ..., poor self-awareness ..., poor self-control ..., low self-esteem ..., and low self-efficacy. ... The inability to effectively deal with stress is one of the strongest predictors of drug craving, relapse, and continued drug use. ... Because these issues are not resolved during the periods of addiction, they resurface during relapse, creating emotional responses that increase the likelihood of relapse. (ibid., 59–60)

Again, this points to a broader basis for alcoholism than is garnered by simply calling it "a neurobiological disease" and leaving it at that. There are numerous factors here forming a web of associated problems, some of which might be treated on the biological level, and many of the others requiring a more personal and integrated psychosocial approach.

Vaccinations for Substance Addiction and Alcoholism

In order to work through this idea, that moral enhancement might be usefully applied to the treatment of alcoholism and substance addiction, it is necessary to see what realistic enhancement possibilities exist and upon what level they can help with the treatment of addiction. To the extent that alcoholism or substance addictions have biological components, then such interventions should be, to some extent, amenable to biologically affective moral enhancement intervention. Let us ask then after the extent to which alcoholism is a biologically related problem and the extent to which a biologically affective intervention might be useful in treating such addiction. "Vaccinations" against such substances present important possibilities that are worthy of extended meditation, for they promise to cut away the very foundation of substance addictions—the effects themselves. If a person is physically unable to get drunk or enjoy the narcotic effects of the substance in question, there is a sense in which the problem is de facto cured. In this sense at least, the removal of alcoholism through purely pharmacological means is a very realistic prospect—and, arguably, this is a threat to the idea presented here that a multidimensional nonreductive approach is essential. Let us look at the various approaches which have been proposed as being able to "fix" or "cure" the addiction in view.

Gene Therapy for Creating "Allergies" to Alcohol

One particularly striking possibility for the treatment of alcoholism has been inspired by a genetic condition suffered by certain Asian populations ("Asian flush") whose livers are incapable of properly metabolizing alcohol. For persons with this condition the consumption of alcohol, even in small quantities, leads to an almost immediate hangover, excessive facial redness, sweating, increased body temperature, and a higher heart rate. Drinking has all of the poisonous effects on the body, but without any of the pleasant stupor that usually makes the substance so attractive.

Using a form of gene therapy, Juan Asenjo (2010) and his team in Chile have created a vaccine of sorts which promises to temporarily generate precisely this disorder in alcoholic patients. To explain this process, it is helpful to know how the liver metabolizes alcohol, a process which can be understood as proceeding in three steps, the first of which is turning the alcohol into acetaldehyde, which is then broken down into acetate, which is then broken down into carbon dioxide and water. Those with the genetic mutation which causes Asian flush have a profoundly impaired ability to perform this second step. Instead, the acetaldehyde, which is a toxic substance, and whose excessive buildup is commonly termed a hangover, results in a rather unpleasant experience for the person who has been drinking.

Asenjo et al.'s gene therapy approach is to temporarily subdue the liver's capacity to carry out this second step, suspending the liver's capacity to express the gene required for acetaldehyde breakdown and thus creating a form of alcohol intolerance. Extreme as such an approach may sound, if the method is successful, any alcoholic who has undergone such treatment would be faced with a terribly unpleasant experience should he or she attempt to continue indulging their addiction. The intervention is delivered by injection and has a life of about six months. Currently, the technique has been successful on rats and is presently undergoing human trials (Gates 2013; Martinez et al. 2010).

Optogenetics to Inhibit Dopaminergic Reward Response to Alcohol Consumption

Another possible avenue for treating alcoholism has been opened up using the optogenetic methods described in part II. Caroline Bass and colleagues (2013) successfully used such optogenetic techniques (introducing viral vectors to alter the DNA of specific classes of brain cells in order to make their function light-sensitive) to gain control over the dopaminergic neurons of the brains of a sample of rats. In this study it was possible to stop the (rat) participants consuming alcohol by simply modulating the reward response produced by the consumption of alcohol. When interviewed, Bass was quoted as saying:

For decades, we have observed that particular brain regions light up or become more active in an alcoholic when he or she drinks or looks at pictures of people drinking, for example, but we didn't know if those changes in brain activity actually *governed* the alcoholic's behavior. (Bass 2014, in Clark 2014; emphasis added)

Co-author Budygin continues:

These data provide us with concrete direction about what kinds of patterns of dopamine cell activation might be most effective to target alcohol drinking ... What optogenetics is enabling, is more precise mapping of the brain so that researchers can identify and plot the *cause and effect* of a whole host of behaviors. (Budygin 2014 in Clark 2014; emphasis added)

Perhaps one ought to be a little suspicious of this approach. As was argued in part II, there is something rather questionable about speaking of agency and the brain, in complex phenomena such as addictions are, using such terms as "cause and effect," or speaking of dopamine "governing" alcoholics' behavior. Indeed, Bass and Budygin's project of "identifying the neural pathways that control alcoholism in humans"[2] (Clark 2014) involves a deeply worrying use of language, which indicates to me a model of addiction which presents persons as passive nonagents. There is no doubt that dopaminergic responses will play a considerable role in certain forms of addiction, but talk of "cause and effect" smacks of a mechanistic outlook on addiction which, as we shall see, neglects the plethora of mediating psychological and cultural influences which contribute to the problem.

Furthermore, I would caution against taking such "rat addiction" studies too seriously. It may well be possible to cause alcohol dependency in rats, but to talk of "alcoholic rats" or, worse, "binge-drinking rats," as some in the popular media have (cf. Clark 2014), is to misconstrue and trivialize alcoholism in a number of significant ways. For example, rats do not inhabit macho drinking cultures; rats do not experience moral conflict; rats do not spend all their money on drinking with their friends and then come home to beat their wives and children; rats do not lose their jobs because they were too wasted to work; rats do not have profound existential despair and use alcohol as a crutch to conceal their anxieties.

The parallels between human addiction and rat dependency are meager, to say the least, and so I would advise caution before getting too excited about this potentiality. Finally, Bass et al.'s approach also has to contend with the fact that persons become alcoholics for various reasons. Not all alcoholism is driven by a reward response. In fact, it can be a consequence of explicitly self-destructive or self-punishing patterns of behavior. Targeting the reward circuits might be less helpful in these instances.

Addiction needs to be recognized as the complex and plural phenomenon that it is. Persons do not become addicts, or remain so, for the same reasons. As Heilig suggests, for "some groups of people—such as long-term addicted individuals or socially anxious addicted people—drinking is not driven by positive reinforcement" (2011, 566). Heilig goes on to suggest alcoholism is an "end-stage disease" which persons come to from a variety of different trajectories and that, in fact, the heterogeneity of alcoholic patients needs to be respected if any meaningful pharmacotherapy is to be pursued (Heilig et al. 2011, 670–1). This is a very different approach from thinking that simply meddling with dopamine responses will be sufficient to deal with alcoholism tout court.

Vaccination to Inhibit Dopaminergic Reward Response to Cocaine Consumption

Another approach worthy of mention, simply by virtue of its incredible ingenuity, is Ronald Crystal's team's attempts to create a vaccine against cocaine use (Wee et al. 2012). Crystal's efforts seem to be a hybrid of Asenjo and Bass et al.'s work, and use gene therapy to affect the dopaminergic response to cocaine use. Already successfully tested in nonhuman primates and presently undergoing human testing, Crystal et al.'s approach has been to link a cocaine-like chemical to a virus, very much like the common cold, and inject it into candidates in order to teach their immune system to recognize the drug as an unwanted invader. This is literally a vaccine against cocaine's effects. Once it has been taught to recognize the drug as a threat, the immune system will itself proceed to neutralize the drug just as it would attempt to neutralize any microbial intruder.

This works in the case of cocaine (which functions as a sort of dopamine reuptake inhibitor) by neutralizing this reuptake inhibition effect, thus maintaining the reward system of the brain's function in the usual non-narcotized fashion. As with any vaccination, booster shots will be required for the intervention to have continued effects, yet when the vaccine is operant, a person can take the drug and no pleasure-stimulation will be produced. It is assumed that the financial cost of cocaine, combined with the fact that it no longer produces such pleasurable effects, will be sufficient to motivate the addict to refrain from its consumption. Crystal is recorded as saying:

The vaccine eats up the cocaine in the blood like a little Pac-man before it can reach the brain. We believe this strategy is a win-win for those individuals, among the estimated 1.4 million cocaine users in the United States, who are committed to breaking their addiction to the drug. ... Even if a person who receives the anti-cocaine vaccine falls off the wagon, cocaine will have no effect. ... I believe that for those people who desperately want to break their addiction, a series of vaccinations will help. (Crystal in Fitzgerald 2013)

Nalmefene: Opioid System Modulation

The final intervention I would like to consider is the most significant for our purposes, primarily because it actually exists, has been tested, and can in fact be prescribed by doctors if deemed an appropriate intervention for the alcoholic in question. Nalmefene (brand name Selincro) functions by blocking a set of the body's opioid receptors. The general idea is that alcoholics can dose themselves with nalmefene on an "as and when" basis. This is thought to offer a means for reducing drinking in alcohol-dependent patients unable to reduce alcohol consumption on their own (Gual et al. 2013, 1432). The various studies conducted into nalmefene suggest that a 60% reduction in "high drinking risk level" (Spence 2014, 1) is possible. Investigators like Mann (2012) suggest:

Nalmefene provides clinical benefit, constitutes a potential new pharmacological treatment paradigm in terms of the treatment goal and dosing regimen, and provides a method to address the unmet medical need in patients with alcohol dependence. (Mann et al. 2012, 706)

The idea of using opioid antagonists to treat alcoholism is not particularly new. To the contrary, there are also other opioid antagonists (e.g., naltrexone) which also have strong research bases (Spence 2014, 1), sharing a similar plasma half-life (8–9 hours) and having a similar list of common side effects (such as loss of energy, depression, and gastrointestinal disturbances; Fudala et al. 1991, 300, 305). The big difference between nalmefene and naltrexone, according to Mann et al., is the manner in which the substances are to be taken. The "as and when" nature of nalmefene treatment means that total abstinence is not required. Instead, naltrexone treatments aim for total abstinence and need to be taken every day—hence "they are rarely prescribed and not acceptable to many patients" (Mann et al. 2013, 706).

The value of nalmefene as a treatment for addiction goes beyond alcoholism. Nalmefene has also been tested on morphine addicts and has been found to result in the diminution of morphine's effects to mere "drowsiness or sleepiness" while presenting its main side effects in terms of "agitation/irritability and muscle tensions" (Fudala et al. 1991, 300)—not particularly surprising symptoms for a substance which removes the desired effects of one's drug of choice. Similarly, nalmefene has been tested on persons suffering from "pathological gambling" ("a disabling disorder experienced by approximately 1%–2% of adults"; Grant et al. 2006, 303). In low doses, the drug was found to be very effective in reducing such obsessive behavior for 59% of participants. However, in high doses the side effects were found to be "intolerable" (Grant et al. 2006, 303).

Again, nalmefene is not a particularly new compound, but rather a refinement of existing opioid system modulators that have been used to a lesser extent for many decades in treating alcoholism. The fact that such compounds have long been in existence, but are largely unheard of, should already raise in commentators certain suspicions with respect to the prospect of such treatments.[3] That said, we do have to accept that nalmefene does have a certain efficacy, in certain kinds of addictive personalities, albeit with a certain profile of negative side effects, and because of the fact that nalmefene is presently already being used in certain contexts (i.e., is not mere potentiality but an actual medical reality for human use), we shall be focusing on this substance specifically.

Do These Interventions Cure Addicts of Their Addictions?

Nalmefene: Elective or Compulsory?

We have just encountered a range of nonfantastical ways of dealing with the problem of alcohol (and cocaine) addiction on the biological level, either preventing the addictive substance from producing a dopaminergic reward response by preventing alcohol agents from bonding with the appropriate opioid receptors, or by making the substance positively repulsive and quite literally sickening to ingest—in either case making the substance entirely unrewarding to consume. Insofar as alcoholism is a moral or morally related problem, we do appear to have some viable proposals here for medical, remedial moral enhancement intervention. Thus, whether our motivation for engaging with such a substance is "hard" or "soft," taking nalmefene as our focus, we have some viable means by which we can look at a particular instance of moral enhancement in a down-to-earth, realistic sort of way.

Throughout the book we have seen that the distinction between compulsory and elective uses of moral enhancement is quite porous. There are all sorts of ways in which persons can be "motivated" to volunteer to take a given intervention. This being so, let us specify two separate instances which might be thought of as more or less voluntary or more or less compulsory. On the one hand, let us take as "elective" the instance in which nalmefene is actually intended—what Mann called "a new paradigm" for treating alcoholism, the "as and when" approach, in which persons simply choose to take the drug on certain desired occasions rather than as part of a total abstinence approach. We can then take as the contrary case, as "compulsory," a situation in which an alcoholic, perhaps one already incarcerated for acts committed while inebriated, is compelled by the courts to engage in a program of nalmefene as a part of their rehabilitation. In this way, nalmefene might even be used as a punitive implement of the judiciary, an alternative to physical incarceration or institutionalization.

Elective Use

Both approaches would have advantages and problems. On the one hand, say nalmefene were to be advocated as a purely elective treatment. In this case, perhaps a person struggling with their addiction would turn around and go to a doctor or relevant healthcare professional—as is presently demanded—and state that they are engaging in a group, like Alcoholics Anonymous, or some other social therapeutic process, and that they would like to be prescribed nalmefene as part of those broader efforts.

This is admirable. Yet there is a rather significant problem lurking in the background. Nalmefene has a relatively brief lifespan (we saw Fudala et al. measure it as approximately 8 to 9 hours).[4] Should the alcoholic in question decide to themselves "enough is enough" and the need to imbibe thoroughly overtake them, all that need happen is a relatively brief wait, after which time the person in question is entirely free to drink once more. In other words, if the moral enhancement here were purely elective, then it is entirely within the alcoholic's own hands whether he or she wishes to take the intervention or not. Leaving such decisions in the hands of addicts has rather obvious consequences. As Fifield (2005) puts it: "It could be safely stated that the inability to 'Just Say No' is a diagnostic indicator of the presence of an addiction" (Fifield 2005, 68). If this is so, that one definitive characteristic of addiction is that the person in question does not have the willpower to resist what they are addicted to (or does not feel him or herself to have such a power, which in a pragmatic sense can amount to much the same thing), then the elective use of nalmefene puts one in a rather absurd situation. For if the addict in question had sufficient willpower in the first place, no enhancement drug would have been necessary.

Now, such an intervention might have a useful placebo effect, it might serve as a helpful psychological crutch for a while, a fallback plan, and the value of such things is not to be diminished or maligned. But the problem with addiction is the long game. Addiction is never overcome but rather managed from day to day, and in the circumstances where an alcoholic is forever "just one drink away from relapse," then psychological crutches can only ever last so long. Should the addict in question reach that "breaking point" where they have psychologically given in, the idea that a personally elective intervention might be of use seems considerably less plausible.[5] Fudala et al. (1991) make this point well. Referring to morphine addiction, they suggest:

Because Naltrexone and Nalmefene ... do not produce morphine-like effects desired by opiate abusers, successful treatment will require individuals who are highly motivated. The lack of morphine-like effects in conjunction with side effects that appear not to be dose related may hinder the therapeutic use of Nalmefene for treatment of opiate addiction. (Fudala et al. 1991, 305)

Precisely the same might be said in relation to the effects of alcohol. High motivation—a strong predisposition of will—is absolutely required, and precisely this is what the addict so often lacks. Indeed, the sometimes rather unpleasant side effects of nalmefene will make things even harder for the addict than they already are. It is little surprise that such opiate antagonists have not met with great success in the past.

Compulsory Use

Now consider another option, nalmefene being made compulsory: imagine that it becomes law that alcoholics who have a track record of domestic violence while intoxicated, or harm to self or others, are mandated by the courts to have the intervention forced upon them. Should they be caught without an effective dose in their system, or caught doing harm while not on this enhancement drug, the punishment would be extremely severe— long custodial sentences in some unpleasant institution or other. This scenario, a decidedly nonelective approach, overcomes the problem raised above inasmuch as it is not left to the addict to choose whether to take the drug.[6] Rather, it is being used more paternalistically.

Some might well advocate such use nonetheless. Drunks who are a menace, consistently violent and dangerous, who have not the least regard for other human beings, might justifiably have such punitive moral enhancement used on them. I am not advocating that at this moment, but rather saying a case might well be made by some for using moral enhancement in this sort of way. Here, the problem that elective moral enhancement poses, that we are putting the choice in the hands of the very addict who has as much as admitted that he or she does not have choice in the first place, is overcome when making it compulsory.

Moreover, given the nature of alcoholism—the role of denial in the problem, in which alcoholics refuse to admit to themselves that they have a problem, and the fact that even those who appreciate that they have alcohol-dependency problems rarely seek medical treatment (it has been suggested that only 2% to 5% of patients with full-blown alcohol dependence actually seek specific treatment; Mann et al. 2013, 706)—relying on alcoholics to volunteer to seek elective treatments seems more than a little optimistic. However, compulsory treatment for alcoholism raises problems of its own. Punitive use of moral enhancement vis-à-vis hard paternalism is likely to provoke a mixed and powerful public reaction—and rightly so.

One implication of compulsory use of nalmefene would be that it makes such an intervention far less likely to contribute to moral growth and the development of a moral identity. That is, such moral enhancement would rarely be "strong." So long as such opioid antagonists are being forced upon persons, there is little sense in thinking this is going to contribute to their growth as a person; though it cannot be ruled out, such growth is less likely where made compulsory. The explicit attempt to use such interventions to develop a moral character based on some explicit quest toward discovering an idea of what it is to be good (Kabasenche 2007; Jotterand 2014) is not terribly congruent with the idea of compelling a person to take a substance which negates the effects of the object of their addiction. And this is true even in a circumstance where an addict is forcibly manhandled, restrained, and injected with such a moral enhancer. Recall that one can never force someone to become a moral agent in this strong sense; one either explicitly wills such a quest and does the work, or one does not—a predisposition of will is the unassailable condition of strong moral enhancement.[7]

In contrast, an elective use of such interventions is much more likely to be congruent with a strong moral enhancement and an explicit attempt to grow as a moral person. Of course, a person could elect to take nalmefene for completely nonmoral reasons—for example, in response to ailing health—but elective use at least suggests the possibility of a starting motivation in more morally generative terms, such as saying to themselves: "It is wrong to beat my wife and children, I only do this when drunk, therefore it is time to stop drinking." In this circumstance, there would be no sense at all in which such use would not constitute "hard" moral enhancement in the strongest possible terms.

Those dead-set against moral enhancement per se will be challenged by such a proposal. For an alcoholic who wishes to refrain from drinking for distinctly moral reasons,[8] appropriating an intervention such as nalmefene (particularly in combination with a larger social or group therapeutic process) would be "hard" moral enhancement, plain and simple. There is nothing futuristic about such a possibility, nothing implausible or speculative at all. Such a possibility is here and now, and, so long as taken within the appropriate integral treatment protocol, manifests as an uncontroversial, desirable example of moral enhancement in its hardest and strongest possible form. Indeed, a strong moral enhancement skeptic would be hard-pressed to justify why an addict wanting to be a better person, already participating in social therapies and seeking nalmefene, should be denied it.

Note that if we are talking about moral enhancement as an explicit project, then the *intent* is one definitive part of the equation. Here is the dilemma raised by using moral enhancement for the treatment of alcoholism. Elective use of such an intervention is

(a) more amenable to being appropriated as explicit moral enhancement in the strongest possible terms, as part of an effort to become a morally better person; but,

(b) elective use relies on an addict being in charge of his own intervention, which, by the nature of addiction is precisely what the addict, on the whole, cannot or will not do.

On the other hand, compulsory or punitive use of such an intervention:

(a) takes away the need for the addict to choose to take it by putting in legislative mechanisms to apply such interventions as a punitive or preventative arm of the judicial system; but

(b) makes it far less likely that the addict in question will appropriate such forced interventions as part of a process of becoming a better person. Compelling a person to take such an intervention changes no attitudes, but likely reinforces them, ensuring later recidivism. Ultimately, in this case, we are just robbing persons of their capacity to get drunk. We are saying, "This person is not capable of drinking responsibly, in fact this person is a menace. We thus revoke his or her capacity to get drunk." The issue of whether a moral development is taking place is put aside. These are public health rather than explicitly moral reasons for compelling such interventions.

Some might not care at all whether compulsory moral enhancement actually contributes to a strong moral development in the addict, but only that the addict in question is no longer creating the misery they were while intoxicated. That is, whether such interventions be moral enhancement or not, some might want them made compulsory all the same. This might be a legitimate response. The point, however, is that one needs to be very clear regarding what one is talking about here. Moral enhancement for addiction raises dilemmas and complexities that are not obvious on the surface. One needs to understand the specific problem one is attempting to enhance, and be prepared to lower expectations accordingly with regard to what such enhancement might be able to achieve.

Looking at alcoholism as a case in and of itself, looking at the practical realities on the ground, raises a number of perplexing issues that cannot be dealt with on the purely abstract level that the sorts of questions "does moral enhancement reduce freedom?" or "should moral enhancement (per se) be made compulsory or voluntary?" can handle. This is why it is absolutely essential for moral enhancement discourse to begin to look at things on a case-by-case basis, to take account of the practical realities involved and work upward from there. When viewed in this light, it simply makes no sense at all to talk of "moral enhancement per se," because there is no moral enhancement per se.

Are Biologically Affective Agents Sufficient to "Cure" Addiction?

Part of the issue we have seen with elective use of nalmefene depends on the limited time frame in which it has its effects. Let us imagine now that the more experimental possibilities we encountered above have been tested successfully and that we have a much longer-lasting intervention available. What needs to be kept in mind is the relatively unique or idiosyncratic nature of this particular proposal for moral enhancement. Such prospects have the capacity to remove altogether the narcotic effects of the agent in question. With perhaps one exception (chemical castration for repeat sex offenders, which robs a person of their sexual urges tout court), there is little parallel between this case of addiction and other objects of moral enhancement. Moral enhancement aimed at, say, enhancing capacities for empathy or compassion, even if they could be devised, would not function in this sort of way.

We have said that no moral enhancement can force a person to do anything, but this case of addiction is slightly different. Because the effects of the alcohol are removed making it essentially pointless to imbibe, the alcoholic can engage all he or she wants, but the gains of such activity are utterly lacking since the drugs will have no effects. The reasons for action here, be they cognitive (e.g., "I drink because the world is a hideous place, and alcohol is the only thing that can take the pain away"); or be they more physiological (e.g., an animal craving for alcohol which transcends any particular propositional reason: "I just need it, and I don't know why"); or some combination of the two, are thwarted, not by forcing a person to do or not do, but thwarted because the reason they engage in such activity can no longer be satisfied by imbibing the substance. So, unlike many other objects for moral enhancement, such vaccinations and biomedical treatments for addiction can realistically stop an addiction dead in its tracks.

Or can it? Is there nothing more to addiction that the taking of the substance itself? We are in a curious position here, for on the one hand the addiction has been de facto cured—the addict is no longer poisoning him or herself and causing mischief due to their alcoholism, but the complex of influencing factors is every bit as present, if not more so, their usual means of satisfaction being so decisively obstructed. So this throws up a rather interesting question: say a person were compelled to take such a long-lasting intervention and stopped drinking altogether (for the duration of the effects of the intervention) because of it. Is this enough to say the addiction has been cured?

There is going to be more than a little controversy in answering this question. It matters a great deal which particular moral theory one cleaves to. A consequentialist may seek to argue that, in looking at the reduced harms done to person and society, when measured against the relatively cheap cost of the intervention to the alcoholic, such an intervention is an overwhelmingly good thing. Of course, there may be more terrible consequences in the longer term—the alcoholic may seek remedy for his addiction by engaging in the abuse of yet worse substances, or simply express his or her destructive and self-destructive behavior in other potentially more harmful ways. So other consequentialists may disagree with the value of such an approach.

An intent-oriented view is going to be more suspicious of the claim that such interventions might constitute a "cure" for addiction. Such longer-lasting vaccines distort the boundary between elective and compulsory use—one may have initially elected to take the longer-lasting intervention based on a sincere intent, and then come to have regretted it later, finding oneself burning for a drink and utterly frustrated in one's capacity to indulge. So intent-based thinking is also plagued with complexities on the issue. This is Sartre's point about "the gambler." The compulsive gambler may well have made a strong decision to give up his gambling ways, but this initial intention is not sufficient: the gambler is forced with having to make the very same resolution every time he passes a gambling table. A strong intent on one day is no guarantee of a strong intent on another day; precisely this is the problem that the alcoholic faces.

As such, the consequentialist has to measure immediate benefits to person and society against the possibility that the addiction will resurface in even worse ways, with worse consequences, further down the line. The intent-based theorist has to deal with the shifting nature of the addict's intentions. The question of whether the addiction is cured by such vaccines is gnarly, and it very much depends on one's point of view. At the very least one can say that, while the addict is "cured" in effect, there is a huge background of larger related issues that simply have not been touched by the introduction of the moral enhancement agent itself.

What may be more valuable than a consequentialist or intent-based approach might be the sort of vision of moral formation that we have hinted at throughout, the more cultural-linguistic approach to ethics which sees moral development (as O'Connell and Kelsey suggested) as being more helpfully understood as the introduction of new relationships, participation in new groups, groups (in this case) in which the habits of sobriety are embedded, inculcated, and valorized. The brand of theological ethics with which O'Connell and Kelsey are aligned seem to me to express a more grounded and realistic vision of how moral formation and moral development actually occurs—one which transcends consequentialist, deontological, and virtue accounts. The essence, if you recall, is that this image of persons making moral decisions by sitting back, reflecting on the good, or detaching themselves from their situation and actually ruminating on moral questions is simply not how things are done in real life. If moral enhancement is to work, then it has to work in real life and has to target the manner in which moral functioning actually happens in real life.

Yet in actual practice, for better or worse, what tends to happen in moral action is that persons get embedded within groups, groups which embody certain values and norms, which act according to long-standing habits and the sharing of common experience. Moral decisions are rarely made by long processes of rumination in isolation, but rather are learned in the way that an apprentice learns a craft, by following and conforming to the habitual actions and kinds of judgment characteristic of the group they are in. Most of the time, this is very much the background framework in which moral action occurs—decisions are pre-made, courses of action are already laid out, and when reflection does come into things, it is rather a matter of thinking through how the implicit habits of the group would likely be applied in the new context one has found oneself in. This is the "generative grammar" of moral functioning as it occurs in groups, the application and improvisation of old group norms in new circumstances. A great deal of the time, morality is largely a matter of unreflective, habitual action shaped by the group or culture one aspires to be a treasured member of.

If this is so, then the question of dealing with addiction is not really one limited to terms of consequences, intentions, or virtues—it will, at the very least, involve finding new environments in which the habits of the group prize sobriety, in which sustained sobriety is rewarded and cherished, in which compassion and encouragement for the addict is an explicit and well-worn norm of the various members of such a group. Without this larger social scaffolding, the notion that addiction can be managed in bio-logical terms only is less tenable. Rather the immersion within a group which values and creates norms for certain behaviors can be an extremely valuable background for motivating, facilitating, and sustaining a genuine wish to change.

At the same time, we should not over-valorize these social-environmental processes. What has to be borne in mind is that, in isolation, such social treatments still do not have an exactly perfect record. Indeed, the long-term recidivism rates of group therapy for alcoholism are somewhat disturbing (Soyka 2014, 580). Such group therapies might well proffer some manner of helpful background, but it must be remembered that society itself proffers a much stronger and more pervasive background which can neutralize the benefits that group therapies can proffer.

Larger societal backgrounds (certainly in the United Kingdom) scaffold drinking cultures which prize alcohol consumption and which generate all manner of pressures against which many persons seek relief through the drinking of alcohol. What we have is really backgrounds within backgrounds—as Kelsey puts it, "host" and "guest" cultures—which have all manner of overlaps and tensions. Even the positive background for sobriety created by the most ideal group therapeutic situation must exist within a larger host culture which is, in part, the source of many of the influences which lead many alcoholics to drink (e.g., macho drinking cultures, depersonalizing work conditions, a "you only live once" party mentality, and so on). The sad fact is that being introduced into groups in which sobriety is valued is often insufficient for many with alcohol dependency. This, bear in mind, is only with respect to that small proportion of alcoholics who do actually seek treatment for their problems rather than languishing in denial or attempting to solve their problems on their own.

In short, social-environmental influences may very well supervene upon the biological aspects of the problem, but if the social-environmental influences are themselves a not-inconsiderable part of the problem, then one is forced into having to attempt to change the entire culture and thinking of the host society (a worthy aspiration, but one not readily or quickly achieved), or seeking to augment and perhaps counter such influences along an alternative means of approach.

As such, one would be utterly remiss to condemn agents such as nalmefene or the other agents that we have been considering if they do have the power to support the group therapeutic protocols. What must be noted is that reducing the solution for alcohol addiction to the level of environment only is actually just another form of reductionism. Overvalorizing the power of scaffolding neglects biological factors, just as biological reductionism neglects the environment. To unnecessarily neglect a perfectly valid, safe, and presently viable means for assisting a person in their aspiration to manage his or her addiction cannot be acceptable.

This suggests, quite obviously, that an intervention which applies both approaches would be best of all, that psycho-social and biologically affective approaches, managed wisely, might be highly complementary in nature. What is required is an effective treatment (the biological approaches, in this case at least, seem to be particularly effective) that can be integrated with a nuanced personal approach—one capable of grasping the deeper nature of why that particular person is an alcoholic in the first place, which offers support, new norms and relationships and experiences to aspire after, and which does not simply sweep the deeper causes under the carpet to manifest elsewhere in other ways.

We want a deeper approach, one which treats the human person as a whole being, a being in an environment, with a history and a culture. I suggest that Markus Heilig, clinical director of the US National Institute on Alcohol Abuse and Alcoholism, summarizes the situation well:

The challenge for alcohol addiction therapy is to find drugs with complementary mechanisms and to work out which patients to treat with which drugs at what stage in the addictive process. (Heilig 2011; 566)

Overly simplistic answers, referral to "dopaminergic responses," simply will not do. Reducing addiction to the level of behavior—saying, as Bass et al. did above, that alcoholics' behavior is "governed" by their brain chemistry, "controlled" by their reward circuitry—is to do a significant harm to the nature of the problem.

Once again, we are caught in that delicate balance of respecting that the problem at hand does indeed have important biological influences without reducing the problem to the biological level of approach or making biological elements primary. The difficulty is finding the right place for such interventions such that a genuinely bio-psycho-social approach can be formed—one which balances most optimally the various biological and social needs of the addict in question, treating the problem of addiction in terms both of its effects as well as its underlying causes.

Pastoral Psychology: "Give Us This Day, Our Daily ... Meds?"

Spirituality has long been recognized by the recovering community as the key to attaining and maintaining a sober life. In the "Big Book" of Alcoholics Anonymous (1976), this principle is clearly proposed: "We are not cured of alcoholism. What we really have is a daily reprieve contingent on the maintenance of our spiritual condition" .... A supportive corollary spiritual principle is that in order to stay sober, the Member must "work" the steps and the program, that beliefs and values alone are not sufficient. The individual must learn how, on a daily basis, to do the "work" of maintaining his or her spiritual progress.
—M. Fifield (2005, 68)

In part III the suggestion was made that moral enhancement discourse, for the most part successfully, has done its utmost to completely ignore the religious dimensions of moral formation, the manner in which billions of persons identify with one religious tradition or another, and the manner in which pretty much everyone in the world has been raised against a religiously shaped background culture. Religious faith cannot simply be bracketed out of moral enhancement discourse, and the issue needs to be redressed, in this case, by extending our meditation on treating alcoholism to show how religious concerns might be married with moral enhancement application.

We are fortunate in this case, for not only do the means for such enhancement exist, as we saw, but there is already a considerable literature relating to the spiritual and religious aspects of treating addiction. These pastoral reflections refuse to ignore, in the way that moral enhancement has tended to, the manner in which religious faith and spirituality can contribute positively (and negatively) to the growth of the individual person as a result of their addiction. Again the call is for a broad-scope, multipronged approach—we need to treat the human as a whole being, a personal and willful agent, a responsible agent, a being in a world. The human is an agent with a biological nature, a social nature, and, as is quite often the case, a consciously spiritual nature, too. All these various elements can and should be appropriated wisely to facilitate the growth of the individual who has been beset by addiction. These must all be explored as essential parts of a more integral bio-psycho-social therapeutic process.

The simple fact is that there is a vast number of persons who self-identify as religious or spiritual. These persons need to be respected, and their particular religious or spiritual outlook has to be examined with respect to the ways it assists or impedes potential treatment. Of course, things differ from country to country, with some countries being more densely theistic (e.g., Andorra), some being more secular (e.g., Denmark), and some countries having a more even balance between theistic, atheistic, agnostic, and "somethingist" outlooks (e.g., the Netherlands). As such, different countries will have to apply different general policies which are appropriate to the range of religious or spiritual outlooks representative of their various citizens.

These are practical, ground-level factors that must be accounted for as part of a serious moral bioenhancement discussion regarding alcoholism. An account of moral bioenhancement for treating addiction which does not broach shaping spiritual influences that may or may not be present in the addict's life, or which is not amenable to their inclusion, will have to be rejected as inadequate for use with respect to the vast number of persons who self-identify in such a way. Existential questions which the addict is used to answering in spiritual or religious terms, the nature of the religious interpretive framework that is applied by the addict, can all manifest in variously adaptive or destructive ways. Similarly, the religious history of the addict, the religious environment and community an addict inhabits are not inconsiderable influences on their behavior, which can likewise be adaptive or destructive. These dimensions overlap with the psychological and social-environmental influences in complex ways. Thus, given the circumambience of religion and spirituality, such dimensions of addiction are influences which need to be accounted for.

Let us start by looking more closely at a theistically driven pastoral approach. Perhaps the most important thing to note is the manner in which something like a Christian pastoral approach to addiction will have a salvatory edge[9]; that is, will have to do with the client growing as a Christian person. A central and unanimously agreed-upon proposition here is that growth as a person—and as a specifically Christian person—is fundamental to the outcome.

As such, pastoral approaches lend themselves toward a "strong" concept of moral development. It is precisely in fostering such growth as a person that the long-term efficacy of pastoral approaches are ideally to be fuelled and ongoing motivation sustained. As in the ideal of virtue formation, the transformation is not complete until, at some deep level, the person in question genuinely desires the good being embodied, or, at least, desires some other good which is inconsistent with being drunk. Thus, dealing with alcohol addiction may not focus on "wanting not to drink," or wanting to be sober, but rather it may focus on wanting to spend more time with one's family, repair destroyed relations, be a healthier person—all of which are incongruent with the alcoholic imbibing their beverage of choice. The idea, then, is to find some larger reason for not drinking, and this can overlap with finding a deeper spiritual purpose for living. As Morgen (2009) suggests:

The counselor's job is to help those who are lost to "find a voice and craft a purpose." I like to think of spirituality as being the same thing—a vehicle for finding a voice and crafting a purpose, no matter how harsh the world may sometimes seem. (Morgen 2009, 2)

Pastoral Approaches for Treating Addiction

There are three main features of the pastoral treatment of addiction that are important for the present discussion—pastoral treatment as involving

(a) a holistic approach;
(b) prayer and meditation; and
(c) a suffering community of support.

The holistic dimension of the treatment of addiction, which includes spiritual dimensions of healing, is a core feature of the treatment of addiction in the United States.[10] The Council for Accreditation of Counseling and Related Educational Programs (CACREP) demonstrates clear recognition that spirituality can be a helpful force in the process of recovery (Section G.3, for example, states that an addictions counselor needs to understand "the assessment of *bio-psycho-social* and spiritual history"; Morgen 2009, 3; emphasis added). This means that addiction treatments are explicitly fashioned in this setting such that they are able to incorporate analysis of all the relevant influences we have hitherto been considering. Even for secular persons seeking treatment, it is important to go into their religious history as one important formative influence. The prevalence of religious influences in the family or in the broader environment can be helpfully brought to the fore, regardless of whether the person is religious.

Under the appropriate conditions, spiritual or religious involvement may be an important protective factor against alcohol and drug abuse (Miller 1998, 984), and apart from the physiological benefits of certain spiritual practices, such as prayer and meditation, spirituality in this context seems to function, using Kenneth Pargament's language, as a vehicle for "benevolent religious reframing" (Pargament 2001, 222). That is, faith contexts can provide a religious orienting system, a system of reinterpretation, which helps the individual in question reframe their situation in positive religious terms (for example, as a test of faith, or as an opportunity from God to grow). Such mechanisms can constitute adaptive coping mechanisms (Pargament et al. 1990, 2001).

What is idiosyncratic about pastoral approaches to treatment is the integration of prayer and/or meditation within the working protocols. In most cases, prayer functions as a reinforcement technique and is usually integrated within counselling sessions as means to establish treatment goals, to "lock in" insights gained, and a means toward becoming mindful of various stressors and triggers in one's environment. Prayer can be used in the course of the session to pray for the healing of hurt, insight, direction, clarity of thought, and peace. The intercession of God is called upon to sustain, to guide, and to comfort. In Norman Wright's words: "Prayer accomplishes several things—it releases the person to God and reminds us that God is the final resource in our lives" (Wright 1987, 128). Finally, growing as a Christian person can be fostered by using prayer to increase a sense of the presence of Jesus or the Spirit working in the lives of the addict. For religiously minded persons, the point is this: "We pray because it is our privilege and a means of communicating with God"—a means by which to be welcomed into God's presence (ibid., 128–140).[11]

Finally, and arguably most importantly, is the notion of the suffering community as a support mechanism—the shared pastoral dimension. Prayer and meditation can offer a "vertical" spiritual or religious channel (i.e., from God to person) wherein a person can refine and explore his or her responsibilities for action, while yet opening up a relationship and receptivity through which grace can be received. The community aspect, in contrast, opens up another kind of channel for grace, a "horizontal" stream, one mediated by the suffering group in their togetherness, wherein the very act of sharing experiences, conversing, and mutual encouragement help sustain and bolster the motivation of participants.

Dangers of Pastoral Treatments

Religious or spiritual aspects of treatments for addiction can be very powerful, and this is precisely why pastoral treatment of addiction needs to be inspected critically, for it can do immense good, or it can be immensely destructive, depending on how such pastoral treatment is deployed. As such, it is important that one not present pastoral approaches as some inherently superior means of treatment. A few examples will suffice to make clear that an uncritical attitude toward pastoral approaches cannot be adequate. One common concern with spiritual approaches to recovery is that "clients can detrimentally use spirituality as a mechanism to avoid responsibility for recovery" (Cashwell et al. 2009, in Morgen 2009, 3). Indeed, the addictions counselor must recognize the many ways in which "spiritual bypass" can hinder the recovery process. Prayer, for example, a staple of pastoral person-centered approaches, is not always an appropriate part of the therapeutic process. Juhnke et al. (2009) note that prayer should be contraindicated by addictions counseling professionals in instances:

Where (a) severely disturbed clients may experience delusions or hallucinations; (b) clients do not believe in God, God's interest in their recovery, or God's willingness to hear and respond to prayer; or (c) clients may have significant impairment resulting from personality disorders, such as dependent personality disorder. (Juhnke 2009, 18)

This underscores the manner in which therapeutic processes are deeply personal in nature, requiring considerable tailoring to individual needs. Similarly, pastoral approaches have to be sensitive to the kind of God-image held by the client. Pargament's (2001) work on religion and coping highlights this point: the kind of idea one has about God can have a considerable impact on the manner in which one interprets one's situation, and can be adaptive or, sometimes, incredibly destructive to the situation at hand.

In this way, a considerable part of the pastoral process might well lie in challenging and realigning distorted and distorting conceptions of God in the client's mind. This is the sort of extraordinarily idiosyncratic sort of conditioning factor that a purely biologically oriented approach could never come close to being able to broach. In the treatment of addiction, the idea of "God-image" is particularly relevant. Ciarrocchi and Brelsford, for example, note the manner in which "the experience of God as a punishing force leads to increased psychological distress" (Ciarrocchi et al., in Morgen 2009, 3). A pitiless and authoritarian image of God combined with the "Just Say No" approach can often produce, through persistent failure, a state of learned helplessness which is particularly difficult to challenge and heal.

Even the community aspects of treatment can be destructive— particularly destructive, in fact—and the power of the community or group of sufferers to help or hinder healing depends on the nature of the group in question. We suggested that denial is a big problem in treating alcoholism; this is a particularly large problem in religious communities. There are many communities that simply refuse to accept that members of their faith tradition, communities, and families are even capable of falling prey to addiction (Solomon 2013, 1). Because of this implicit refusal with respect to admitting that problems can exist among practicing religious members of a community, many such alcoholics are less likely to come forward for treatment or feel comfortable attending support group meetings in places where they worship (Solomon 2013, 1). Communities such as these not only offer no practical support to the addict, they make things worse by encouraging denial or stigmatization and ultimately have no understanding of how the process of addiction is best managed (religiously or otherwise). Such communities are not helpful forces in spiritual healing.

Lastly, perhaps the biggest factor in whether a pastoral treatment is likely to be successful is the counselor him or herself. The previous chapter ended by noting the paramount importance of the quality of the practicing care professional in question. This is very much a concern in pastoral approaches. Clergy are not perfect, and are sometimes in need of pastoral guidance themselves (Wright 1987, 19). A pastoral approach can thus be helped or hindered by the quality, professionalism, and values of the counselors involved. For example, one common but very unhelpful attitude taken by the clergy in the past was to focus on the "Just Say No" approach. While personal responsibility is without doubt an essential aspect of addiction recovery, focusing only on personal responsibility has poor long-term results. Like a purely biologically reductive approach, it takes no account of the complex background causes influencing the particular person's addiction. Inasmuch as this method, through repeated failure, often results in generating more shame, guilt, and acting out, it is clearly an unhelpful way of dealing with addiction—particularly when couched in religious terms. Fifield (2008) writes:

Historically, the Christian Church's approach to helping an individual who was held in the grip of addiction was to repeatedly challenge the person to "Just Say No." When the suffering person was repeatedly unable to comply, the Church condemned him or her further and increased the amount of shame and guilt in a mistaken effort to obtain compliance. This misguided approach rested on two faulty assumptions about the nature of addiction, assumptions that continue to be made by some Christians today. Simply stated, the Church assumed that addiction was the result of individual choice and, therefore, the individual could choose to stop if she or he wished to do so. Because the addict did not stop, the Church concluded that the person would not stop because he or she did not want to stop. This Christian attempt at helping the alcoholic failed because the alcoholic could not stop. Many individuals, who began with a seeker's heart for God in the Christian faith, left their church of choice with a broken heart in order to find a way to survive. (Fifield 2005, 68)

This is a perfect example of how pastoral approaches can distort and hinder the alcoholic's wish (albeit a divided wish) for recovery. As such, pastoral approaches to treatment have to be presented as fallible, and certainly not above critique. At the same time, applied appropriately, the religious and spiritual elements of addiction treatments arguably have the most potent capacities for healing of all means for shaping and reshaping the outlook, purpose, meanings, and values with which the addict views themselves and the world. These are considerations which, once genuinely embodied, can have an incredibly powerful effect on the addict's thinking and behavior.

Just as we have sought throughout to reduce addiction not to biology, nor to social influence, we have to resist any spiritual forms of reductionism too. We have to resist many of the easy, and false, pseudo-spiritual answers that are prevalent within our culture—that addiction can be healed through prayer alone, or by "visualization" or "affirmation," that God is simply going to come riding over the hill with the cavalry and solve all of one's problems without effort from oneself. Thinking of addiction as a purely spiritual problem is completely unacceptable if one misperceives its treatment as something which does not involve active engagement. For it is precisely in active engagement, as much as in appropriate surrender, that the spirit is roused and the channels of grace can open up. In contrast, a superficial spiritually reductive approach is much more problematic than treating addiction in biological terms only, or relying on social or group therapies in isolation. *We need some means of integrating all three modes of approach.*

Treating Alcoholism: Mental Health Problem or Moral Enhancement?

Sleight of Hand?

Addiction is rarely considered as being a form of "madness," but has often been judged as morally blameworthy. And thus a form of "badness." However, ...

substance abuse does find a place in classifications of mental disorder, such as DSM-IV and ... ICD-10. Furthermore, according to most American psychiatrists, it should be understood not just as a "disorder," but as a disease.

—C. Cook (2006, 148–149)

We have considered various biological, cultural, and spiritual forces involved in addiction. But what about the basic idea of the addict taking responsibility for their own actions? The question of personal responsibility is, in part, a culturally mediated one. While psychiatrists in the USA are more likely to accept a disease model, UK psychiatrists are more likely to understand addictive disorders as a form of learned behavior (Cook 2006, 149). In the end, the treatment protocols are the same regardless of whether one believes alcoholism to be a disease or behavioral issue (and for now it is a matter of definition, since the currently available evidence is very much open to contrary interpretations). In either case:

The focus of treatment will still be upon counselling group work and an emphasis upon the need to take personal responsibility for addressing the problem. (Cook 2006, 149)

Addiction is positioned then in an uncomfortable middle ground as both:

(a) a problem requiring personal responsibility (and thus as something more readily thought of in moral or morally related terms); and
(b) a problem that has medical treatments available for it (and thus as something fitting within the medical or mental health domain—whether one defines "disease" in terms of physiological deficit or social convention).[12]

While the idea that something can be treatable while yet being a matter for personal responsibility is perfectly sound, the concern is that if one defines addiction as a "disease," this involves implicitly sanctioning the view that addicts are no longer responsible for their actions and for seeking their recovery. This concern is significant but not a necessary consequence of thinking of the problem in "disease" terms, for it assumes that medical or mental health practitioners can only ever treat matters over which persons have no power of responsibility. Obesity can be treated using medical means (gastric bands, liposuction), but this does not mean there is no personal responsibility involved.

So, again, what we have here is an uncomfortable overlap with moral bioenhancement and medicine and mental health treatment—addiction is both a matter of personal responsibility having substantial moral dimensions, and something apt for medical or mental health treatment. In other words, addiction is both a moral issue and a medical or mental health issue.

This presents a rather large problem for the moral bioenhancement debate. It means that treatments, often long-standing treatments, that are usually considered mental health or medical in nature (for that is what they are) can also be thought of as moral bioenhancement.

This is a problem. One particularly astute commentator has made a powerful case for suggesting that the label "moral enhancement" is highly dubious in such cases of clinical intervention (such as treating serious addictions, or something like antisocial personality disorder or psychopathy, particularly in violent individuals). One is simply taking a standard mental health practice, say, using SSRIs in the treatment of those with antisocial personality disorder, and then—as if from nowhere—simply calling it something else: "moral enhancement."

This overlap between that which is moral and that which is in the purview of mental health practice makes drawing the line between moral enhancement and simple mental health intervention particularly difficult in these kinds of serious remedial cases. In this case, simply giving the label "moral enhancement" to what already exists as a mental health intervention is nothing more than a sleight of hand. We are just taking something which is properly a clinical intervention and rebranding it to give the illusion that we have created something new. This would then be a case of the emperor's new clothes.

How is one to make sense of moral enhancement for serious remedial issues when questions of morality and mental health overlap so extensively? If we have treatments available for matters which are also at least in part matters for personal responsibility, where does "mental health" end and "moral enhancement" begin? How does one avoid this rebranding problem? As we saw in the previous chapter, the trend in present mental health diagnosis is one of generating an ever-expanding purview. Thus we are faced with a dual assault here. On the one hand, the field of that which constitutes mental health problems is increasing all the time, disturbingly so, such that it is beginning to include behaviors that are well within the realm of normality (e.g., smoking, as "tobacco use disorder"). On the other hand, certain "moral sickness," as we saw in the previous chapter, can manifest to such pathological extents (e.g., extreme violence combined with lack of empathy), that it appropriately steps into the realm of mental health. Particularly in light of this increasing tendency to determine any behavior which falls even slightly outside the norm as a "disorder" to be treated in a mental health context, this dual migration of morality into mental health, and mental health into morality, makes things such that there are very few clear and distinct lines to be drawn between morally related problems, such as alcoholism is, and the domain open to mental health practice.

If we are talking about serious remedial cases, moral enhancement and mental health treatment must be understood as overlapping. This is to some extent inevitable, and in some cases potentially desirable: having treatments to assist an existing predisposition of will is precisely what moral enhancement should be concerned with. But what then are we to do with the important and powerful observation raised above, that it is deceitful to simply take an existing mental health intervention and, all of the sudden, simply relabel it "moral enhancement"? If the term "moral enhancement" adds nothing to what was already there, is it not entirely gratuitous?

Responding to the Charge of Gratuity: "Hard" and "Soft" Forms

One potential response to this charge is to suggest that in using the term "moral enhancement," we are not hoping to add to anything new, but merely recognizing the possibility that "soft" moral enhancement has existed and been applied for quite some time. This points back to the observation with which we toward the end of part I—that moral functioning is very rarely an explicit activity, but rather diffuse. Moral functioning is diffused throughout our daily activities, a dimension of pretty much everything that one does throughout the day. It is not a separate part of daily existence, as if one puts aside half an hour a day to be moral and then goes on with the rest of one's day in a completely amoral or nonmoral fashion.

Moral functioning is spread throughout almost everything that we do, and most of these things have moral dimensions which can be projected upon them, be that something as simple as holding a door open for someone, refraining from assaulting someone who has been particularly offensive, sparing a thought for someone other than oneself, not indulging a vice that one desperately wishes to indulge, watching one's tongue and being tactful with another's feelings, or casually dispensing some pocket change to a homeless person on the street. In all the miniscule and obvious ways in which moral functioning blends with daily existence (and this is particularly visible in relational contexts), we have to note that moral functioning proceeds almost invisibly on a continual and ongoing basis. Little wonder, then, that mental health and moral functioning cannot be readily separated: this is a mere amplification of the manner in which daily living and moral functioning are so thoroughly intermingled.

One possible response to the charge that talking about pharmacological treatments for addiction as moral enhancement is gratuitous may simply be to recognize this fact that moral functioning overlaps with almost everything anyway. Calling a mental health intervention for addiction "moral enhancement" is only gratuitous if one is attempting to posit the existence of some extraneous element to the intervention. If one is merely recognizing that certain mental health interventions have moral dimensions, that they always have and will, and thus that such interventions have been, in a sense, "soft" moral enhancement all along, is not to posit some new extraneous dimension of the treatment. Perhaps it is simply to disclose something that was implicit within it the whole time. This is to remind us that we have already had moral enhancement "by another name" in subtle forms for quite some time. Insofar as (a) any given problem is couched in biological terms, understood as a biological problem; and/or (b) the problem is treated of using biomedical means, then, in such cases, we have arguably had soft moral bioenhancement going on under our noses for quite some time.

Alternatively, returning once more to the introduction of the present book, we have to remind ourselves that so much of the matter comes down to how we define our terms. Throughout this section, we have been dealing with moral enhancement in a distinctly soft sense. This soft sense is very broad by design, and it has been very helpful in opening up discussion to highlight the larger issues within which moral bioenhancement is necessarily embedded. Yet those purists who wish to define moral enhancement only in the hardest and strongest terms, as something which is explicitly used by persons as part of the process of them attempting to develop morally, or to sustain and facilitate moral growth, character, or moral identity, are likely to be considerably unimpressed with this inclusion of the whole of pharmacologically based addiction treatment within the remit of moral bioenhancement.

That is something that needs to be acknowledged. Using the most stringent possible definition for moral bioenhancement (in this case, looking at nalmefene and alcoholism), such an intervention would be moral enhancement if and only if such persons were using the substance to overcome their addictions on explicitly moral grounds. If a person, as we earlier indicated, recognized him or herself as an addict and saw this as problematic, not just for health reasons, but because of all the terrible things they perpetrate while inebriated, and as such used a pharmacological intervention to deal with their issues, *then there is no sense at all in which such an intervention is not moral bioenhancement in the hardest possible way of understanding the term.*

This then provides another response to the charge of gratuity raised above. In this case, one is not just taking a mental health intervention and relabeling it "moral enhancement" (one is doing that, but at least there is more going on than purely gratuitous rebranding). The mutation of the mental health intervention into something at once moral enhancement and medical is something created by the intent of the person taking the intervention. It is the moral intent here which does all the work in determining whether a particular mental health intervention is also hard moral enhancement. As such, pharmacologically assisted therapies for dealing with addiction would then not be moral enhancement of themselves, but only made so by the intention of the patient to engage with such treatments on explicitly morally formative grounds.

This being so, the charge of gratuity still retains a cautionary force. However one conceptualizes the relation between moral functioning and mental health, one must be cautious, on purely pragmatic grounds—given the punitive and covert manner in which medical and mental health interventions routinely move into value-laden territory—regarding the temptations to disguise what may be invasive interventions for perceived social diseases in the more benign terminology of mental health intervention.

Moreover, to bring the discussion back to the complex distinction between voluntary and compulsory moral enhancement, as we argued earlier, there may very well be defensible ways of arguing that such anti-addiction interventions be made compulsory in cases of dangerous addicts. It is an unfortunate reality that persons subject to certain forms of substance addiction can be a menace. Persons intoxicated with, or in a state of severe withdrawal from methamphetamine, for example, have been known to perpetrate crimes too despicable to warrant mention, displaying behaviors and lack of judgment of a truly bestial nature. For the good of society and for their own good it may well be justifiable, if such interventions could be made to exist, which in light of this chapter seems entirely feasible, to compel such persons into vaccination programs—whether we mention moral enhancement or not.

Even the dangers of biological reductionism, which we have highlighted throughout, may pale in terms of the human cost of allowing such behavior to continue. William Burroughs (1993), speaking from lengthy and entirely personal experience, described the dehumanizing effects of heroin addiction as follows:

Junk is the ideal product ... the ultimate merchandise. No sales talk necessary. The client will crawl through a sewer and beg to buy. ... The face of "evil" is always the face of total need. ... Beyond a certain frequency need knows absolutely no limit or control. In the words of total need: "Wouldn't you?" Yes you would. You would lie, cheat, inform on your friends, steal, do anything to satisfy total need. ... A rabid dog cannot choose but bite. (Burroughs 1993, 8)

Burroughs characterizes this state as one of "being possessed," as "total possession." And in the face of such dire and often hopeless conditions, particularly for those living in abject poverty (whether that be in dark recesses of Manchester, UK, or the ghettos of Cape Town, South Africa), there may very well be a case to be made for a decisive compulsory project of vaccination to be enacted. Such a program would then be compulsorily "passive" ; that is, applied only to those who display such addiction. Recall, the active/ passive distinction, when applied for compulsory intervention, refers to the difference between an intervention applied to everyone, all the time (active), and one which applies to all persons, but only activates when they have transgressed some moral boundary or other (passive), like addicts who have become lost in the belly of some dangerous substance or other, and have become dangerous themselves. There may yet be cases—realistic, and presently attainable—in which the expression of a highly limited compulsory moral enhancement be legislated for certain groups as one potential arm of the judiciary. Not to prevent the possibility of world annihilation, not to be applied unilaterally to all persons, but simply as an extension of the measures states routinely do apply to that small minority of individuals presenting a serious and predictable threat to others and to themselves due to their addictions. In such cases, nalmefene or any correlative intervention like those elaborated above might well be justifiably mandated as a part of a court-directed program of rehabilitation. As with so many of the suggestions made throughout this chapter and the last, such a move would likely blend in imperceptibly with what is already being carried out.

All the same, this is very dangerous territory. The application of moral enhancement solely to certain marginalized groups is a recipe for disaster. The marginalized are, by definition, not accorded the same concern and checks as those more adept at conforming to mainstream social expectations. We have to be vigilant that in enacting moral enhancement, regardless of how soft that enhancement is, regardless of how seamlessly it blends into what we already have, we do not do so in ways that are themselves morally problematic. Using biomedical interventions on the marginalized has so often proven to be a passport for the perpetration of outright evil. That being so, if an appropriately humane supervisory institution is in place, there may yet be extreme cases (in this case, the radically dehumanizing nature of certain illicit substances and the unspeakable harms committed by persons addicted to them)—circumstances which must remain very limited, focused, and constrained in nature—that may demand a heavier hand in their resolution, for the public good as well as for the individual in question, even if the word "morality" never comes into the equation at all.

A Final Word on Remedial Moral Enhancement

It could very well be that something as pedestrian as nalmefene, applied as an adjunct to person-centered therapy and within a larger mental health context, represents the ideal remedial moral bioenhancement ought to aspire to. It is nonreductive—it does take biological influences seriously, and draws on methods to tackle biological aspects of the problem. At the same time, it does not presume the biological dimensions of the problem at hand are "all there is to it," nor does it even assume that these biological aspects are primary. In fact, most of the "moral work," as it were, is done in the face to face of group therapy, or the face to face of the ongoing patient-doctor conversation. The biological agent here is simply a facilitator for this work to occur. Such an approach takes seriously the biological aspects of the given problem at hand and meets them on their own terms without reducing the matter thereto.

And is not this the real truth that I have been attempting to get at all the way through? Moral functioning is travestied when the biological components of such functioning are taken to be the sole, primary, or dominating forces, regardless of the number of influences that are recognized. Such a pedestrian approach to hard moral enhancement as found with nalmefene in present practice may then be at the limits of what is desirable with pharmaceutical agents. If that is so, we ought not cry about it. It is better to have something pedestrian that actually works, and works appropriately alongside a deeper, more integrative approach, than some reductive travesty of an intervention, or some fantastical aspiration with no grasp on practical reality. Of course, some issues are going to be more or less manageable by pharmaceutical agents. Yet it is to be noted that we have been considering here interventions which are, in principle, capable of removing the surface difficulty altogether. In this way, we have been talking about the absolute peak potency for pharmaceutical approaches to moral bioenhancement. Even this has been found wanting. Indeed, the suggestion could be that the more thoroughly the drugs in question cover up the surface of the moral issue, the more dangerous they become. Thinking of addiction as a purely biological problem is a terrible mistake, just as with thinking that violent crime is a problem of faulty amygdalae. Too few people are asking *why* it is that such persons' amygdalae are "malfunctioning." Too few people are asking why it is that the addict's dopamine system is causing problems. The wonderfully vague answer we have encountered twice, from Hughes and Savulescu, that "our genes" are "about 50%" of the problem, tells us less than nothing. It does not tell us how each of the various parts of this supposed 50% interact, and it is *precisely* in the interaction that the actual reality of things manifests and coheres into the particular concrete reality that it is. Such paradigmatic lazy thinking needs to be avoided at all costs, for it misdirects us from the real issues that need to be remedied.

Too few are looking toward the psychology at hand. When we ignore the interpretive frameworks through which a person understands himself and his world and go straight to "the dopamine system," to "the genes," to "the amygdalae," we ignore one of the most fundamental parts of the problem—the way that individual is relating to his world, and the way that individual understands his relation to his world and the people in it. This is the complex answer, and thus it is the undesirable answer. It is difficult to understand a person's psychology, or the way we live our lives, and it is even more complex to try and actually deal with all this. Would it not be wonderfully convenient if such problems could be the fault of a malfunctioning dopamine system? Of course, we want to believe that. But what needs to be kept firmly in mind is that the *existential* part of the problem, which is overwhelmingly complex, cannot be dealt with on the biological level except in the most superficial way, such as nalmefene does, which is precisely why the real work goes on in the therapeutic context. If this therapeutic work fails, as it often does, it is because existential problems are massively difficult to remedy. If biological agents like nalmefene might smooth the flow of such existential examination in the face to face between the doctor, patient, and the suffering community, then we may already have to hand the best that can be hoped for in this remedial, hard domain.

V Conclusion

10 The Limits and Potential of Moral Bioenhancement

Critical Edge: Bursting the Bubble

"Whoever lives there," thought Alice, "it'll never do to come upon them this size: why, I should frighten them out of their wits!"
—L. Carroll (1994, 65)

This book has been a story about two main kinds of moral enhancement, those which are done explicitly to improve upon moral functioning, and those more subtle, covert forms which touch upon profoundly morally related functions but which do not necessarily mention morality at all. The former kind, insofar as we are thinking in terms of what can be politically realized by a given state, have been presented as a nonstarter in a democratic world. No politician could get elected calling for a program of explicit biomedical moral enhancement. This still leaves over a range of hard options that might be yet be possible, in principle at least, and we will come to these in a moment.

In contrast, the latter, soft kinds of moral enhancement have been presented as embodying prospects that are already upon us—prospects that have been with us, and will inevitably continue to be so enacted. This latter project is harder to detach from the state because, for the most part, it is itself nothing other than the action of the state and the many various institutions that comprise it. Indeed, the irony with moral enhancement is that, while no politician could get elected calling for an explicit form of biomedical moral improvement, it is entirely feasible that a politician would be quite popular if he or she called for very similar enterprises without mentioning morality as the rationale. A politician calling for nonemergency surgery to be withheld from smokers and the obese "because they lack the virtues of moderation, self-discipline, and self-restraint," would not do so well as a politician calling for the same measures "because such persons put undue strain on the taxpayers and health authorities." The final truth may be that moral enhancement is nothing other than a matter of branding.

Yet, philosophically, these two modes, hard and soft, proffer a set of prospects for moral enhancement that are so radically different in nature that they genuinely merit splitting the entire discourse in two and separating out questions relating to either the one set of prospects or the other. For this reason (and many others), all talk of "moral enhancement per se" must be curtailed. Actually, the idea of moral enhancement can be construed in such a plurality of ways that we really need a much better-fragmented discourse to accommodate the various possibilities available. We need a whole range of conceptual paradigms to wrest the conversation free from the unwarranted dominance of the "hard" state-sponsored paradigm, which, along with its apocalypse-based rationale,[1] unfortunately is what instantly springs to commentators' minds when the idea of moral enhancement is raised.

If anything, I hope this book has shown how narrow, superficial, abstract, and impractical such a paradigm is. That approach, one which looks toward the state as a possible vehicle for enacting a version of hard moral enhancement (that we might be "saved from ourselves"), is unrepresentative of the realistic and desirable potentialities that the field of moral enhancement holds for us at the present moment. It is because of the dominance of that paradigm that so many commentators think of moral enhancement as merely speculative, future-fantastical nonsense (looking solely at that paradigm, they are right to think so), or outright evil, and completely ignore the presently realizable, and already realized, dimensions that moral enhancement portends.

I hope that some of the specific case-meditations provided at length here have been sufficient to show that moral enhancement is a very intriguing prospect and the subject needs to be taken seriously. I hope the present book has been able to proffer a strong new approach to the subject. We have been following the discourse through the lenses of

(a) the various practical realities involved (political realities, biological realities, institutional realities, and cultural and faith-based realities, at the very least, are relevant here). Such realities must be appealed to in order to shape any plausible conversation on the subject given the limits such realities place on the enactment of any given instrument of moral enhancement. Practical reality limits things severely, and, as we have said, "ought implies can." It is all very well making the claim that "we must create beings with moral standing superior to our own" (Rakić 2015, 58), but if there is no realistic way of doing this, then we have not really said anything at all;

(b) the complexities involved in realistically managing various moral problems (the complexities of issues like global pollution, poverty, terrorism, institutional violence, racism, subjugation of women, and so on), without trivializing them into crude, offensive, biologically reduced caricatures of themselves, and without massively overestimating the contribution biological alterations might make to such massively complex problems, which are not really appropriately construed in biological terms at all;

(c) the gnarly idiosyncrasies raised by each particular intervention and target proposed (where each prospect has issues, limits, and ground-level realities specific to itself, all of which need to be considered precisely as particularities rather than in the senselessly vague terms of "moral enhancement per se"), that we may be profoundly sensitive to the different needs of different kinds of individuals with different kinds of moral issues, which might be treated in very different kinds of integral ways; and,

(d) an interdisciplinary approach as the medium for optimal conversation. Given the interwoven nature of moral functioning, the manner in which moral living traverses so many domains of human experience, any account of moral enhancement that does not engage deeply with a range of disciplines must be decidedly limited. Particularly with respect to moral living, a multidimensional approach is required to gain any kind of foothold on the subject matter. In this way, interdisciplinary dialogue is a mirror, a microcosm of sorts, which echoes the interwoven, multidimensional nature of moral living itself.

I hope the focus on these kinds of specifics has been sufficient to show that the term "moral enhancement" embraces a set of prospects going far, far beyond the very limited terms that the default "hard, state-sponsored paradigm" allows.

Moving the discourse beyond the above paradigm has been one of the core hopes of the book, though my worry is that the dominance of this paradigm has done irreparable damage to the conversation. The hideous visage of Persson and Savulescu's *Unfit for the Future* (2012), the thesis of which is, literally, morally enhance or die, has really made a joke of this domain and presents an easy target for anyone who wishes to dismiss outright the many and various prospects involved in moral enhancement. It is truly my hope that the above paradigm be abandoned by commentators completely, leaving nothing over, and that it never be spoken of again.

Dreaming and Reality: Where to Moral Enhancement?

Philosophical Reflection

One of the key propositions developed throughout the book is that practical reality is the most severely limiting factor when it comes to moral enhancement. Just as we have seen that the scientific studies carried out in "the moral laboratory" do not adequately model real-life moral functioning and arguably leave our understanding worse off than before, so too we have discerned many ways in which the philosophical machinations have gotten lost, becoming divorced from the reality they purport to be describing—and often producing accounts which trivialize the tremendous significance of the problem at hand thereby.

There are too many factors in reality, a "freakonomics" (Levitt and Dubner 2007) of moral enhancement if you will, wherein factors too oblique or innocuous to warrant consideration get swept over when it is precisely such minutiae that are decisive for purposes of our conversation. Such a freakonomics creates a much more difficult and uneven ground for the enactment of any given moral enhancement intervention, all the time raising issues idiosyncratic to itself.

The section on pharmaceutically based moral enhancement illustrated this complexity. On the one hand, there are massive research costs, the length of time for testing, and the profit-based realities that pharmaceutical companies must face. This makes it highly unlikely that specific morally affective drugs will be created. On the other hand, in principle, side effects of existing drugs might have some morally affective benefits in certain circumstances. We saw that nalmefene is one realistic prospect, particularly since mental health and moral enhancement concerns overlap. But then again, testing and marketing would still be an issue, and at present, as we saw, the psychiatric research departments of many labs are being downsized, if not outright closed. This is a very multifarious picture, the many intricacies of which must be taken account of before we just sally forth and make grand claims about what might be possible here. How such intricacies (and all the intricacies that we cannot yet see), are liable to manifest in actual reality is simply too difficult to predict.

Indeed, if we take actual economic reality as our measure, it is clear, to the extent that persons indicate their will with their wallets, so far from seeking pharmacological remedies to develop "positive" moral traits—empathy, generosity, altruism, and the like—currently men and women are lining up to obtain precisely the opposite. Prescriptions for hormone replacement therapies have exploded. People want more aggression, more sexual intercourse, and they want the energy and lifestyle of youth, not the wisdom of age, and certainly not the traits of saintliness. Even if interventions to foster such traits were to be developed, few would buy them. Testosterone, aggression, heightened sexual urges, a great head of hair—this is what persons are paying for. That is reality.

These sorts of reality-based considerations, which on the surface at least may appear like small considerations, unworthy of philosophical attention, turn out upon reflection to be indicative of more decisive impediments. These are factors not always readily visible at the giant overview-level of analysis upon which much philosophical enthusiasm exists. It is of course necessary to make overview generalizations, but these have to be leavened by, and arise out of, pertinent details and realities which are incredibly hard to discern without extensive expertise within the domains under consideration.

All this being so, as I hope the present book has illustrated, an approach to moral enhancement that places practical realities at the forefront is possible. In addition to the above stated lenses through which the present book's hoped-for "strong, new approach" has been proffered, it can be added that the book has aspired toward showing how an approach is possible which takes a critical eye toward the grand scientific claims that are made regarding "the moral brain," without simply accepting them wholesale. I hope this book has shown how an approach is possible which is sensitive to the narratives undergirding our empirical endeavors, one which is able to evaluate the strengths and failings of such narratives, how they distort or enrich our enquiry. I hope it has been shown that an approach is possible which sees moral enhancement as a tremendously diverse plurality, one which appreciates that there are a range of subtle interventions which might be desirable and brought to pass on a smaller, more piecemeal scale. I have hoped to show that an approach is possible which looks for existing parallels within extant, well-functioning institutional measures as potential entry points for a realistic moral enhancement, while being constantly vigilant for potential dangers and distortions that might arise at the institutional level. Finally, above all, an approach is possible which is integrative from the outset, which sees morally formative influences as manifold, interwoven (and understands moral enhancement as having to respect this interwoven quality), which sees the human person as a whole being, not a passive nonagent or biological device; an approach which refuses to be biologically reductive, instead seeing biological dimensions of moral functioning as a lesser, but not insignificant, partner in the larger quest to provide appropriate means for moral education, respecting matters of both personal agency as well as broader cultural specificity. All these are my prescriptions for philosophers in the moral enhancement domain—and I very much hope that this present book has gone some way toward showing that such a very intricate and demanding approach is in fact possible. Then, I believe, moral enhancement discourse can rescue itself from the dominant, highly superficial, and massively fantastical paradigm within which it is presently mired.

Neurobiological Accounts of Morality

So it goes for the philosophical enthusiasts. What then of those who think that moral functioning can be appropriately described in primarily neurobiological terms? The suggestion has been that such de facto reductionism involved in the methods used, and their subtending assumptions, are outmoded, inappropriate, misleading, and false. Insofar as any moral enhancement claims are based on uncritical reading of such research, these claims are likely to become dinosaurs before they have even got off the ground. While our biological and neuroscientific models are still conforming to atomistic and reductionist paradigms, we are going to have a hard time going between the empirical findings gained in "the moral laboratory" and the actual reality that such science is purported to describe. For even where the present science is good, as we have seen, the assumptions and narratives about the human person embedded within such science are more questionable. Certainly, biology and neuroscience are incredibly important areas of investigation, there can be absolutely no doubt about that; so long as the disciplines do not overstep their legitimate boundaries, such biomedical research holds incredible promise for continuing to benefit humankind into the future.

But there are boundaries to the usefulness of such research, and it is the reliance on such domains to describe social and relational matters of immeasurable complexity, realities which are so mercurial and interwoven, that is the problem here. At the very least, the biomedical domain has no primacy with respect to questions of moral functioning. Instead, such primarily social and relational realities such as moral functioning comprise, while they can certainly be enriched by empirical biological and neurological investigation, are best empirically approached through a more human lens. It is the more human sciences, psychology, sociology and the like, that are actually far better equipped for taking the lead with the empirical examination of the human side of moral functioning. This is a point made recently by Mihai Avram and James Giordano (2014)—though their observation must be qualified by noting that the present psychology of morality, as it stands, does leave an awful lot to be desired and is shaped by precisely the same narratives which undergird the neuroscience of "the moral brain."

That being so, the human sciences do at least attempt to straddle the line between objective, empirical study and the fact that the phenomena in question are precisely human realities and thus have unassailable subjective dimensions. This balancing act is advantageous, in principle, when it comes to the exploration of something like moral functioning. For moral functioning is precisely like that: it has primarily human dimensions and is better off understood less in overly abstract, objective terms, but rather in terms of the lived reality of such phenomena. These are the most promising empirical avenues for investigation, then—qualitative methodologies from within the human sciences—that can facilitate appreciation for the complexity of moral functioning.[2]

Understanding moral functioning primarily in the atomistic and mechanical terms that the present neurobiology of morality takes as standard involves the assumption of too passive a model of human agency. There is something very worrying about a discourse which sells, through the popular media, the ideas which implicitly contain the proposition that the human being is essentially a slave to the chemicals in his or her brain. Certainly the general public have been voracious in eating up such claims; the self-moving forces of bad faith which wish above all to abrogate themselves of having to make decisions and, worse, of having to take responsibility for those decisions or being held to account for them, are all too eager to accept as "objective scientific truth" the grounding assumption of reductionist discourse that "we are our brains," that "my neurons made me do it." But this is a philosophical assumption, not an objective scientific truth. It is a narrative which has directed the research from the outset, one embedded in the very nature of reductive discourse. The idea of the human person determined by his or her brain was of course enshrined in the objective scientific results because *those were the guiding presuppositions going in.*

This is important. One of the key reasons that such an assumption is dangerous can be understood in the quite simple terms of the "Pygmalion" or "Golem" effect (Rosenthal and Jacobson 1968, 1992). That is, quite often persons tend to rise only as high as they are expected to. Of course this is quite simplistic, but there is more than a grain of truth here too. One tells the general public that they are no longer responsible for their actions, that it is their brain chemistry that is making them impatient, aggressive, explosive; that it is their neurons that make them untrustworthy, weak, incapable of discipline; and one should not be too surprised by the outcome.

This is not mere speculation. As soon as the studies into oxytocin were made the subject of intense media attention, whole swathes of incarcerated criminals launched court appeals against their incarceration on the grounds of "diminished responsibility" because of "lack of oxytocin in the brain" (e.g., Hans Reiser, convicted killer; Hayward 2012). Serial killers have had reduced sentences simply because their lawyers held up a colorful fMRI scan before the jury, whose perception of the quality of such data has been shaped by precisely the science that we have been considering. Indeed, there is an ever-growing academic literature discussing the use of neuroscience in criminal defense cases, use of which has skyrocketed of late (e.g., Koebler 2012).[3]

In other words, the very science that is being touted as offering moral enhancement potential could itself be a morally de-enhancing power. Would it not be ironic then if the very science which is used to fuel moral enhancement discourse, the very study into "the moral brain," was itself an instrument for lowering the public standard of moral responsibility? Such science feeds the general public a view of the human person as a marionette, dancing on the strings of "nature," of "brain chemistry,"[7] of "faulty machinery" in Joshua Greene's words, as a passive receptor of biological signals which are faithfully translated into bad deeds that no one is any longer responsible for, except that they can take their medication to resolve it. The consequences, as measured in the courts of law, should not have been too surprising.

In contrast, continental philosophy offers a deeper framework for grasping the dynamic nature of the human person and the complexities of personal responsibility. Indeed, focusing on figures more of the order of Kant and Merleau-Ponty helps us see how grasping that the mind actively participates in constructing its own experience is an insight which runs exactly parallel to the problem of biological causation in moral functioning as construed by the various investigators considered. We are not mere antenna picking up our biological signals and faithfully representing them in behavior and judgment; we actively shape how those biological impulses manifest at various levels of consciousness.

At the very least, we consciously and preconsciously shape the sense given to a biological impulse and how it manifests. It is similar to the experience of hearing one's alarm clock while still asleep and dreaming that a fire-engine is passing by. Both conscious and preconscious forces participate in the manifestation of the biological "causation." A biological predisposition toward aggression may "cause" a person to become a boxer, but it might just as well "cause" that person to become a soldier, or perhaps a terrorist or a criminal; such a biological disposition might "cause" him or her to become introverted and reflexive, turning the aggression inside against him or herself; it may manifest as suicidal impulses, depression, self-harm; it might "cause" that person to become a saint, a person of restraint and extraordinary self-control, a creature of duty whose integrity would not even be possible without some measure of introverted, self-directed aggression. Persons may embrace their aggressive nature, they may fight it—what then is there to be said about this idea of direct biological "causation" in such complex matters? It is pure nonsense.

It has to be impossible to reliably predict what a biological impetus toward aggression is going to do to a person by looking at biological impulses only. It is impossible to reliably predict how such persons are going to shape themselves in relation to it, or be shaped by their peers and surrounding environment. What then is left of this idea of direct biological causation? To think that one can fashion a significant moral enhancement by doing something as radically vague and simplistic as, say, giving someone SSRIs for their aggressive outbursts relies on so shallow an account of the nature of the problem, and of the human person, that we should be very worried whenever we hear such noises—be that in academia or in the public press. Biological influences on moral functioning are only influences as shaped. While this shaping cannot be divorced from biological influence, it must not be limited to it either.

There is a belief in our contemporary times that philosophy, as a discipline, is either outright defunct, or that it exists now only to bolster science and to refine it, as if it is now merely some nagging grandma, ranting away at the scientists, offering unsolicited warnings, or at best some decrepit sage that scientists ought really seek for helpful clarifications here and there (even though they rarely do so). But in this domain, at least, philosophy must still be our leading light. A capacity to disclose and interrogate the distinctly philosophical images of the human person, and for examining the appropriateness of the metaphors which guide such neurobiological research, is fundamental to the process of scientific advancement. Grasping that such objective science exists first in philosophical space is extremely important when considering an area of biomedical investigation which might be used to alter the moral nature of the human creature, and I think it is entirely appropriate to have rebuked the various philosophical enthusiasts for merely kowtowing to the neurobiological studies without subjecting them, and the metaphors at their foundations, to the least amount of critical thought.

Alternative Biological Paradigms: Emergentism and Systems Biology

Finally, then, the "science of morality," currently in its infancy, is likely to remain in precisely such an infantile condition until a broader, more "systems-oriented" approach to biological functioning is made mainstream. If we are to use biology to supplement our understanding of moral functioning, we have to look for alternative biological paradigms to help develop a more complex, richly human set of accounts. The problematic biological paradigms that I have been railing against above are not at all without alternatives. Within the biological sciences there have long existed a considerable plurality of interpretive lenses through which the various biological findings have been understood. Yet the atomic, reductionistic, and deterministic interpretive lenses seem to be the only ones nonbiologists are at all familiar with, despite their taking up only a relatively small proportion of the biological domain as a whole. This has shaped the philosophical expression of moral enhancement, much to its detriment. What is more, there is currently a resurgence of these long-established "systems" traditions within the biological sciences, which are much less reductionistic, and which look at biological creatures, particularly with reference to their genetics, in a much more ecologically interwoven fashion.

Many theses and commentators within the neurobiological community (e.g., the "nonreductive physicalism" of Warren Brown and Nancey Murphy 1998 is a paradigmatic exemplar of this approach; or the more philosophical Philip Clayton's 2006 *Mind and Emergence*), are taking up the long-talked-about "emergence" position. A brief gloss on the subject will be sufficient to show the significance of taking alternative biological paradigms when looking at the complex matters we have been examining.

"Emergence" is a view which suggests that certain phenomena, once sufficiently complex, take on active, self-organizing properties of their own which cannot be understood in terms of the lower-level elements which make them up. An example might be, well, life itself, which is made up of the very same building blocks as inanimate objects, which when organized differently, and sufficiently complexly, suddenly literally take on a life of their own. Life is then emergent from the same physical building blocks which make up everything inert in the universe. When they are sufficiently complex, this inert matter somehow becomes conscious, and when sufficiently more complex this life becomes self-conscious, and so on. Life would then be an emergent property, a self-organizing mechanism over and above its physical building blocks, and not graspable purely by looking at those building blocks alone.

Another example might be the human mind. Brown and Murphy (1998), for example, suggest that the mind is such an emergent phenomenon, one which displays properties not fully graspable in terms of the neurons which create it. If this is the case, then even a full neuroscientific grasp of the brain will be inadequate to encapsulate the self-organizing properties of the mind. This emergence view, then, if true, would do considerable damage to the reductionist theses which want to see complex functioning, in this case moral development or moral functioning, in either purely biological or neuroscientific terms. Rather, moral functioning could then be understood as a larger emergent property of a self-organizing neurobiological system, but one not adequately captured in either biological or neuroscientific terms only. A more emergence-based biological paradigm would then proffer a much richer and more complex framework for approaching the biological bases of moral functioning and would require reference to a higher-order mode of description to help account for it.[4]

In sum, the biological assumptions and frameworks that have been drawn upon by the moral enhancement enthusiasts, assumptions with reductionist and deterministic undertones concealed within them and taking inappropriate, over-simplistic biological frameworks as their basis, may actually serve to completely undo the conclusions drawn in the overwhelming majority of moral enhancement discourse. That is, relying upon such false premises is going to result in the conclusions so garnered having very short-lived tenure before their decisively built-in obsolescence becomes unavoidably transparent. Again, this has precious little to do with "the science" or scientific method per se, which, like logic, cares little for the truth of its premises but only about the validity of the connections between its hypotheses and the conclusions drawn.

In other words, it is not so much that "the science" is the problem here; rather, it is the assumptions which are driving the use of such science. If such assumptions—humans as passive receptors for their neurobiology—are suspect, then the conclusions drawn therefrom are likely suspect too, regardless of the validity of the method used to ascertain them. Thus, such moral enhancement discourse, and the "science of morality" that it is based upon, may very well have to be entirely rewritten, sooner rather than later, in light of the resurgence of the systems approach, emergentist frameworks, and epigenetic insights. The important point is that this "systems" paradigm has not yet filtered sufficiently well through to the public, nor to the philosophers in the debate. Yet the asteroid of systems biology is hurtling fast toward us. If the entire edifice of moral enhancement is to avoid going the way of the dinosaur, going extinct before it has even got off the ground, then it needs to adapt, and quickly, to this new tendency toward ecogenomic and integrated systems biological thinking.

Constructive Conclusion: What Is Possible, Desirable, and Worthwhile in Moral Enhancement?

There is no question that my rebuke of moral enhancement discourse has been severe. Yet, for all of this, none of these criticisms are sufficient to say that the entire discourse is to be rejected outright. I hope the approach taken here has shown that a more realistic, fine-grained approach is possible and worthwhile. Just because the subject matter has been dealt with poorly for the most part, and superficially, does not mean that we should not be talking about it. It has been seen throughout that there are, plainly and simply, opportunities for biomedical augmentation of larger personal, social, and institutional efforts to influence behavior and thought. Let us conclude with two very different possibilities, based on the two dominant means of construing moral enhancement utilized throughout the book—"hard" and "soft" approaches to moral enhancement.

"Soft" Moral Enhancement and Light-Touch Paternalism: A Potential Public Good

There is always an extent to which measures are appropriately put in place by states and regimes to protect the citizenry that compose them. There are more or less forceful ways of doing this. One very clear and direct way of doing this is making, say, the wearing of seatbelts in cars a law, or crash helmets for motorcyclists (Dworkin 2010). This is a very clear measure introduced by the state to help protect persons from others and from themselves.[5] There are also more oblique ways of exerting such influence. Over the past decade a profound change in attitude toward the smoking of tobacco has been actualized. A combination of more indirect methods such as exorbitant taxation on tobacco, with the direct ban of smoking in public places, has resulted in a genuine change in attitude on the part of the general public toward smoking and the harm it can do to self and others.

As such, there is certainly room for some optimism with respect to political and legal methods for guiding and manipulating behavior and judgment. Similar techniques are used on a tremendous number of levels, be it in advertisement, or the way in which supermarkets present groceries to the passing public, or the myriad ways in which institutions and businesses attempt to seize attention and shape it. A great deal of thought has gone into how the minds and actions of the general public can be "nudged" into taking courses of action or coming to certain desirable conclusions, sometimes as simply as by manipulating the order in which information is presented to them.[6]

In short, there are means for encouraging certain behaviors, and in this respect, a very soft paternalistic moral enhancement should also be possible. This might rely on some combination of psychological techniques (which have had some established success) and biomedical means. We suggested that mandating some biomedical interventions for addicts, particularly for dangerous addicts, as part of an integral treatment program may very well make for a desirable and responsible form of soft moral enhancement. Morality would not be mentioned, but the strong moral dimensions of issues of addiction would find a humane, integral treatment context through which the moral dimensions of addicts could be managed all the same.

In this way, it is not only possible, but, in principle, desirable and good that certain very clearly defined prospects for soft moral enhancement be enacted, in certain circumstances. It would be a very radical person indeed who wanted to suggest that the government had absolutely no right whatsoever to interfere with persons in any way, shape, or form. It is in this respect that a very, very limited, focused (and preferably "passive," as we saw with addiction) account of moral enhancement might be justified. Such interventions are arguably necessary, and in certain cases unambiguously desirable and good. In this sense a soft, nonexplicit moral enhancement might well have some point of entry by means of its overlap with justifiable paternalistic control.

It is in this sense that soft moral enhancement is likely not at all futuristic, but actually inevitable and already occurring. We have suggested that the increasing encroachment of definitions of mental disorder into what most persons think of as normal human weakness is an example of how an indirect, soft, nonexplicit moral enhancement already exists by means of its overlap here with clinical psychology's encroachment into domains which have clear moral dimensions. We have also seen that present medical practice, which often uses definitions of "health" that are distinctly socially constructed and heavily laden with value judgments, involves the implications that, sometimes, the very treatment of disease can be understood as an oblique and indirect form of moral enhancement.

It is very difficult for us to perceive how we distort the boundaries between illness and moral misbehavior in our own culture, because we do not have sufficient social distance. But looking back at the way we have treated marginalized groups in medical terms, such tendencies have a sufficiently long and pervasive history for us to be concerned that scapegoating "social undesirables" by mischaracterizing them in inappropriate biomedical terms is a tendency to be constantly vigilant against. In any case, medical practice has to deal with moral issues every day and make moral stands in practice (as we saw with doctors refusing to inform parents from certain backgrounds about the gender of their child, because of the number of gender-based abortions associated with that culture).

Given that soft moral enhancement, by the present definition, avoids mentioning morality altogether, the rubric throughout would likely be "promoting the public good," and this would probably be its primary viable justification. Even then I would contend that such interventions be extremely constrained and "passive" in nature; that is, applying only to a small range of particularly dangerous individuals or well-focused "problem groups." Even a soft moral enhancement, if it involved biomedical means, could be very, very disturbing if applied unilaterally to the population. Psychological, manipulative "nudge" techniques are routinely applied to entire populations, as are social incentives, like tax breaks for certain behaviors. If, however, we get to the point where entire populations are being prescribed biomedical interventions as a means of covert moral enhancement, we might well have grounds for concern, whatever the justification applied. This is, of course, assuming that the general public would even stand for such measures; likely they would not. Given the extent of the public outrage that manifested over the mere idea of mandating flu jabs, it is highly unlikely that any strongly morally conditioned concerns could be sufficiently well dressed up to be mandated on "public health" grounds (in the absence of extreme emergency conditions, at least).

So, if we are talking about soft moral enhancement as a compulsory measure or a socially incentivized measure, we would be likely talking about a "passive" version, which applied only to certain "problem populations," like addicts or the pathologically violent, and then, only on an "as and when needed" basis. We have talked throughout of those addicted to dangerous chemicals, predatory sex offenders, and so on, but it should be noted that we have not yet even been specific enough—it is not as if members of these categories can be universalized and treated as homogenous entities. In any case, such interventions would be justifiable, arguably, because these persons' behavior has severe (and it would have to be severe) consequences for self and society. Just so long as we are talking about "soft," nonexplicit, morally related issues, then a case could well be made for such an oblique and indirect moral enhancement to be forwarded.

We must be vigilant, though. As we have seen, these socially constructed visions of health which overlap with treating perceived social diseases can be monstrous. Not even forty years ago doctors in the United States were still sterilizing Native American women without their knowledge or consent, and doing so with the absolute conviction that this medical intervention was both a medical and moral good to be endorsed and pursued with the fullest vigor (Rutecki 2010; Temkin-Greener et al. 1981; Wagner 1977; Dillingham 1977). There is no doubt then that medical paternalism and the soft moral enhancement it overlaps with have tremendous potential for abuse. The problem, though, is that paternalism and paternalistic medical and psychiatric practice, which overlap with moral judgments, are both largely inevitable and largely invisible.

Yet potential for abuse exists with everything. Rather, we must ensure that we have institutions, watchdogs, and various parties involved in keeping an eye on the practices involved. This is not a question of "bogus science"—though that can indeed be a problem. It is rather the interpretations and application of such science in disturbing value-laden ways that create the problems. It is these foundations, assumptions, and interpretations that need to be continually interrogated. Admittedly, this is not easy to do when government bodies, institutions, and certain high-ranking academics have financial incentives to continue their positions and legacy by quelling critique—particularly when large institutions have sufficient funds to tie down dissenters through expensive litigation that individual researchers cannot hope to pay their way through, regardless of the validity of their critiques.

Dangers for abuse exist, and they always will. The key question must then be one of negotiation—negotiating the limits and manner of such state influence on a case-by-case basis. Given that, as has been contended, soft moral enhancement is an already existing inevitability, mere critique will not do. What one needs are ways of working with the already existing system and means. Equally, what one needs is conversation about the matter, that we might become aware of all the subtle ways in which we are getting moral enhancement "by another name" through invisible and taken-for-granted social processes occurring right before our eyes. As long as we can become aware of the issues here, and have responsible institutions and care practitioners, then we have some room for at least negotiating where a nonreductive moral enhancement might fit in for the public good. We at least have a place where we can start.

Let us conclude this meditation with a possibility, perhaps shocking given our modish faith in biomedical enhancement potential, that biomedical approaches may very well be the *least* promising means to moral enhancement that can be made available to us. In other words, the biomedical angle may very well have been a red herring, except within limited constraints, from the outset. We may simply be looking in the wrong direction by seeking out biological manipulators for moral functioning when psychological manipulators and nonbiological technology have been proven to do a particularly effective job.

We mentioned James Hughes' optimism that it is "quite likely" we will one day have neurodevices to read and direct our various impulses. Persson and Savulescu also produced a similar idea as a thought experiment, something called the "God machine"—a device that can read all thoughts and intervene to prevent serious harms being committed. But is this, on reflection, really such an absurd possibility? Could it not be said that in a very real sense the "God machine" is already upon us? The idea of a fine-tuned way of reading all our thoughts and intentions directly from our brains may well be very, very far into the future (if it is possible at all), but the simple fact is that anyone who has an e-mail account has his or her every online thought read, analyzed, and actioned by companies such as Google, Facebook, Yahoo, and a range of other data-mining companies, which then collate, analyze, and sell your personal information to whomever is willing to pay for it.

On the rubric of "helping us provide services more helpful to you," that is, for purposes of targeted advertising, every single e-mail, every last attachment, every single blog entry, every single website visited, every video and picture that is posted in our increasingly more pervasive online lives is analyzed and put into predictive profiling programs by extremely sophisticated machinery. There is no question that such intrusions have been used to (a) notify the security services when a terrorist or security threat is spoken of, and (b) identify harmful pedophiles for arrest and detainment (Evans 2014).[7]

In this sense, the God machine already does exist and is an already operational arm of state control. Such monitoring can (and does) read our thoughts, quite literally, because we are increasingly in the habit of writing our every thought down in our various online communications. The expectation of privacy is ever diminishing. Samsung's recent Smart TV has a warning issued alongside it, for families to not discuss private business while watching it, since it is passively listening to and analyzing all words spoken while the voice recognition is switched on, and potentially sending that information "to partners and third parties" (which is usually just another way of saying: "*anyone at all*, just so long as they are willing to pay us money to get hold of your information") (BBC, 2015a). Samsung insists that this information is not being sold, but public trust of bodies collecting their private information is at present, understandably, at a very low ebb. LG's Smart TV has already been accused of perpetrating illegal data breaches by collecting details about viewing habits even when users' privacy settings had been activated (Kelion 2013). Microsoft's new Kinect technology is reputed to be able to detect emotional cues and is capable of lip-reading its users. Planes (2012) writes: "It is already accurate enough for Microsoft to use it to create a mall advertising system that changes content based on who it thinks you are. … The only way to "opt out" of Micro-soft's individualized ads is to avoid places that use it." Analysis of such data, particularly by Google and the like, can and does lead to interventions to prevent extreme harm. So the big conclusion for moral enhance-ment enthusiasts may then be this: that the biomedical aspect may be the least promising and least worthwhile avenue to pursue. If we have "soft" moral enhancement in mind, we already have powerful, powerful means for enacting such paternalistic influences and surveillance. The question then, is not "is moral enhancement possible?" nor "should moral enhance-ment be allowed?" When we have a soft vision in mind, the question in fact becomes: how do we ensure that moral enhancement, which is already here, does not get radically out of control?

"Hard" Moral Enhancement for Individual Moral Development: A Private Good

As far as hard moral enhancement goes, the primary realistic potential comes from the intention of the user. That is, if it is just too politically untenable to think that explicit moral enhancement be enacted in the public space, then what is left over are those persons who would elect to engage a biomedical intervention, for their own purposes, in an attempt to improve upon their own moral functioning. This is a specific sense of "voluntary" moral enhancement; it is not "compulsory versus voluntary" in any political or institutional sense whereby the state offers interventions for moral improvement which are then forced upon you, or left for you to choose to take. This would be a purely self-chosen activity—taking a given substance, known for having certain morally related side effects as part of a personal project to become more moral along a certain self-chosen trajec-tory. In the abstract such a possibility cannot be ruled out, though whether any such thing would ever happen is a different question. What can be ruled out is the idea of people being "forced to be moral."

It was argued in chapter 1 that our definition of moral enhancement can benefit by being loosely construed. The section above demonstrates why it is important to have a definition of moral enhancement which goes far beyond the hard-line "strong" account (which sees moral enhancement as existing only if a person's character or moral identity has been developed). As we saw, a looser construal opens debate to include much more serious and contemporary issues that are nonfantastical and sometimes deeply disturbing in nature, and not to be ignored.

That being so, we do have to respect that there are "hard," explicit, and "strong" character-based ways of defining the term which are deeply important too. Can we do justice to a "hard" and "strong" vision of moral enhancement, and find some way of giving a positive appraisal of the possibilities therein? I think so. The core problem with "hard" moral enhancement seems to be that of the limited manner in which biological or technological agents can realistically influence moral functioning. As we have maintained throughout, complex moral problems are not predominantly biological in nature to begin with, and thus biology can only ever be a very limited way of approaching such complex problems. The way around this problem, something explored throughout this book, has been to envisage such interventions as being part of a larger, more integrated bio-psycho-social approach, one which includes biological dimensions but which is also necessarily embedded—perhaps within a larger person-centered therapeutic approach, a pastoral context, family setting, or, at the very least, having social and relational dimensions broadly defined. Above all, we have contended throughout that there will always be a limiting factor to these types of hard biomedical interventions, namely the need for a pre-existing disposition of will.

There is no way such an account can be accused of mincing words. We are talking here about an explicitly moral project to become an explicitly morally better person (along one or more moral dimensions), to develop an explicitly moral identity, and as a part of an ongoing, multi-angled (i.e., bio-psycho-social), self-generated effort to become a more moral person. There need be no fear, as Francis Fukuyama and others have been interpreted as saying, that "if we take drugs to help us improve ourselves we lose the pride of self-sufficiency"; the fact is that moral enhancement will not do its own work for us. There is no magic, no hocus pocus here. Even with such interventions, we will still have to instigate, we still have to put in the work ourselves, we still have to morally reflect. As such, hard moral enhancement, construed this way—as offering a helpful "nudge" to those already putting in the efforts—is a pretty hard thing to find fault with. If, as has been contended here, the peak penetration of biologically targeted moral enhancement has already been reached, we need not fear any future fantastic nonsense. What we already have is indicative of the quality and potency of what will be available. The main problem here, the practical reality, is whether anyone in the world would actually use or develop such interventions.

So, in terms of this hard, voluntary enhancement, one would need to put in the efforts, and in relation to those efforts, use pharmacology and technology as a crutch to *lighten the burden*, as a hydraulic suit helps a factory worker lift burdens beyond his muscular powers. Such a moral enhancement is absolutely dependent upon the efforts and will of the person so enhancing. As with sports enhancement, there is no profit in giving anabolic steroids to a nonathlete. The steroids are not going to go running for you, the steroids are not going to lift the weights by themselves—steroids are there to augment already pre-existing efforts, for persons already trying their hardest.[8] This, at the very least, as has been argued throughout, is the precondition for explicit, voluntary forms of hard moral enhancement. No intervention can make one be moral in this strong sense, nor can any such intervention prevent someone intent on being immoral from being so, but with respect to someone already putting in the efforts, making the movements, and living (preferably) inside a social-environmental context which facilitates and treasures such behavior, then we at least have some ground to work with.

Yet even this is too generalized. In each case we will still have to look at all the ground-level conditioning realities before we are able to endorse any given use of a biomedical intervention in explicit moral context. These factors cannot be bracketed out. Thus a requisite of a pre-existing disposition of will is not the only preconditioning factor here, but at least we have some bedrock, a fulcrum against which to leverage some potential for intervention.

What we have in mind is a specific individual, with a specific goal that has been conceptually shaped by that specific individual, reflected upon and emotionally invested in as part of an identity that is salient for that individual, and which they have put in determined efforts to manifest in the specific contexts of interest. This is a completely different circumstance to much of what we have been presented with as moral enhancement— it is a very limited prospect, and perhaps not something to get massively excited about. We are talking about a relatively pedestrian sort of intervention of the order that one might purchase over the counter in a pharmacy. But this at least, pedestrian though it may be, sits squarely within the realm of actual reality.

Such "hard" and "strong" moral enhancement must be understood then as (a) limited, (b) generalized, (c) integrated, and (d) directed.

It is (a) limited inasmuch as we have to reign in our expectations regarding what is possible through biomedical means, even in combination with other approaches. It is (b) generalized in the sense that moral enhancement cannot be fine-grained enough to make specific changes; rather, it must rely on creating a "biological context." Moral enhancement must be understood as (c) integrated in the sense that biological or technological means of influencing persons cannot be relied on in isolation, but must be applied as part of an embracing bio-psycho-social approach which brings to bear as many different motivational angles as possible. The precise manner for such an integration must be constantly dynamic—it must be determined on a case-by-case basis and be reassessed and adapt to change as the process of moral development progresses.

Finally, such hard moral enhancement must be understood as (d) directed, once again, because it relies on a number of conditioning features to turn this limited and generalized, integrated phenomenon into something practicable and workable. Such interventions must be directed by the individuals themselves, in conjunction with any medical or clinical staff that might be involved in such a process and with the supervening environmental dimensions which help shape moral goods into precisely the moral goods that they are, such as religious or spiritual conceptual backgrounds which are necessary for giving appropriate shape to the moral good being formed. It must be understood that even traditional means of moral formation occur within a context and absolutely rely on a background to give them meaning and direction. The values formed through such formation are not idle, they are not cogs unattached to the larger machine—they are integrated, and they make sense with respect to the larger moral horizons of the culture and the various determinations of the individual's reflective conscience.

All this may, in real terms, sound like it has priced moral enhancement out of the market completely. If the terms for a realistic form of hard moral enhancement are made so demanding, this can be another way of saying that such hard moral enhancement cannot exist? Tempting as it may be to come to such a conclusion, we have already given at least one example of how such a hard moral enhancement—one which meets all these criteria and which is, again, entirely plausible, if not already being enacted—can be said to already be in practice. A person taking nalmefene for explicitly moral reasons, in a group therapeutic context, fits the bill entirely and can be said to be hard moral enhancement without any prevarication or dispute. The person here is volunteering to take a substance to manage his or her addiction for explicitly moral reasons, and is doing so within an integrative treatment paradigm. There is nothing about this that is not hard moral enhancement, plain and simple. Those who think moral enhancement is impossible, or too speculative, or too fantastical and futuristic, will be challenged by such an example. The only difficultly would be that raised in chapter 9, the problem of gratuity, the possibility that we are simply taking an existing mental health intervention and calling it something else.

But such hard moral enhancement is mischaracterized when understood as "making persons more altruistic," "making persons show more empathy," and the like. These are vagaries which involve the fiction of direct biological causation in complex behaviors. What kind of empathy? What kind of altruism? These are the questions we need to ask. Within what context, and shaped by which vision of the good? What are the appropriate standards of excellence by which to recognize an act of altruism and empathy? These are questions that require a thinking person's engagement. Brute effects can never be sufficient, and the personal, human touch will always be required to make such a prospect work.

One last time, then, for this is perhaps the most important point of the entire book: let us repeat that a credible hard moral enhancement will rely for the greatest part on the active participation of the individual in moving toward and shaping the desired good. Without this active participation, only the crudest possibilities for moral enhancement remain—or arguably none at all. Throughout we have spoken of biological influences as well as the supervening effects of culture and society. It is pleasing then to finish here by recentering things back upon the individual and his or her personal responsibility. It is the individual and his own efforts which ultimately make moral enhancement into moral enhancement—for without his willing participation, there is no intervention on earth, nothing feasible for the imagination to even conceive, which can make a person more moral in the most stringent, explicit sense. For all the biomedical possibilities that may or may not be made to exist, for all the biological bases of our moral action, for all the scaffolding that we inhabit, finally it is up to the individual precisely as the individual moral agent to take responsibility for his or her own moral progress. For, if we treat the individual as a responsible moral agent and expect the individual to be so, instead of appealing to the machinery of "the moral brain" to whose conflict and domination he or she is supposedly subject, then perhaps we will be surprised by what we might find.

Afterword: The Goodness of Badness

We live in a culture where everything is replaceable; if something has the least blemish we abort it. Enhancement fits right in here. Yet there is value in the strangest of things, though for us to see it requires not enhanced intelligence or augmented "moral sense" but vision and imagination. Since no intervention can force a person to be moral in the strongest sense, the responsibility lies with us to learn first how to apply what we already do have to the best ends. St. Paul wrote that God turned his weakness into strength. It was not that God "made Paul strong"; it was more subtle than that. Paul was observing how his very weaknesses were the things that ended up being his strengths. It was his weaknesses that allowed him to conquer. The God of surprises is highly perverse like this.

Nietzsche, if the juxtaposition is not too curious, wrote something not too dissimilar, that one of our core goals as self-making individuals should be to "brew our poison into milk." The goal is not to diminish our various impulses; rather, it is to learn to sublimate and redirect our various urges toward targets which can be taken as valuable goods. This is the fundamental ambiguity that we have been talking about throughout the entire book. Those impulses that moral enhancement seeks to diminish or enhance, in the overwhelming majority of cases, admit of tremendous ambiguity and diverse potential for those in whom a shaping, directing mind is present. Though not without tremendous effort, if the creative, imaginative will is there, such impulses can be shaped into something far greater than the watered-down, homogenized, anemic blueprint for "goodness" that undergirds the moral enhancement discourse we have been exploring. It may be the case that many of the prospects for moral enhancement that have been suggested by the enthusiasts would make us less as a species, and not more.

Goodness and badness are inextricable. It is not just that certain vices, taken in moderation, actually enrich our lives in ways that simply living within the constraints of traditional virtue cannot (i.e., that badness is itself a good, a means for enriching our lives when appropriated within reason). But we still need to ask: what it is that makes the difference between evil expressions and heroic expressions of these very same impulses? In the end, it is the shaping that makes this difference. The deeper point, then, is not just that good and bad impulses are inextricable, nor even that they might be identical in some cases. The deeper point is that what makes the difference between evil, heroic, and morally neutral expressions of the same impulses is not to do with biological manipulation to begin with; that is, the imaginative, creative will of the person involved, ideally with social support, to shape their impulses into something good transcends talk of biology.

Violence can be expressed as the need to protect, hatred can be directed at something worthy of hatred, an utterly self-absorbed narcissist might be more suitably motivated to create a better world out of selfish, self-glorifying impulses than a duty-bound or extremely well-intentioned moral agent through long-sustained efforts. As an approach, this focusing on shaping, the capacity to use the bad precisely as good, to let our weaknesses be our strengths, must be richer, more authentic, and superior to the sort of "castration" ideal found throughout moral enhancement—the mindless "off with his head" way of dealing with any impulse deemed unsavory. Nor is any of this easy, this call for imaginative self-possession, not easy by any stretch, but the individual's imaginative and creative will, vision, and social support, along with mass movements—powers and phenomena absolutely not appropriately talked about as biological phenomena—will always constitute the limits of moral enhancement. That is, we need to give a shape to our impulses that we might turn what is most primal in ourselves toward our various visions of the good.

We have seen that certain small-scale visions of moral enhancement might very well be desirable and worth exploring, but we must be careful to ensure that moral enhancement, however we go about construing it, does not simply become a case of cutting off an offending impulse as if it were a gangrenous limb, of making us less, but rather contributes toward and works within a context that fosters an increase in the richness of moral living. Our goal should not be to water ourselves down, nor to reduce ourselves to "the mean." Instead, our goals should be to welcome and accept difference and weakness, and badness and evil, and develop in ourselves creative and imaginative ways of turning them to the good. We need to learn how to shape ourselves within the constraints of what we already do have, for this skill is fundamental and, in the end, more valuable to the living of a good life than any biomedical moral enhancement could hope to be. To face and accept this idea that human weakness simply cannot be resolved, nor our nature "perfected," but only managed—understanding that there will always be risks, and learning to deal with weakness and evil with compassion and justice—requires more than enhancement can ever give us. It requires wisdom, and for this no technology or pharmaceutical can avail us in the least.

Notes

Chapter 1

1. See Harris (2007) and Savulescu (2001, 2005), according to whom we need not fear such "positive" eugenic prospects. The horrors of the recent past, indeed of the entire span of human history, the horrors that engulf the world even now outside what is arguably a precarious soap bubble of Western security, need not trouble us in the least. Such thinkers suggest that we need not fear backslide to a Nazi-style abuse of eugenics. Sparrow (2012) observes the senselessness of the optimism of Harris' fervor. He writes: "The cover illustration of the hardcopy edition of this book—a muscled male arm, dressed in what appears to be Superman's blue costume, with the rising sun behind it—suggests that Harris is not unduly concerned about the historical resonances of his philosophical program" (Sparrow 2012, 40). At least Harris' position is coherent. Savulescu advances claims that we are both (a) sufficiently trustworthy to disregard fears over the terrible possibilities of abuse of moral enhancement technologies required to make such significant changes to humans' biologically based moral nature, yet, curiously, also maintaining the view that we are (b) so much of a threat to ourselves that we are in desperate, *urgent* need of moral enhancement to save us from ourselves and all our manifold, ingrained destructive tendencies. Fortunately, a defense of moral enhancement in certain contexts does not rely on making big claims about how evil or problematic human moral psychology is. All the same, it is worth noting that Savulescu's contradictory claims raise a big problem that cannot be swept under the carpet: if human creatures are desperately in need of pervasive moral enhancement then, by implication, those same creatures cannot be morally responsible enough to handle the technologies required to remedy such moral imperfections. More constrained, case-sensitive accounts of moral enhancement that do not rely on apocalyptic rationales to justify pervasive state compelled moral enhancement are not subject to this decisive flaw.

2. Biomedical, as distinct from something like moral education, which will be the subject of chapters 6 and 7. It is important to note that the terms "moral enhancement" and the more clunky term "moral bioenhancement" will be used as synonyms throughout this text. While moral education might be thought of as a form

of moral enhancement, for present purposes we are talking only about enhancement that utilizes biomedical means as at least a part of the process used.

3. An example of this up close and personal, case-by-case analysis would be my exploration of the claim that SSRIs might be used as potential moral enhancers for treating reactive aggression (Wiseman 2014b). The idea is to focus in on a very specific case and explore it as the specific, idiosyncratic case that it is. Another example can be found in chapter 9, examining the use of nalmefene in a moral enhancement context.

4. The validity of this term "chemical imbalance" then became an urban legend of sorts, pervading the public consciousness and becoming unquestioned public coin (see Kirsch 2009). We will be discussing throughout how ridiculous it is to reduce complex psychological and relational problems to something like the monocausal gremlin in the machine that goes by the name "chemical imbalance."

5. I will assume that the reader already understands moral functioning is a plurality, that the various kinds of moral functioning are quite unlike each other in many ways (e.g., truthfulness is not the same as faithfulness, which is not the same as benevolence, and so on, since each of these uses very different kinds of processes). This is borne out in the neuroscientific studies; cf. Sinnott-Armstrong's observation that there is "no general wrongness detector in the brain" (2012) simply because the parts of the brain used in the various kinds of moral functioning are used in numerous other functions too. There is no "morality spot" in the brain, no region which is dedicatedly responsible for "doing morality."

6. Note this is not a matter of choosing the hard policies for the sake of doing hard things. There is a nuanced discourse in the enhancement literature about where it is legitimate to take shortcuts and where it is not (see Schermer 2008, 2013). That is, there is sometimes a case to be made for certain shortcuts. But the call here for disregarding lazy approaches is not a call for choosing hard means just because they are hard. As will be argued throughout, the lazy answers and lazy approaches will turn out, in the long run, to be self-defeating. By travestying the problem and dealing with it only on the most superficial level, we will end up merely filling a hole by digging an even bigger hole. Thus, we need to engage with the hard answers simply because, in the long run, only the hard, complex solutions have any hope of producing an enduring and nonsuperficial set of solutions to the complex moral difficulties that face us.

7. Indeed, we are never quite sure of the extent to which the "biological causes" of our problems are in fact manifestations of these larger social issues. We have a "chicken and the egg" situation between culture and biology very similar to that found in the neurobiological discourse on love: do we love persons because of the chemical cascade in our brains, or do we have a chemical cascade in our brains precisely because we love that person? Which is cause and which is effect? Or is the whole question based on a hopelessly superficial construct of the object in question

(a question particularly pertinent for the biological study of moral functioning)? Where do biology and environment begin and end? Biology and society shape each other, and this is precisely the mélange problem—the boundaries between biological causes and larger social concerns are porous, and atomizing them is not always the appropriate way to understand the nature of the situation.

8. Much of Crockett's (2010, 2012a, 2012b, 2013) work on serotonin in modulating moral judgment, for example, has a very dubious relation to moral judgment, and arguably bears no relation to moral judgment at all (see chapters 4 and 5 for more).

9. See Grubin (2010) for an analysis of recent instantiations and justifications for use of antiandrogenic drugs and physical castration in Poland and the Czech Republic (where such treatments are legally mandated for certain classes of sex offender under the rubric of "protective treatment"), the various effects of such treatments, the larger nonbiological scaffolding in which such treatments are given, and Grubin's argument as to why "Doctors should avoid becoming agents of social control" (Grubin 2010; 340).

Chapter 2

1. Indeed, preventing such attacks may well require security services to behave in precisely the unpleasant ways that moral enhancement proponents are seeking to prevent. As Dame Eliza Manningham-Buller, previous head of British Intelligence MI5, once stated of her officers: "they are no Angels."

2. There is a certain sleight of hand going on here, for this very loaded word "myopic" is being given as describing what is essentially "communitarian" moral thinking (i.e., it is permissible to give greater moral weight to our kin than to those beyond our community), which is incidentally a competing moral system to Savulescu's (2006) own "welfarist" vision (i.e., that we should value the maximization of the welfare of all persons everywhere equally). Calling man's psychology "myopic" is a way of denigrating a competing moral system by projecting it onto our biology and then posing it as a world-threatening ideology.

3. Rakić (2014a, 2014b, 2014c), for example, has argued that while the idea of compulsory moral enhancement is repugnant, the idea of preventing ultimate harm is a good rationale all the same, and thus proposes a voluntary form of moral enhancement in its stead. As time has gone on Rakić has come to distance himself from the ultimate harm rationale. This is a most welcome move, and fortunately for Rakić it is possible to entirely excise the idea of ultimate harm without it undermining the core foundations of his argument. Moreover, I would suggest that the extent to which Rakić felt the need to engage with the idea of ultimate harm, and to appeal to it in his various arguments, is a symptom of the powerful negative influence that Persson and Savulescu's dominance in the domain has had on the conversation. In any case, voluntary or compulsory, the suggestion here is that moral enhancement,

however envisaged, is not capable of even slightly diminishing the potential for such decisive, self-caused, malevolence-based destruction.

4. Some commentators may want to appeal to a "balance of probabilities" argument at this point, the suggestion being that the smallest chance that moral enhancement may diminish the potential for ultimate harm should motivate us to seek such interventions. This might seem reasonable given the way Persson and Savulescu have framed the issue. However, part of the intent of the present book is to shift the very way these issues have been framed, to destabilize the assumptions dominant in the literature which have been set in place by Persson and Savulescu. The balance of probabilities argument in this context is readily destabilized by showing how its basis, the concept of ultimate harm, was a red herring to begin with, an unfortunate misdirect, drawing attention away from the practical realities that should have been driving and shaping the discourse from the outset, rather than merely forwarding overgeneralized and massively overdramatized abstractions for debate. In the end, the balance of probabilities argument is fallacious here because the tangled complexity of the causality involved in such world-destructive problems cannot be circumvented by giving someone a tablet to make them more compassionate—even if such an intervention existed, such problems are simply too pervasive and systemic. Thinking that world-destructive problems like global pollution can be remedied by biomedical intervention is as realistic as a snake-handler thinking that the poisonous bite of the viper in his hand can be remedied by prayer, or that the solution to the melting polar icecaps is for the general public to collect rainwater, put in the freezer, and post it to the Arctic. *It is entirely the wrong approach, plain and simple.* The solution is utterly inappropriate to the nature of the problem which it utterly misrepresents by framing it in such a distorted way. And when one is dealing with such a misguided approach (which is precisely what we are dealing with in thinking that systemic world-destructive problems are appropriately thought of in even partially biomedical terms), it makes no sense to talk of "the balance of probabilities." Rather, the best response is to admit that the entire way of framing the issue was fundamentally inappropriate from the outset, to go back to the drawing-board, and start again elsewhere. This is what must be made absolutely clear—on this point we are not talking about how coherent Persson and Savulescu's argument is, we are saying that the *entire framework* is wrong from the bottom to the top. After nearly a decade of pushing the ultimate harm line of argument it is little surprise that Persson and Savulescu expend so much time and effort, as Michael Hauskeller (2015) put it, on "the art of misunderstanding critics." The entire framework needed to be abandoned from the outset.

5. Pope Benedict XVI did declare pollution to be a sin (Mar 2008). Wilkinson writes in the *LA Times*: "In this age of expanding globalization, the Vatican is telling followers that sin is not just an individual act but can also be a transgression against the larger community. An offense against God, said senior Vatican official Msgr. Gianfranco Girotti, 'is not only stealing or coveting another man's wife, it is also destroying the environment'" (Wilkinson 2008).

6. Though I would not be too quick to associate Nietzsche with the sort of vision proposed by Hughes. As Nietzsche's Zarathustra proclaims in *Of the Virtue that Makes Small*: "I go among this people and keep my eyes open: they have become *smaller* and are becoming ever smaller: *and their doctrine of happiness and virtue is the cause.* For they are modest even in virtue—for they want ease. But only a modest virtue is compatible with ease" (Nietzsche 1969, 189).

7. The Big 5 model analyzes the personality into five sliding scales (Bloom 2007a): (a) openness to experience (curious vs. cautious); (b) conscientiousness (efficient vs. easygoing or careless); (c) extraversion (outgoing and energetic vs. solitary and reserved); (d) agreeableness (friendly and compassionate vs. analytic and detached); and (e) neuroticism (sensitive and nervous vs. secure and confident).

8. Some might want to suggest that long causal chains might still allow for fine-tuning. That may very well be the case in certain circumstances, but it is not possible that it likely applies here. When the degree of approximation, such as Hughes draws upon, is so broad, cavernous in fact, the range of potential unintended side effects (which include the very opposite effects of those intended) gets multiplied considerably. As we will see in chapters 4 and 5 with respect to SSRIs for aggression, even in cases where SSRIs have been shown to correlate with lowered explosive aggression, a by-product of lowered explosive aggression is an unpredictable increase in premeditated aggression. One merely swaps one ill for the threat of another.

10. For example, recent work on epigenetics (i.e., genes' capacity to change expression through environmental influence), significantly complexifies the whole question of biological determinism.

11. Though it is to be noted that this immediacy may well be how moral functioning *actually* works—even if ideally we should be more circumspect and reflective in our judgments. If moral functioning does actually work de facto on the more immediate level, to some not small extent in terms of gut responses to contexts, inertia created by habits, and conformity to surroundings or circumstances, that would only strengthen Douglas' proposal.

12. Talking about moral enhancement in terms of very limited "nudges" is far more realistic than talking of moral enhancement for the prevention of global annihilation, or its contribution to the democratic potential of a nation, or the creation of certain specific kinds of personality and character.

13. There are more issues with Douglas' rationale for moral enhancement, not least of which is the manner in which he uses the word "motive" to describe mere manipulation of persons' most basic animal impulses and drives. The suggestion is that moral enhancement should be about changing persons' "motives," when by this account it is only animal impulses that are in sight. There is nothing going on here other than attempting to drug away mischievous animal impulses. A change incidentally which would last only just so long as one remains on the drugs, and for no longer (i.e., is the moral equivalent of sweeping the problem under the carpet; see

Wiseman 2014b). This is not even to mention the disputed philosophical and neuroscientific question of whether there even exists such a thing as a "noncognitive" emotion or impulse.

Chapter 3

1. In fact there is a growing trend within neuroscience, and the philosophy thereof, to attempt to rediscover universal bases for moral judgment by locating morality "in the brain." There are better and worse ways of doing this. Unfortunately, some contributors (e.g., Sam Harris) want to tell us that we can not only derive morality from the brain, but that we can actually tell what human beings should value by looking at how the brain works. Such attempts are rightly treated with suspicion. They tend to amount to little more than unconscious attempts to back up personal convictions by appealing to (bad) science (as we will see in chapter 5) in order to give the illusion of "objective truth." Such attempts to derive values from the brain have the appearance of validity when applied to very unambiguous moral situations—for example, "looking at the brain" one can tell that it is not good for human welfare to be tortured or subjected to sadistic actions (i.e., examples where no one needs neuroscience to tell us such things are undesirable anyway). However, as soon as one moves beyond these obvious cases into more complex moral questions which require nuanced answers, the neurological approach falls apart (Earp 2011): looking at the brain offers no clear answers to complex moral problems, no concrete answers to do with which values are worthwhile and which are not, but always relies on prior convictions and value judgments imposed by the commentator him or herself. Appealing to neuroscience in such crude ways represents an abuse of science, wherein matters of personal and cultural morality are projected outwards (or in this case, into the brain), and the authority of science is hijacked in order to give credence to one's own personal value judgments. This crude universalism is radical perspectivism's opposite number, substituting the bottomless fractured abyss of being unable to say anything fully objective about morality for the imposition of a parascientifically defended objectivized absolutism—moral values which are objective, apparently, because they are "in the brain," but which just so happen to correspond to those held by the persons who are positing the values' supposed scientific objectivity. The provisional and embryonic nature of the science appealed to is usually obfuscated by such commentators, and both the significance and credibility of such studies is then dramatized to absurd degrees in much the same way one finds the likes of Hughes, Persson and Savulescu, Zak, and Harris doing. A sustained analysis of the quality of such studies with respect to their moral enhancement potential is desperately needed. Please refer to chapters 4 and 5 for an attempt at such an analysis.

2. Precisely this was the battle between Socrates and the poets over two millennia past—which is better to motivate action, emotively driven exemplars of heroic

action, or disciplined reason and debate? It is one of philosophy's great ironies that Socrates' victory of reason over the poets was achieved through an emotively driven poetic retelling of the story of his execution.

3. Those pro-enhancement individuals who respond to the charge that science is often abused in horrendous ways, for example by Nazi doctors, by merely shrugging and saying "that could never happen again" need to pay close attention to the existence of massive genocide and terrible crimes being perpetrated on the world stage to this very day and our own Western participation in them.

4. This argument about the unwillingness to actively engage with the processes of moral education leaves aside those who, so far from refusing to moralize themselves, instead actively engage in evil deeds, perpetrating crimes their institutions allow. That is, not only are some people unwilling to pursue a moral education, some people actively seek out immoral forms of life.

5. This is not to diminish the importance of aggregate benefits, but only to show how unrealistic it is to claim that such aggregate benefits can produce very specific, fine-tuned results of the order that our most demanding social ills demand.

6. Moral eugenics. Or, as it is more popularly called, eugenics.

7. Of course, the lack of investment is not an "in principle" argument against pharmaceutically based moral enhancement, but that misses the point. We are talking about practical realities here, practical impediments in the here and now. Such impediments will not be overcome without sweeping change, the motivation for which is entirely lacking.

8. In his appropriately titled lecture, "The Illusion of the Technological Moral Fix" (Wallach 2012).

9. Jotterand is clear that such methods would not make persons "more moral," but rather only contribute to their capacity to undergo a project of moral development through reflection on the good.

10. This ignores entirely the adaptively successful (i.e., noncriminal) psychopaths, and the social forces which actively encourage and reward their behavior.

12. Others, e.g. Rob Sparrow (2011), have also made a strong case that the "new eugenics" of Harris and Savulescu is not at all different, in effect, from the old eugenics that were so despicable.

13. Guinness writes: "Studies on the effect of child abuse and neglect, particularly in the first year of life, produce some very disturbing findings. There is a lasting imprint on the neuronal networking of the brain, to a great extent irreversible, and producing a permanently damaged personality. ... The Maudsley Romanian Orphan Adoption Study is demonstrating that severe stimulus neglect (infants left 24 hours per day unattended in their cots) results in mental retardation, induced autistic

defence and attention deficit disorder, and failure of language formation as well as attachment disorder. ... The worst result was profound imbecility in spite of normal genetic potential. ... John Bowlby described 'affectionless psychopathy' in the resulting adults who had never had the opportunity as children to bond with a primary attachment figure. As Bowlby said, the infant needs a 'besotted caretaker' to interpret what it is to be human" (Guinness 2006, 34–36).

14. Savulescu's (2010) position is that genetics "may hold the key" to influencing behavior. Apparently there is "no reason why human moral behavior and its biological underpinning could not be in principle altered," that "through bioprediction" or "genetic testing" we can identify genes that increase risk of sociopathic disorder and select them out, segregate them, and increase surveillance over them (Savulescu 2010). We know that Savulescu is guilty of the crudest possible form of genetic reductionism from his lecture title: "Unfit for Life: Genetically Enhance Humanity or Face Extinction" (2010). Unless this title was chosen purely for shock value, what can it mean other than that Savulescu thinks we are sufficiently reducible to our genes to make their manipulation the difference between human survival and its destruction? Savulescu argues that, just as dogs have had their nature domesticated through years of selective breeding, we can do this "over a single generation" with human persons. Apparently placing dogs and humans in a comparable relationship, in the context of genetic manipulation, for purposes of "domestication," raises no eyebrows at all.

15. This is not even to mention the manner in which genes interact with each other and the very complex way we are now learning they interact with the environment (epigenetics) in ways which can switch certain genes structures on or off depending on the living creature's environmental context and personal endeavors. As Benson and Proctor's (2010) work has indicated, even something like meditation, regularly performed, can alter the expression of a person's genetic profile, switching on or off certain gene structures accordingly. This point about epigenetics is incredibly important for those that think morality can be improved by targeting genetics—it suggests that moral behavior is going to be influential in shaping one's genetic profile as much as one's genetic profile is in shaping one's behavior. Before we start suggesting that segregation or eugenics is the appropriate response to "problematic" genetic profiles it might be worth exploring the extent to which immoral activity influences our biology rather than just assuming that biology is the explanatory and causal bedrock here.

Chapter 4

1. Given the truly prodigious amount of empirical work surrounding the effects of oxytocin—as of August 2014, there are approximately 10,000 papers with the word "oxytocin" in their titles in journal publication website Pubmed alone—a mountainous body which only appears to be increasing, certain strands of the research

will have to be ignored for present purposes. As such, this strand of the research attempting to relate the generalized oxytocin levels of the population of a whole country and the economic success of that nation shall be left to one side. Suffice to say, these studies attempt to show that oxytocin can be isolated as a causal factor, on the national level, in promoting good economic conditions. That is, free, liberal democratic states with good economic conditions correlate with increased oxytocin levels, trust, and happiness on the part of the general population (Zak and Fakhar 2006).

2. Is there not a poorly disguised misogyny implicit in associating oxytocin with goodness and testosterone with the opposite? Actually, Churchland and Winkelman's (2012) critique of oxytocin as morally affective is simply to reduce oxytocin to the level of a "de-stresser"—that is, oxytocin is a mild anti-anxiety medication, and any moral goods that may arise would then be merely side effects of the anxiety-reducing effects of oxytocin, rather than being direct improvements on human moral functioning per se.

3. One which, happily, and quite coincidentally, seems to correlate with precisely the broadly liberal-democratic political structures we already have. Zak's equation then is simple: more oxytocin = more social interactivity = more moral goodness = more commerce and economic prosperity = more happiness.

4. Throughout part II, it is worth keeping in mind Mary Midgley's critique of the so-called selfish gene idea, inasmuch as it refers to the inappropriate anthropomorphisation of the basic biological constituents of persons and their presentation as external agents—evil hobgoblins affecting us from without—and to which human persons are then posited as no more than passive slaves. What would make for a more appropriate view, as Midgley rightly asserts, and a point I have drawn on heavily throughout, is that justice must be done to balancing appreciation that we are both (a) influenced passively by various factors in our world and biology (our facticity, to use the existential term), as well as (b) agents with psychology and personal responsibility, creatures able to rise above the so-called culturally, genetically, and biologically deterministic forces which do surround us, but which absolutely do not enslave us. Throughout we will see oxytocin, serotonin, dopamine, and various brain regions talked about as if they are homunculi, themselves capable of independent agency and thought, which alternately advise or cause havoc in the merely passive receptacles that we call human persons—in this case, speaking as if oxytocin were itself prosocial, as if oxytocin were itself in possession of capacities for discernment which enable it to indicate to its "carrier" (i.e., you and me) whether or not persons in such a context should be trusted (see Zak, 2012, and his talk of the "gyroscopic" capacities of the oxytocin molecule).

5. Prosocial behavior is not necessarily the same as "moral behavior," but there is some overlap between facilitating social harmony and some of the many relational moral goods that occur in social contexts.

6. It is useful to note here, as Gendler (2012) and Mihai and Giordano (2014) do, that these are very loose ways of construing the meanings of the words "deontological" and "consequentialist." It is certainly worth noting that the projection of this dilemma onto the deontological and consequentialist schools of thought (of which there are a surprising variety, with much internal complexity) represents at best very crude simplifications of the views involved. We shall not dwell on this point too much here, even though it is important (for if the empirical studies do not even accurately model the school of moral thinking that they claim to, then what is left over?), because the entirety of the next chapter will be devoted to arguing that the methods used for such empirical research as one finds in Crockett and Zak's work involve such radical simplification that the idea these studies bear any credible relationship to the reality they are meant to be modeling is too dubious to take seriously.

7. Of course, the possibility that there might be more important and rationally defensible goods to the responder than merely accepting the small amount of money on offer never occurs to those designing the experiments. I would suggest that these assumptions tell us more about the experimenters and their values than they do about the participants.

8. Which seems to be, actually, more of a shorthand for a constellation of terms such as "anger," "impulsivity," "aggression," "the need to punish and harm."

9. Indeed, Takahashi et al. are keen to make this point clear: "Understanding the complex role of 5-HT in aggression requires consideration of multiple factors, including (1) the type of aggressive behavior, its topography and function, (2) the genetic background of the individual and the typical phenotype for this background, (3) that trait characteristics of the human subject or the mouse ... and (4) the situational conditions under which aggressive behaviors have been engendered" (Takahashi et al. 2011, 201).

10. Anyway, it is hard to see why SSRIs and the focus on serotonin have produced so much excitement in the enhancement literature. If all one wishes to do is to reduce aggressive impulses and behavior, plenty of substances already exist to do so; for example, Lithium, anticonvulsives (e.g., carbamazepine), benzodiazepines, some beta-blockers, and some atypical antipsychotics (e.g., clozapine, which is "a highly effective drug with which to treat aggressive behavior"; Prado-Lima 2009, 61–62). Whether simply drugging people will adequately count as "moral enhancement" is another question.

Chapter 5

1. Such a possibility was also explored by Gruber and Dickerson (2012) and Hook and Farah (2013). Both sets of authors found the images alone did not necessarily have as much impact on the nature of the evaluations of the science in question as imagined.

2. This assumes that such methods are genuinely "the best we have," that other more integral and realistic approaches are not presently possible, and that the other qualitative elements of the human sciences might not also be more appropriate ways of going about scientific investigation here.

3. Indeed, Zak's confidence in the reliability of the method used in his oxytocin studies seems all the more curious when one considers his own reservations regarding precisely the same methods used in examining the effects of testosterone. Zak et al. (2009) write: "These correlations should be viewed with caution as (testosterone) is highly dependent on a variety of environmental conditions. ... The inability to control experimental subjects' behaviors before they enter the lab, and the high degree of variability in basal (testosterone) indicate that correlational studies can only be considered provisional findings" (Zak et al. 2009, 1). Why Zak does not extend the same crucial concerns to his work on oxytocin is quite mysterious.

4. Once again, there is no objective standard of "cost" or "liberality in giving." Of course, at the extremes everyone can recognize certain acts as involving a loss or of being particularly generous, so there is, no doubt, some objective standard by which most would agree altruism or generosity has taken place. Yet this is not what is being measured here. No extreme acts are being observed or looked for, and instead what are presented can only be understood as readily disputable boundary cases.

5. It seems the preference, with Zak at least, is to attempt to apply incredibly generalized notions of the trait he is investigating. The same might be said, for example, of his examination of trust, something he calls, in one study, "generalized trust" (Zak 2006, 413). Generalized trust is defined by Zak as "the likelihood that two randomly chosen individuals will trust each other in a given environment" (ibid., 413). The measure he uses to assess trust is the proportion of those in each country who respond affirmatively to the question: "Generally speaking, would you say that most people can be trusted, or that you can't be too careful in dealing with people?" (ibid., 416). But how much sense does it make to speak of "generalized trust"? And how useful is it to speak of "generalized trust" in the context of a scientific study which tries to relate the population of *an entire country* to their oxytocin levels, to whether they answer in a self-report questionnaire affirmatively to the very vague question "can people be trusted?" (Note that the question asks whether people can be trusted, and not whether the participant actually does in fact trust people.) But trust people to do what? To respect your property and person? To tell you the truth, even if it hurts? To not betray you if it is in their own interests? Or something else? The construct is hopelessly vague.

6. It is to be noted that even the human sciences have to deal with the problem of simplification regarding definitions of the virtues (see Peterson and Seligman's [2004] classification of the "character strengths and virtues" for a wonderfully superficial account of a range of moral goods). Likewise, such human sciences often apply crude and quantitative measures for study rather than the more appropriate qualitative research appropriate to moral functioning. The former mode of study produces

wonderful pie charts, but not much else. Commentators such as Avram and Giordano (2014, 23) have noted that it is important for those conducting such neurostudy to engage more closely with psychological insights into moral functioning ("moral psychology"). In principle this is an excellent suggestion. The problem is that current moral psychology is in a pretty dire state. In fact, much the same critique that has been raised here can be applied directly to moral psychology—it deals with cavernously vague, generic constructs of extremely fine-grained moral phenomena which it then reduces to massively oversimplified measures in order to make them amenable to the production of neat and colorful pie charts. Ideally, the neurological and psychological study of morality should indeed be better integrated—but both disciplines have a long way to go before they are able to take an approach fine-grained enough to make their output meaningful and relevant to the embodied reality of moral living. Again, a more qualitative set of approaches would be most welcome.

Chapter 6

1. With the exception of Hughes (2013) and his colleague Andrew Fenton (2009), who speak of enhancement in the context of Buddhist faith, and how the various "paramitas," or disciplines, might be developed by pharmacological and genetic means.

2. It is true that there are plenty of nontheological philosophers who have emphasized the importance of the social environment in moral formation (e.g., MacIntyre 1981). What these thinkers do not offer, with regard to our present concerns anyway, is a very close inspection of the implications that the ground-level idiosyncrasies of religious faith are likely to have on how moral enhancement will need to be grasped in specifically religious communities. Yet even in choosing Christian theology as the primary voice, such theology is so diverse and conflicting even among itself (not even considering the differences that exist between theology and actual practice), that it is not really possible to give a full account of even "the" Christian voice anyway—there being no such thing as "the" Christian voice in any unified sense. In keeping with the overall mind-set of the book, the attempt has been to distil a number of core Christian theological points (some of which are considered "nonnegotiable" even among the various particularist voices) into a picture which amounts to a Christian virtue ethical theology.

3. O'Connell writes: "Making disciples: that is our vocation; that is our responsibility ... that is the whole purpose of the Church" (O'Connell 1998, 13).

4. This accountability to "the Kingdom" as a standard for Christian moral enhancement is going to cause tremendous problems. At the very least, there is the epistemological question of whether human persons are even capable of articulating that end point with great enough acuity to justify cementing such visions into our

natures by biomedical means. As history moves on, our visions of what the Kingdom should look like change. Enshrining present visions of the Kingdom, "locking in" our present interpretations and visions by biotechnological means, could end up causing more problems than it solves.

5. In this sense, Christian practical wisdom involves, in part, a testing out of a range of possible improvisations on Christian themes, which aim to be true to both the original scriptural meanings, the history of Christian practice, the constraints of the present world, and the desired future to come.

6. This Christian outward-facing emphasis on turning away from self stands against at least two self-interested positions. There are the more superficial self-centered doctrines (such as those found in consumerism or brutish social Darwinist ethics), but there are also more philosophically credible doctrines which are concerned with the interrogation of self. Existentialist-based theses which recognize anxiety as the necessary condition of having an authentic self-chosen self are not so easily swept under the carpet by conflating them with the more superficial forms of self-centric ideology.

7. There is much talk of late by "situationists" regarding the idea that virtue or character is actually an illusion (see Alfano 2013 in Van Zyl 2013) using studies which attempt to show that persons behave in very different ways depending on the context in which they find themselves. Now, of course, persons do behave differently in different contexts—yet one must be careful to not over-read such studies, since they do not take into account the fact that virtue is a process, the work of a lifetime, and that there are differences between virtue "novices" and "experts" which are not represented in these studies (Annas 2011).

8. Kelsey goes on to suggest the following "powers and dispositions," split into three kinds, embodied, intellectual, and affective: "respect, prudence, and justice as virtues of intentional bodily action; truth, vision, discernment, and mastery of Christian concepts ... as intellectual virtues; delight's joy and love as affective virtues" (Kelsey 2009, 355–6).

9. The 1960s and 1970s were rife with such scandal; see Richard Baker and the San Francisco Zen Temple in 1984 (Lachs 1994, 2002).

10. How one gets from this enlightened love of knowledge, which is quite literally a form of worship, to the state where Islamists are executing school children is impossible to fathom.

11. Andrew Fenton (2009) has made the case for endorsing concentration enhancers in a Buddhist mindfulness context.

14. As Ford notes: "Without this practical immersed element, rules and structures, and anthropological theory, as explanations are radically impoverished and inadequate" (Ford 1999, 141).

15. Indeed, O'Connell argues that it is moral theology's failure to tackle the empirical question of how persons come to hold their values that has held back its capacity to motivate an interest in taking on the project of discipleship (O'Connell 1998, 2–4). Similarly, Fraser Watts notes that theological anthropology can be, at times, a little too abstracted from the empirical reality of the human condition—that the human sciences offer resources for grounding a more empirically justifiable picture of the human being. He writes: "claims are often made in theological anthropology just because they fit in with other theological positions, and without theologians feeling any need to check empirically whether their claims are correct. ... If theological anthropology were more inter-disciplinary, and conducted in dialogue with an empirical discipline like psychology, the result would be much more satisfactory" (Watts 2010, 196).

16. Traditional theological means of communal moral formation usually involve ideas of participatory formation through practices of worship, particularly Eucharistic or liturgical celebration; the telling and retelling of essential Christian stories and revitalizing the principal Christian narratives; and embracing the past traditions in an ever-changing present, toward a hoped-for Kingdom that we are attempting to bring about. Being a Christian is understanding yourself as part of an ongoing story which began at creation, through to Christ's crucifixion and resurrection, and into the present, where we are working. This development of Christian identity is an essential communal component of "forming together," a formation which includes embracing the moral dimensions of Christian living as an essential and inextricable part. In short, moral formation is both directed by and energized by the stories its members identify with. These stories add meaning and generate motivational power for taking on the task of moral development. Such stories mean that moral formation is not some extraneous task, like a hobby, not at all "one task amongst others," but rather is crucial to the continuation of one's place within that story. Thus, moral formation has an integrated sense, a salience which it would not have were it not caught up within the terms of the story itself. Of course, it matters which story we tell ourselves—the postmodern and pessimistically fractured story that many contemporary persons tell themselves leaves precious little room for sustained moral development. Certainly such stories do not place moral formation at their center, and no overarching telos is provided to motivate such centrality.

17. It is important at this point to note that while focusing on the larger scaffolding is crucial, primary even, just focusing on scaffolding is also a form of reductionism. While group norms are important, it still requires individuals to do the hard work of "putting on the virtues." Individual responsibility is still at the very heart of the matter.

Chapter 7

1. It is important to make clear that enhancement and theology are not natural enemies, as is commonly held. In fact, there are plenty of pro-enhancement

theological thinkers in both conservative and liberal Christian theological domains; see Lustig (2009). For example, it is possible, in principle, for a theological commentator to step forward and suggest that such moral biotechnology might in principle constitute a form of the mediation of the will of God to a broken world, thus serving a reparative social function. It should be made clear that I am not myself endorsing that view; what I am doing here is not itself theology, and I am not espousing any particular theological view on moral enhancement. There is an important difference to be had between presenting a theology of moral enhancement on the one hand, and what is being attempted here: discussing the fact that moral enhancement will necessarily be enacted in a religious thought world. The latter merely accepts that a religious thought world is an inextricable part of the context in which moral enhancement is expected to be enacted.

2. Such a prospect brings to mind the film *The Matrix* (an inoffensive but utterly nonsensical yarn), in which abilities to do skilled and complex procedural tasks, such as riding motorcycles or piloting a helicopter, can be simply "downloaded" into the brain.

3. Though it is to be noted that grand biological accounts will always involve a not-inconsiderable amount of storytelling and worldview creation, too (Midgley 2004).

4. While Zak has not explicitly said oxytocin should be used as a moral enhancement intervention, others in the literature have (see part I).

5. Kelsey stresses the finitude of human life, which he describes in this context as "dying life lived on borrowed breath" (ibid., 402). He continues: "Lived wisely in faith, it flourishes as the glory of God" (ibid., 402).

6. This is to generalize. Of course, there are certain pagan outlooks which favor vegan lifestyles.

7. It is important to note that even within the same faith tradition numerous modes of interpretations are possible, that there are cultures within cultures, and traditional modes of interpreting moral goods are "underdetermined"—that is, they leave plenty of room for generating alternative interpretations. There have been plenty of examples of religious zealotry coming from within the dominant faith traditions, persons who self-select to engage in acts of terrorism to support what they perceive to be their traditional moral viewpoints—for example, pro-life Christian activists who have murdered doctors and bombed abortion clinics, and, of course, the recent wave of terror produced by ultra-hardline militant Islamist groups like ISIS.

8. Retired New York Teacher of the Year Gatto writes: "In 1926, Bertrand Russell said casually that the United States was the first nation in human history to deliberately deny its children the tools of critical thinking; actually Prussia was first, we were second" (Gatto 1994, 280).

9. It is to be taken as read that Christian history is at least as bloody as any of the other faith traditions, and precisely the same points made with respect to the

religious acts focused upon here (which are but a tiny proportion of the whole) can be also made with respect to Christian history too. Jim Jones' *Blood that Cries out from the Earth* (2008b), cited above, is generally recognized as the definitive text on the psychology of religious terrorism and religious violence through history.

10. Though such action does not always result in murder: "often burning or scarring with acid are the preferred weapons of the men committing such crimes" (Jones 2008a).

11. Though perhaps it is better to say that any devised biological intervention, regardless of how fine-grained, will be superficial by default.

Chapter 8

1. For example, it is morally responsible for governments to legislate for the wearing of seatbelts, for not allowing driving while intoxicated, for not allowing persons freedom to destroy their bodies through endless cosmetic surgery or use of hard drugs (Dworkin 2010).

2. In fact, conservative religious views on enhancement are much more varied than Hughes indicates, with both liberal and conservative voices offering positive and negative theological appraisals of the validity of enhancement projects with respect to God's creative will (Lustig 2009).

3. Norman Daniels gives the example of "the running-away disease of slaves" as a socially constructed and highly convenient way of diagnosing undesirable behaviors (Daniels 2000, 318).

4. I admire Harris' confidence here. It seems to me that such a claim would be rather hard to prove.

5. It is worth noting that the treatment/enhancement distinction is not always actually used in current practice anyway. As Sparrow suggests: "Many existing medical interventions provide individuals with capacities that are beyond those typical of the species. Insofar as one thinks that it is at least permissible (cosmetic surgery) and arguably obligatory (vaccination, the Pill) to make these technologies available to people, existing practice does not locate the distinction between permissible and impermissible interventions at the therapy/enhancement boundary" (Sparrow 2010, 118). By way of example, the FDA "recently approved the use of human growth hormone in very short children with no other medical problem" (Dees 2007, 377).

6. Reviews suggest that between 20% and 50% of depressive patients will fail with the use of pharmacotherapeutic methodologies. There are accounts of neural pacemakers being implanted into depressed patients' brains, curing life-long melancholia in ways no medication or therapy have hitherto been able to manage (Karas et al. 2013; Chivers 2013).

7. For example, the line between "cosmetic" and "essential" features of our nature is in many cases just as sketchy as that between therapy and enhancement. Some give the example of memory enhancement as an example of a cosmetic alteration (that genetic enhancement of memory is "unproblematic because the intervention merely increases storage capacity and has 'little or no effect on the essence of the person'" (Lustig 2008, 49). In contrast, "genetic interventions aimed at reducing aggressiveness in boys would involve a 'fundamental alteration' of the future persons they would become" (ibid., 49). Yet, from looking at Alzheimer's patients, it seems fairly evident that changes to a person's memory capacities can indeed have profound effects on their personality, their identity, or what is considered essential about their personhood. The other example, height as a cosmetic alteration for children, is also tenuous, if a person's esteem and future prospects are in part defined by a severe lack of stature, the long-term effects of this cosmetic change might be significant enough to manifest as an essential change to their identity or personhood. This is not at all to suggest that the ideas of "identity" and the like are not useful lenses for exploring the question of enhancement. Rather, it is to note that the boundaries here are a little more muddy than first sight would suggest.

8. Sparrow is one of many quality contributors to the moral enhancement discourse that have been all but excluded from the present text. The main reason for this is that Sparrow's most relevant work, for the most part, concerns egalitarian ethics, which have little place in the present argument, and more importantly, his arguments only really work against the sort of enhancement project that Persson and Savulescu have in mind. Without at all diminishing the quality of Sparrow's arguments, which have had a harder time than they deserve, my point is that the moral enhancement debate is travestied if we take Persson and Savulescu as the whole of the matter. There is more to moral enhancement than can be found in Persson and Savulescu's philosophy. While Sparrow's polemics against Persson and Savulescu are both incisive and decisive, the suggestion is that articulating a moral enhancement *beyond* Persson and Savulescu is one of the greatest contributions to the field one can make right now. Indeed, when it comes to the likes of Sparrow and Rakić, their real sin is simply staying within the confines of the paradigm offered by Savulescu and the like and arguing only from within that framework, instead of developing some alternative paradigm, or a clearly different way of going about thinking through the subject from the outset. The art of getting moral enhancement wrong may well consist in not much more than remaining within the bounds set forth by Persson and Savulescu.

9. More significantly, though, what was found with Ritalin was the appropriation of mental health language and interventions which are then dispensed on moral or (morally related) grounds, to resolve behavior perceived as disobedient or bad. A similar controversial example can be discerned in recent British health services attempts to refuse nonemergency surgery for those who are morbidly obese and persons who smoke (BBC News 2014). The initiative is defended as a "cost-cutting

measure," but there is a clear morally related dimension going on beneath. It is not a coincidence that obesity and smoking have been singled out here. The implication is clearly that "such persons have not looked after themselves, so why should the taxes of more responsible citizens be wasted carrying the weight of those who refuse to control themselves or moderate their behavior." Indeed, not a month seems to pass without some new measure of this sort being passed and with such rationales being presented. Another proposal to deny benefits to the obese and those with alcohol problems has also been set forth by UK Prime Minister David Cameron (BBC News 2015b), who has been quoted as saying:

It is not fair to ask hardworking taxpayers to fund the benefits of people who refuse to accept the support and treatment that could help them get back to a life of work. (BBC News 2015b)

Whether we agree with this policy or not, the point is that there is a distinct moral substructure to decisions being dressed up as "cost-cutting measures" and denial of benefits articulated in terms of fairness (i.e., justice). In the case of the "work-shy" alcoholics and obese people mentioned by Cameron above, perhaps a whiff of the Protestant work ethic can be discerned as an ingredient in the rhetoric used to persuade persons to accept such measures?

In the case of Ritalin, we are treating disorderly conduct as ADHD, qua mental illness. We have unruly children, misbehavior, treated as a mental illness. The Victorian sense of children needing to be "seen and not heard" can be sensed hanging around in the background. Even the "cure" of this apparent mental illness is cast in moral-sounding terms. One news correspondent for the Telegraph reports of Ritalin that it "does not turn children into 'robots'" but rather "frees them to make the right moral decisions" (Adams 2012).

10. The British Psychological Society responded by criticizing these diagnoses as "clearly based largely on social norms, with 'symptoms' that all rely on subjective judgments ... not value-free, but rather (reflecting) current normative social expectations" and suggesting that "clients and the general public are negatively affected by the continued and continuous medicalization of their natural and normal responses to their experiences ... which do not reflect illnesses so much as normal individual variation" (Allan 2011).

11. While these points are important, one needs to ask whether it is right to simply reduce such cases to matters of organic deficit only. There are mediating factors which are significant too. As was argued in chapter 2, it is not as if processes of socialization and conditioning cannot to some extent ameliorate these deficits. In extreme cases, such environmental dimensions might not be sufficient to compensate for such deficits, which are indeed problematic and not something to be demeaned in significance. All the same, it is important to note that appropriate socialization is still possible for those with such deficits, and not something to be ignored.

12. Healy refers to this as "shadow science" (2004, 109): ghost-writing performed by public relations and communications agencies and the writing and publication of articles for purposes of marketing newly approved drugs (ibid., 112). Healy points the reader to the following further reading on the topic: Wright (2002); Rampton and Stauber (2001); and Mundy (2001).

13. Except within certain preestablished niches, voices themselves being sold to a small but hungry anti-reductionist market.

14. A quick note regarding the appropriation of thinkers such as Szasz and Breggin is warranted. There is little doubt that many of the authors drawn upon in the present chapter are outliers, considered by many as mavericks within their field. It is important to note the extent to which one can draw from the important insights they have elaborated without necessarily being drawn into a full commitment to the larger "package deal" their work represents. For example, regarding Szasz, it is perfectly reasonable to agree with his views that more emphasis on personal responsibility needs to be forwarded with respect to the moral dimensions of clinical care, and that there are clear dangers inherent in talking about disturbances that are more appropriately understood as mental or agency-based in nature in physiological or biological terms. This being so, we do not need to commit ourselves wholesale to the view that mental illness is a myth simply because of the inappropriate use of language. Similarly, it is perfectly reasonable to be in agreement with Breggin insofar as he is talking about the existence of two subtending philosophies of human nature undergirding the practice of psychiatry. Indeed, this need for greater awareness of the hidden narratives which subtend and give shape to various elements of medical science (and science more generally) is an incredibly important point for which we find wide support. Breggin's articulation of these two contrasting philosophies represents a very important distinction that deserves to be presented at length. We can accept this call for greater awareness of hidden, dehumanizing narratives within medical science; we can also accept this distinction between two kinds of philosophies of human nature. Yet we do not have to accept that this is an "either/or" position. It is possible to treat people more or less like passive machines. It is possible to use drugs in a psychiatric context while still treating the human person as a responsible subject. The approaches can add to each other, in the right hands. Certainly such use has to be determined on a case-by-case basis, but in principle one can say that the use of drugs and treating persons as responsible human beings are not mutually exclusive (indeed, the following chapter exploring alcoholism and drug abuse is a perfect example of how drugs and person-centered therapy might work alongside each other in a mutually supportive relationship). These qualifications are all massively significant with respect to our current focus on moral enhancement. The attack on reductionism and its ever-present siren song is not at all an exhortation to abandon moral enhancement, but to indicate as clearly and as vehemently as possible that unless moral enhancement is understood in a way which can do justice to the interwoven complexity of the biological, and the many,

many nonbiological influences—personal, social, and so on—then there is no moral enhancement at all. What we would have then would be a dehumanizing caricature, something ultimately self-defeating.

15. Recently there has been a call for UK doctors to (literally) prescribe "Weight Watchers" classes to obese persons.

Chapter 9

1. Appealing to "the future" offers no recourse here; moral functioning is and will always be an interwoven and contextual business with strong conceptual shaping involving habits and norms which cannot be biologically reduced or reproduced. That is, one can develop whatever future technology one desires—these fundamental problems of moral development are the same now as they have always been.

2. Or, in Bass et al.'s more technical language "these data demonstrate ... the first causal evidence demonstrating that tonic activation of VTA (ventral tegmental area) dopamine neurons selectively decreases ethanol self-administration behaviours" (Bass et al. 2013, 1).

3. The possibility has been raised by Danish Health authorities that nalmefene, traded as Selincro, may create in some persons psychiatric symptoms such as confusion, hallucinations, and dissociation, and should be used with caution in patients with moderate to severe depression. It is suggested that "patients prescribed nalmefene should receive continuous psychosocial support on adherence to treatment and reduction of alcohol consumption" (*Reactions Weekly* 2014).

4. The problems that this shortness of effect raises might be dealt with by creating a longer-lasting agent, like Asenjo's, which lasts about six months. This complexity distorts the line between compulsion and election considerably. If one changes one's mind four months into the treatment, one is yet tied into the commitment to abstain. This presents a curious case which is at once elective and compulsory. It was elective when one took the intervention; now it is compulsory because of that elective choice made by oneself back then, and one has now changed one's mind.

5. This is not to say that alcoholics do not overcome their addictions, or rather learn to manage them; it happens every day.

6. Though in neither case would it prevent a person intent on doing such harm from perpetrating it. An addict may be force-fed a vaccine which renders their substance of choice ineffectual, but they can still take it; thus the freedom of choice and the ontological "freedom to fall" has not been removed.

7. However, this does not rule out the possibility that compulsory moral enhancement later turns into a strong vision of moral enhancement. A person who was forced into a program of such intervention might think to themselves later on that

being forced off alcohol was "the best thing to ever happen" to them and take the program as an opportunity to grow as a person.

8. Even if such a reason is very vaguely defined or articulated—say, an alcoholic one day taking a good hard look at him or herself and simply saying "No, this is wrong."

9. It is worth observing that the root of the word "salvation" means "to heal"; salvation is considered the fullest and most complete kind of healing.

10. As Juhnke et al. put it: "Spirituality is in vogue" (Juhnke et al. 2009, 16).

11. In addition to these uses of prayer, meditative prayer, or meditation more generally, is a staple part of the spiritual aspect of recovery. According to Young, Armas, and Cunningham:

Meditative practices are used in a variety of health care settings (Baer 2003) and have been acknowledged as important components of addiction recovery. ... Meditation is specifically referenced in AA's 11th step, which reads, "We sought through prayer and meditation to improve our conscious contact with God as we understood Him, praying only for knowledge of His will for us and the power to carry that out" (AA 1976, p. 71). Moreover, meditation is a tool that is frequently used in drug and alcohol treatment centers. In a national survey on the frequency of prayer, meditation, and holistic interventions in addiction treatment centers, Priester et al. (2009) found that 58% of 139 addiction treatment programs used meditation as a component of treatment separate from prayer alone. Meditation has also been shown to help increase anger management skills (Vannoy 2005), strengthen recovery motivation, bolster detoxification, manage stress, increase self-confidence, and boost personal well-being. (Young, Armas, and Cunningham 2009, 59)

Given then that meditation is such a low-cost and effective supplement to the recovery process, it is hardly surprising that it has received widespread endorsement as a positive adjunct to the treatment process.

12. Physiologically, it is hard to dispute that certain biological factors are influences with respect to addiction; for example, childhood abuse can lead to neurobiological deficits which can contribute to addiction. At the other extreme, defined in social terms, alcoholism or substance addiction is a maligned condition with respect to which treatment options are available. As such, defined objectively or subjectively, there are grounds for defending the idea that addiction is fittingly described as a disease, though whether this is best understood metaphorically or literally is again a question for debate.

Chapter 10

1. It is essentially none other than the use of a "shock doctrine" approach (Klein 2008), the academic equivalent of "disaster capitalism," invoking apocalyptic consequences in order to push through a rationale for a hideously undesirable project of

pervasive state control, all in the name of "safety." This is a pure hard sell—be afraid, and then accept our solutions.

2. Though such methods are not at all without their problems either, such human sciences often apply crude, quantitative measures for study rather than the qualitative research more appropriate to moral functioning.

3. Koebler (2012) examines the case of teacher in Virginia convicted of child molestation, who upon being examined was found to have a brain tumor, who upon removal of said tumor no longer displayed pedophilic tendencies. Koebler also notes the double-edged nature of using brain-scan imagery, usually presented in the attempt to absolve the culprit of responsibility; it can backfire and be presented as showing the person in question to be a "natural-born killer," someone at profound risk of reoffending, and thus in need of ongoing monitoring and treatment.

4. This gloss on emergence is useful for articulating for those unfamiliar with biological paradigms that credible alternatives to biological reductionism do exist and have existed for a considerable amount of time. That emergence is not so well heard of is more to do with the popularizers of science rather than the quality of the paradigm. The incredible black-and-white simplicity of reductive accounts make them far easier to communicate to general audiences and makes for very grand headlines.

5. Another more presently controversial prospect is the linking of GPS to the speedometers in our cars to ensure that no one ever breaks the speed limit, thus making traffic deaths less likely. Even this seems a little tame next to recent Russian driving restrictions. A 2015 BBC news article reads:

Russia has listed transsexual and transgender people among those who will no longer qualify for driving licences. Fetishism, exhibitionism and voyeurism are also included as "mental disorders" now barring people from driving. The government says it is tightening medical controls for drivers because Russia has too many road accidents. "Pathological" gambling and compulsive stealing are also on the list. Russian psychiatrists and human rights lawyers have condemned the move [saying] some people would avoid seeking psychiatric help, fearing a driving ban. (BBC News Europe, 2015)

6. A recent article in *The Psychologist* stated:

The House of Lords Science and Technology Committee has held sessions of evidence to hear follow-up information to the 2011 Behaviour Change Inquiry. The original inquiry looked into how policy interventions and "nudge" techniques could affect change among the population. ... Oliver Letwin MP, Minister for Government Policy, was among those to speak about the effectiveness of the work of government so far. He told the select committee that behaviour change had become one of the best-evaluated sectors in government. ... Owain Service, Managing Director of the Behavioural Insights Team ... gave some examples of departments where nudge techniques had been used: ... HMRC estimates now that these trials alone in one area of HMRC have helped to bring forward about £200 million in revenue to the Exchequer. They include things like using social norms to encourage people to pay their tax on time. So we now know for example that telling people that nine out of ten people pay their tax on time encourages more people to pay their tax on time. (Rhodes 2014, 572)

Imagine how unfeasibly complex it would be to develop and test a drug that could do such a thing, the decade of research, the billions in costs, when all that was required was the addition of a single sentence in a letter to the taxpayers.

7. *Telegraph* crime correspondent Martin Evans writes: "Technology giant Google has developed state of the art software which proactively scours hundreds of millions of email accounts for images of child abuse ... Last month the National Crime Agency (NCA) announced that more than 600 suspected paedophiles including doctors, teachers, and care workers had been arrested in a major crackdown on the trade in images of abuse. ... The system operates automatically and nobody working for Google is able to see any of the images being examined" (Evans 2014).

8. The parallel with sports enhancement also invites a similar parallel with the related question of "cheating"—I will leave it to the reader to reflect upon whether they feel such moral enhancement, were it to be effective, would also count as cheating, and whether or not they think this would be a problem.

References

Adams, S. 2012. Ritalin Does Not Turn Children with ADHD into "Robots" but Frees Them to Make the Right Moral Decisions, Researchers Claim Today. *The Telegraph.* October 15. http://www.telegraph.co.uk/news/health/news/9605455/Ritalin-doesnt-turn-ADHD-children-into-robots.html.

Agar, N. 2013. A Question about Defining Moral Bioenhancement. *Journal of Medical Ethics* 40 (6): 359–360. doi:.10.1136/medethics-2012-101153

Allan, C. 2011. British Psychological Society Response. http://apps.bps.org.uk/_publicationfiles/consultation-responses/DSM-5%202011%20-%20BPS%20response.pdf.

Amnesty International. 1999. Pakistan: Honor Killings of Girls and Women. http://www.refworld.org/docid/4a1fadcdc.html.

Anderson, D. 2013. Your Brain Is More Than a Bag of Chemicals. http://www.ted.com/talks/david_anderson_your_brain_is_more_than_a_bag_of_chemicals/transcript.

Annas, J. 2011. *Intelligent Virtue.* Oxford: Oxford University Press.

Arnett, J. 2008. The Neglected 95%: Why American Psychology Needs to Become Less American. *American Psychologist* 63 (7): 602–614.

Avram, M., and J. Giordano. 2014. Neuroethics: Some Things Old, Some Things New, Some Things Borrowed … and To Do. *American Journal of Bioethics Neuroscience* 5 (4): 23–25.

Baer, R. 2003. Mindfulness Training as a Clinical Intervention: A Conceptual and Empirical Review. *Clinical Psychology: Science and Practice* 10 (2): 125–143.

Barraza, J., and P. Zak. 2009. Empathy toward Strangers Triggers Oxytocin Release and Subsequent Generosity: Values Empathy and Fairness across Social Barriers. *Annals of the New York Academy of Sciences* 1167 (6): 182–189.

Barraza, J., M. McCullough, S. Ahmadi, and P. Zak. 2011. Oxytocin Infusion Increases Charitable Donations Regardless of Monetary Resources. *Hormones and Behavior* 60 (2): 148–151.

Bass, C., V. Grinevich, D. Giola, J. Day-Brown, K. Bonin, G. Stuber, J. Weiner, and E. Budygin. 2013. Optogenetic Stimulation of VTA Dopamine Neurons Reveals That Tonic but Not Phasic Patterns of Dopamine Transmission Reduce Ethanol Self-Administration. *Frontiers in Behavioral Neuroscience* 7 (173): 1–10.

Baumeister, R., E. Bratslavsky, M. Muraven, and D. Tice. 1998. Ego Depletion: Is the Active Self a Limited Resource? *Journal of Personality and Social Psychology* 74 (5): 1252–1265.

BBC News. 2014. NHS Devon Surgery Restriction for Smokers and Obese Plan Revealed. December 3. http://www.bbc.co.uk/news/uk-england-devon-30318546.

BBC News. 2015a. Not in Front of the Telly: Warning over 'Listening' TV. February 9. http://www.bbc.co.uk/news/technology-31296188.

BBC News. 2015b. Sickness Benefit Review to Consider Obesity and Drug Problems. February 14. http://www.bbc.co.uk/news/uk-31464897.

BBC News Europe. 2015. Russia Says Drivers Must Not Have "Sex Disorders." January 8. http://www.bbc.co.uk/news/world-europe-30735673#.

Beck, B. 2014. Conceptual and Practical Problems of Moral Enhancement. *Bioethics* 29 (4): 233–240. doi:.10.1111/bioe.12090

Benson, H., and W. Proctor. 2010. *Relaxation Revolution: The Science and Genetics of Mind Body Healing.* New York: Scribner Book Company.

Bertsch, K., M. Gamer, B. Schmidt, I. Schmidinger, S. Walther, and T. Kästel. 2013. Oxytocin and Reduction of Social Threat Hypersensitivity in Women with Borderline Personality Disorder. *American Journal of Psychiatry* 170 (10): 1169–1177.

Bezlova, A. 2002. China Faces Music for Psychiatric Abuse. *Asia Times.* August 21. http://www.atimes.com/atimes/China/DH21Ad03.html.

Blasi, G., L. Bianco, P. Taurisano, B. Gelao, R. Romano, L. Fazio, A. Papazacharias, 2008. Functional Variation of the Dopamine D2 Receptor Gene Is Associated with Emotional Control as well as Brain Activity and Connectivity during Emotion Processing in Humans. *Journal of Neuroscience* 29 (47): 14812–14819.

Bloom, P. 2007a. Why Are People Different? Differences. http://oyc.yale.edu/psychology/psyc-110/lecture-13#transcript

Bloom, P. 2007b. A Person in the World of People: Morality. http://oyc.yale.edu/psychology/psyc-110/lecture-15#transcript

Bolt, I. 2007. True to Oneself? Broad and Narrow Ideas on Authenticity in the Enhancement Debate. *Theoretical Medicine and Bioethics* 28 (4): 285–300.

Boyden, E. 2011a. Enhancing the Brain, Past, Present and Future. http://syntheticneurobiology.org/videos

Boyden, E. 2011b. A Light Switch for Neurons. http://www.ted.com/talks/ed_boyden

Breggin, P. 1993. U.S. Hasn't Given Up Linking Genes to Crime. *New York Times*. September 18.

Breggin, P. 2003. Psychopharmacology and Human Values. *Journal of Humanistic Psychology* 43 (2): 34–49.

Breggin, P. 2008. *Medication Madness: The Role of Psychiatric Drugs in Cases of Violence, Suicide and Crime.* New York: St. Martin's Griffin.

Breggin, P., and D. Greenberg. 1972. Return of the Lobotomy. *The Washington Post.* March 12.

Brewer, M. 2000. Research design and issues of validity. In H. Reis, and C. Judd, eds., *Handbook of Research Methods in Social and Personality Psychology*, 3–16. Cambridge: Cambridge University Press.

Brigandt, I. 2010. Beyond Reduction and Pluralism: Toward an Epistemology of Explanatory Integration in Biology. *Erkenntnis* 73 (3): 295–311.

Brink, W., H. Aubin, A. Bladström, L. Torup, A. Gual, and K. Mann. 2013. Efficacy of As-Needed Nalmefene in Alcohol-Dependent Patients with at Least a High Drinking Risk Level: Results from a Subgroup Analysis of Two Randomized Controlled 6-Month Studies. *Alcohol and Alcoholism* 48 (5): 570–578.

Brown, W., N. Murphy, and H. Malony. 1998. *Whatever Happened to the Soul? Scientific and Theological Portraits of Human Nature.* Minneapolis: Fortress Press.

Bryant, R., and L. Hung. 2013. Oxytocin Enhances Social Persuasion during Hypnosis. *Public Library of Science One* 8 (4): e60711. doi:.10.1371/journal.pone .0060711

Burroughs, W. 1993. *Naked Lunch.* London: Flamingo.

CACREP (Council for Accreditation of Counseling and Related Educational Programs). 2009. Standards. http://www.cacrep.org/2009standards.html

Carrillo, M., L. Ricci, G. Coppersmith, and R. Melloni, Jr. 2009. The Effect of Increased Serotonergic Neurotransmission on Aggression: A Critical Meta-Analytic Review of Preclinical Studies. *Psychopharmacology* 205 (3): 349–368.

Carroll, L. 1994. *Alice's Adventures in Wonderland.* London: Penguin.

Carroll, R. 2014. Experimenter Effect. http://skepdic.com/experimentereffect.html

Carson, R. 2000. Review of Erik Parens (ed.), Enhancing human traits: Ethical and social implications. *Theoretical Medicine and Bioethics* 21 (6):613–616.

Casebeer, W. 2002. Moral Cognition and its Neural Constituents. *Nature Reviews: Neuroscience* 4:841–846. doi:.10.1038/nrn1223

Casebeer, W. D. 2012. The Promise and Peril of Neuroscience Technology (with a Hopeful Coda). http://bioethics.as.nyu.edu/object/bioethics.events.20120330 .conference

Casebeer, W., and P. Churchland. 2003. The Neural Mechanisms of Moral Cognition: A Multiple-Aspect Approach to Moral Judgment and Decision-Making. *Biology & Philosophy* 18 (1): 169–194.

Cashwell, C., P. Clarke, and E. Graves. 2009. Step by Step: Avoiding Spiritual Bypass in 12-Step Work. *Journal of Addictions & Offender Counseling* 30 (1): 37–48.

Chan, S., and J. Harris. 2011. Moral Enhancement and Pro-Social Behavior. *Journal of Medical Ethics* 37 (3): 130–131.

Chivers, T. 2013. Putting a Brake on Depression with Deep Brain Stimulation. *The Telegraph*. September 22. http://www.telegraph.co.uk/health/10327129/Putting-a -brake-on-depression-with-deep-brain-stimulation.html

Churchland, P. 2012. Braintrust: What Neuroscience Tells Us about Morality. https://www.youtube.com/watch?v=9Bv4k8CJnuc

Churchland, P., and P. Winkelman. 2012. Modulating Social Behavior with Oxytocin: How Does it Work? What Does it Mean? *Hormones and Behavior* 61 (3): 392–399.

Ciarrocchi, J., and G. Brelsford. 2009. Spirituality, Religion, and Substance Coping as Regulators of Emotions and Meaning Making: Different Effects on Pain and Joy. *Journal of Addictions & Offender Counseling* 30 (1): 24–36.

Clark, L. *Wired*. 2013. Neuroscience Is about Discovering Our Limits, Then Hacking to Get around Them. October 13. http://www.wired.co.uk/news/archive/2013-10/ 17/molly-crockett

Clark, L. *Wired*. 2014. Optogenetics Used to Stop Rats Binge Drinking. January 6. http://www.wired.co.uk/news/archive/2014-01/06/alcoholism-cure-rats

Clayton, P. 2006. *Mind and Emergence: From Quantum to Consciousness*. Oxford: Oxford University Press.

Conrad, P. 2005. The Shifting Engines of Medicalization. *Journal of Health and Social Behavior* 46 (1): 3–14.

Cook, C. 2006. Personal Responsibility and its Relationship to Substance Misuse. In M. Beer, and N. Pocock, eds., *Mad, Bad, or Sad? A Christian Approach to Antisocial Behavior and Mental Disorder*, 148–160. London: Christian Medical Fellowship.

Coplan, A. 2011. Will the Real Empathy Please Stand Up? A Case for a Narrow Conceptualization. *Southern Journal of Philosophy* 49 (s1): 40–65.

Cosgrove, L., and L. Drimsky. 2012. A Comparison of DSM-IV and DSM-5 Panel Members' Financial Associations with Industry: A Pernicious Problem Persists. *Public Library of Science Medicine* 9 (3): 1–5.

Crockett, M. 2008. Your Inner Jekyll and Hyde. http://www.youramazingbrain.org .uk/2008_Reseacher_runner-up_Molly_Crockett.pdf

Crockett, M. 2011. Drugs and Morals. http://www.starttreatment.com/ drug-addictions-facts/tedxzurich-molly-crockett-drugs-and-morals/

Crockett, M. 2012a. Moral Enhancement? Evidence and Challenges. http://bioethics.as.nyu.edu/object/bioethics.events.20120330.conference

Crockett, M. 2012b. Morphing Morals: Neurochemical Modulation of Moral Judgment and Behavior. http://bioethics.as.nyu.edu/object/bioethics.events .20120330.conference

Crockett, M. 2013a. Moral Bioenhancement: A Neuroscientific Perspective. *Journal of Medical Ethics* 40:370–371.

Crockett, M., L. Clark, M. Hauser, and T. Robbins. 2010. Serotonin Selectively Influences Moral Judgement and Behavior Through Effects on Harm Aversion. *Psychology and Cognitive Sciences* 107 (40): 17433–17438.

Daniels, N. 2000. Normal Functioning and the Treatment-Enhancement Distinction. *Cambridge Quarterly of Healthcare Ethics* 9 (3): 309–322.

De Dreu, C., L. Greer, M. Handgraaf, S. Shalvi, G. Van Kleef, M. Baas, F. Ten Velden, E. Van Dijk, and S. Feith. 2010. The Neuropeptide Oxytocin Regulates Parochial Altruism in Intergroup Conflict Among Humans. *Science* 328 (5984): 1408–1411.

DeGrazia, D. 2013. Moral Enhancement, Freedom, and What We (Should) Value in Moral Behavior. *Journal of Medical Ethics* 25 (3): 228–245.

Dees, R. 2007. Better Brains, Better Selves? The Ethics of Neuroenhancements. *Kennedy Institute of Ethics Journal* 17 (4): 371–395.

Dillingham, B. 1977. American Indian Women and IHS Sterilization Practices. *American Indian Journal* 3 (1): 27–28.

Domes, G., M. Heinrichs, J. Glascher, C. Buchel, D. Braus, and S. Herpertz. 2007. Oxytocin Attenuates Amygdala Responses to Emotional Faces Regardless of Valence. *Biological Psychiatry* 62 (10): 1187–1190.

Domes, G., M. Heinrichs, A. Michel, C. Berger, and S. Herpertz. 2007. Oxytocin Improves "Mind-Reading" in Humans. *Biological Psychiatry* 61 (6): 731–733.

Domes, G., A. Lischke, C. Berger, A. Grossmann, K. Hauenstein, M. Heinrichs, and S. Herpertz. 2010. Effects of Intranasal Oxytocin on Emotional Face Processing in Women. *Psychoneuroendocrinology* 35 (1): 83–93.

Douglas, T. 2008. Moral enhancement. *Journal of Applied Philosophy* 25 (3): 228–245.

Douglas, T. 2011. Human Enhancement and Supra-Personal Moral Status. *Philosophical Studies* 162 (4): 473–497.

Douglas, T. 2013. Enhancing Moral Conformity and Enhancing Moral Worth. *Neuroethics* 7 (1): 75–91.

Douglas, T. 2013. Moral Enhancement Via Direct Emotion Modulation: A Reply to John Harris. *Bioethics* 27 (3): 160–168.

Dworkin, G. 2010. Paternalism. http://plato.stanford.edu/entries/paternalism/

Dwyer, J. 2012. For Detained Whistle-Blower, a Hospital Bill, Not an Apology. *New York Times.* March 15. http://www.nytimes.com/2012/03/16/nyregion/officer -adrian-schoolcraft-forcibly-hospitalized-got-no-apology-just-a-bill.html?_r=0

Earp, B. 2011. Sam Harris Is Wrong about Science and Morality. http:// blog.practicalethics.ox.ac.uk/2011/11/sam-harris-is-wrong-about-science-and -morality/#more-2334

Evans, M. 2014. Paedophile Snared as Google Scans Gmail for Images of Child Abuse. *The Telegraph.* August 4. http://www.telegraph.co.uk/technology/news/ 11012008/Paedophile-snared-as-Google-scans-Gmail-for-images-of-child-abuse.html

Fenton, A. 2009. Buddhism and Neuroethics: The Ethics of Pharmaceutical Cognitive Enhancement. *Developing World Bioethics* 9 (2): 47–56.

Fifield, M. 2005. Spirituality in the Therapeutic Community. *American Journal of Pastoral Counseling* 8 (1): 67–72.

Fitzgerald, K. 2013. Cocaine Vaccine Close to Human Clinical Trials. *Medical News Today.* http://www.medicalnewstoday.com/articles/260468.php

Ford, D. 1999. *Self and Salvation.* Cambridge: Cambridge University Press.

Freud, S. 1991. *The Essentials of Psycho-Analysis.* London: Penguin.

Fudala, P., S. Heishman, J. Henningfield, and R. Johnson. 1991. Human Pharmacology and Abuse Potential of Nalmefene. *Clinical Pharmacology and Therapeutics* 49 (3): 300–306.

Galton, F. 1869. Hereditary Talent and Character. *Macmillan's Magazine* 14:322.

Gates, S. 2013. World's First Alcoholism Vaccine to Begin Preclinical Trials in Chile. *The Huffington Post.* January 29. http://www.huffingtonpost.com/2013/01/29/first -alcoholism-vaccine-chile-preclinical-trial_n_2569033.html

Gatto, J. 1994. The Tyranny of Compulsory Schooling. *Educational Forum* 58 (3): 276–281.

Gendler, T. 2012. Feeling Good about Feeling Bad: Moral Aliefs and Moral Dilemmas. http://bioethics.as.nyu.edu/object/bioethics.events.20120330.conference

Getty, T. 2014. Christian broadcaster: Ebola could cleanse US of atheists, gay people, and sluts. *Raw Story.* August 7. http://www.rawstory.com/rs/2014/08/ christian-broadcaster-ebola-could-cleanse-us-of-atheists-gay-people-and-sluts/

Gilbert, F., A. Harris, and R. Kapsa. 2014. Controlling Brain Cells with Light: Ethical Considerations for Optogenetic Clinical Trials. *American Journal of Bioethics Neuroscience* 5 (3): 3–11.

Gillman, K. 2006. A Review of Serotonin Toxicity Data: Implications for the Mechanisms of Antidepressant Drug Action. *Biological Psychiatry* 59 (11): 1046–1051.

Giordano, J. 2011. On the Implications of Changing Constructs of Pain and Addiction Disorders In the DSM-5: Language Games, Ethics, and Action. *The International Journal of Law, Healthcare and Ethics* 7 (1). https://ispub.com/IJLHE/7/1/3592

Giordano, J. 2012. Neuromorality: Implications for Human Ecology, Global Relations and National Security Policy. http://bioethics.as.nyu.edu/object/bioethics.events.20120330.conference

Goldacre, B. 2012. What Doctors Don't Know about the Drugs They Prescribe. https://www.ted.com/talks/

Goldberg, C. 2014. Beyond Good and Evil: New Science Casts Light on Morality in the Brain. www.commonhealth.wbur.org/2014/08/brain-matters-morality

Goldman, M., A. Gomes, C. Carter, and R. Lee. 2011. Divergent Effects of Two Different Doses of Intranasal Oxytocin on Facial Affect Discrimination in Schizophrenic Patients With and Without Polydipsia. *Psychopharmacology* 216 (1): 101–110.

Grammaticas, D. 2013. "Airmageddon"; China Smog Raises Modernization Doubts. January 31. http://www.bbc.co.uk/news/world-asia-china-21272328

Grant, J., M. Potenza, E. Hollander, R. Cunningham-Williams, T. Nurminen, G. Smits, and A. Kallio Antero. 2006. Multicenter Investigation of the Opioid Antagonist Nalmefene in the Treatment of Pathological Gambling. *American Journal of Psychiatry* 163 (2): 303–312.

Greene, J. 2009. Dual-Process Morality and the Personal/Impersonal Distinction: A Reply to McGuire, Langdon, Coltheart, and Mackenzie. *Journal of Experimental Social Psychology* 45 (3): 581–584.

Gruber, D., and J. Dickerson. 2012. Persuasive Images in Popular Science: Testing Judgments of Scientific Reasoning and Credibility. *Public Understanding of Science (Bristol, England)* 21 (8): 938–948.

Grubin, D. 2010. Chemical Castration for Sex Offenders. *British Medical Journal* 340, c74. doi:.10.1136/bmj.c74

Gual, A., L. Torup, W. van den Brink, and K. Mann. 2013. A Randomized, Double-Blind, Placebo-Controlled, Efficacy Study of Nalmefene, as-Needed Use, in Patients with Alcohol Dependence. *European Neuropsychopharmacology* 23 (11): 1432–1442.

Guastella, A., P. Mitchell, and M. Dadds. 2008. Oxytocin Increases Gaze to the Eye Region of Human Faces. *Biological Psychiatry* 63 (1): 3–5.

Guinness, E. 2006. Childhood Influences on Antisocial Behavior. In M. Beer, N. Pocock, eds., *Mad, Bad, or Sad? A Christian Approach to Antisocial Behavior and Mental Disorder,* 16–67. London: Christian Medical Fellowship.

Haan, N., E. Aerts, and B. Cooper. 1985. *On Moral Grounds: The Search for Practical Morality.* New York: New York University Press.

Haidt, J. 2006. *The Happiness Hypothesis.* London: Arrow Books.

Harris, J. 2007. *Enhancing Evolution: The Ethical Case for Making Better People.* New Jersey: Princeton University Press.

Harris, J. 2011. Moral Enhancement and Freedom. *Bioethics* 25 (2): 102–111.

Harris, J. 2012. On Our Obligation to Enhance. In E. Sargent, ed., *Superhuman: Exploring Human Enhancement from 600 BCE to 2050,* 38–39. London: Wellcome Trust.

Harris, J. 2013. Moral Progress and Moral Enhancement. *Bioethics* 27 (5): 285–290.

Hassan, Y. 1999. The Fate of Pakistani Women. *New York Times.* March 25. http://www.nytimes.com/1999/03/25/opinion/25iht-edhass.2.t.html

Hauerwas, S. 2007. Carving Stone, or, Leaning to Speak Christian. http://www.ptsem.edu/uploadedFiles/School_of_Christian_Vocation_and_Mission/Institute_for_Youth_Ministry/Princeton_Lectures/2007_Hauerwas_Carving.pdf

Hauerwas, S. 2013. *Hannah's Child.* London: SCM Press.

Hauskeller, M. 2015. The Art of Misunderstanding Critics: The Case of Persson and Savulescu's Defense of Moral Bioenhancement. http://www.academia.edu/10099292/The_Art_of_Misunderstanding_Critics_the_Case_of_Persson_and__Savulescus_Defense_of_Moral_Bioenhancement

Hayward, G. 2012. Reiser Attacks Slain Wife's Character in Wrongful Death Case. *CBS.* July 11. http://sanfrancisco.cbslocal.com/2012/07/11/reiser-attacks-slain-wifes-character-in-wrongful-death-case/

Healy, D. 2000. Emergence of Antidepressant Induced Suicidality. *Primary Care Psychiatry* 6 (1): 23–28.

Healy, D. 2004. *Let Them Eat Prozac: The Unhealthy Relationship between the Pharmaceutical Industry and Depression.* New York: New York University Press.

Heilig, M. 2011. Trial Watch: Nalmefene Reduces Alcohol Use in Phase III Trial. *Nature Reviews: Drug Discovery* 10:566. doi:.10.1038/nrd3518

Heilig, M., D. Goldman, W. Berrettini, and C. O'Brien. 2011. Pharmacogenetic Approaches to the Treatment of Alcohol Addiction. *Nature Reviews: Neuroscience* 12:670–684. doi:.10.1038/nrn3110

Henrich, J., S. J. Heine, and A. Norenzayan. 2010. The Weirdest People in the World? *Behavioral and Brain Sciences* 33 (2–3): 61–135.

Hick, J. 1990. *Philosophy of Religion.* 4th ed. New Jersey: Prentice Hall.

Hoge, E., M. Pollack, R. Kaufman, P. Zak, and N. Simon. 2008. Oxytocin Levels in Social Anxiety Disorder. *Neuroscience & Therapeutics* 14 (3): 165–170.

Honigsbaum, M. 2011. Oxytocin: Could the "Trust Hormone" Rebond our Troubled World? *The Guardian.* August 21. http://www.theguardian.com/science/2011/aug/21/oxytocin-zak-neuroscience-trust-hormone

Hook, C., and M. Farah. 2013. Look Again: Effects of Brain Images and Mind-Brain Dualism on Lay Evaluations of Research. *Journal of Cognitive Neuroscience* 25 (9): 1397–1405.

Hookem-Smith, K. 2012. Experts Recommend a "Love Pill" to Save Marriages. *Yahoo News.* May 3. https://in.news.yahoo.com/love-pill-save-marraiges-relationships-couples.html

Hopkins, P. 2012. Moral Disease. http://bioethics.as.nyu.edu/object/bioethics.events.20120330.conference

Hughes, J. 2004. *Citizen Cyborg: Why Democratic Societies Must Respond to the Redesigned Human of the Future.* Cambridge: Westview Press.

Hughes, J. 2006. Virtue Engineering. http://ieet.org/

Hughes, J. 2012a. Morality in a Pill? http://ieet.org/index.php/IEET/more/hughe20121009

Hughes, J. 2012b. The Benefits and Risks of Virtue Engineering. http://bioethics.as.nyu.edu/object/bioethics.events.20120330.conference

Hughes, J. 2013. Using Neurotechnologies to Develop Virtues: A Buddhist Approach to Cognitive Enhancement. *Accountability in Research: Policies and Quality Assurance* 20 (1): 27–41.

Hurlemann, R., A. Patin, O. Onur, M. Cohen, T. Baumgartner, S. Metzler, I. Dziobek, J. Gallinat, M. Wagner, W. Maier, and K. Kendrick. 2010. Oxytocin Enhances Amygdala-Dependent, Socially Reinforced Learning and Emotional Empathy in Humans. *Journal of Neuroscience* 30 (14): 4999–5007.

Hyman, S. 2013. Psychiatric Drug Development: Diagnosing a Crisis. *Cerebrum.* Mar-Apr (5). http://www.ncbi.nlm.nih.gov/pmc/articles/PMC3662213/?report=reader

Hyman, S. 2014. I Hope That We Are Not Living in a Post-Fact World. *American Journal of Bioethics Neuroscience* 5 (3): 1–2.

IJzendoorn, M., and M. Bakermans-Kranenburg. 2012. A Sniff of Trust: Meta-Analysis of the Effects of Intranasal Oxytocin Administration on Face Recognition, Trust to In-Group, and Trust to Out-Group. *Psychoneuroendocrinology* 37 (3): 438–443.

IJzendoorn, M., R. Huffmeijer, L. Alink, M. Bakermans-Kranenburg, and M. Tops. 2011. The Impact of Oxytocin Administration on Charitable Donating Is Moderated by Experiences of Parental Love-Withdrawal. *Frontiers in Psychology* 2 (258): 1–8.

IJzendoorn, M., R. Huffmeijer, L. Alink, M. Tops, and M. Bakermans-Kranenburg. 2012. Asymmetric Frontal Brain Activity and Parental Rejection Predict Altruistic Behavior: Moderation of Oxytocin Effects. *Cognitive, Affective & Behavioral Neuroscience* 12 (2): 382–392.

IsHak, W., M. Kahloon, and H. Fakhry. 2011. Oxytocin Role in Enhancing Well-Being: A Literature Review. *Affective Disorders* 130 (1–2): 1–9.

Jones, A. 2008. What Is Gendercide? www.gendercide.org

Jones, H., R. Johnson, P. Fudela, J. Henningfield, and S. Heishman. 2000. Nalmefene: Blockade of Intravenous Morphine Challenge Effects in Opioid Abusing Humans. *Drug and Alcohol Dependence* 60 (1): 29–37.

Jones, J. 2008. *Blood That Cries Out from the Earth: The Psychology of Religious Terrorism.* New York: Oxford University Press.

Jotterand, F. 2011. "Virtue Engineering" and Moral Agency: Will Post-Humans Still Need the Virtues? *American Journal of Bioethics Neuroscience* 2 (4): 3–9.

Jotterand, F. 2012. Enhancing Criminal Brains. http://bioethics.as.nyu.edu/object/bioethics.events.20120330.conference

Jotterand, F. 2014. Psychopathy, Neurotechnologies, and Neuroethics. *Theoretical Medicine and Bioethics* 35 (1): 1–6.

Jotterand, F. 2014. Questioning the Moral Enhancement Project. *American Journal of Bioethics* 14 (4): 1–3.

Juengst, E. 1998. What Does Enhancement Mean? In E. Parens, ed., *Enhancing Human Traits: Ethical and Social Implications*, 29–47. Washington: Georgetown University Press.

Juhnke, G., R. Watts, N. Guerra, and P. Hsieh. 2009. Using Prayer as an Intervention with Clients Who Are Substance Abusing and Addicted and Who Self-Identify Personal Faith in God and Prayer as Recovery Resources. *Journal of Addictions & Offender Counseling* 30 (1): 16–23.

Jung, C. 1959. Face to Face with Carl Jung: Interview with Richard I. Evans. http://www.openculture.com/2012/07/face_to_face_with_carl_jung_man_cannot_stand_a_meaningless_life.html

Kabasenche, W. 2007. Emotions, Memory Suppression, and Identity. *American Journal of Bioethics* 7 (9): 33–34.

Kabasenche, W. 2012a. Engineering for Virtue? Towards Holistic Moral Enhancement. http://bioethics.as.nyu.edu/object/bioethics.events.20120330.conference

Kabasenche, W. 2012b. Moral Enhancement Worth Having: Thinking Holistically. *American Journal of Bioethics Neuroscience* 3 (4): 18–19.

Kahneman, D. 2012. *Thinking, Fast and Slow.* London: Penguin.

Kamarck, T., R. Haskett, M. Muldoon, J. Flory, B. Anderson, R. Bies, B. Pollock, and S. Manuck. 2009. Citalopram Intervention for Hostility: Results of a Randomized Clinical Trial. *Journal of Consulting and Clinical Psychology* 77 (1): 174–188.

Kamarck, T., M. Muldoon, S. Manuck, R. Haskett, J. Cheonge, J. Flory, and E. Vellag. 2011. Citalopram Improves Metabolic Risk Factors Among High Hostile Adults: Results of a Placebo-Controlled Intervention. *Psychoneuroendocrinology* 36 (7): 1070–1079.

Karas, P., C. Mikell, E. Christian, M. Liker, and S. Sheth. 2013. Deep Brain Stimulation: A Mechanistic and Clinical Update. *Neurosurgical Focus* 35 (5): e1.

Kelion, L. 2013. LG Investigates Smart TV 'Unauthorised Spying' Claim. November 20. http://www.bbc.co.uk/news/technology-25018225/.

Kelsey, D. 2009. *Eccentric Existence: A Theological Anthropology.* Kentucky: WJK Press.

Kéri, S., I. Kiss, and O. Kelemen. 2009. Sharing Secrets: Oxytocin and Trust in Schizophrenia. *Social Neuroscience* 4 (4): 287–293.

Kierkegaard, S. 1985. *Fear and Trembling.* Trans. A. Hannay. London: Penguin.

Kirsch, I. 2009. *The Emperor's New Drugs: Exploding the Antidepressant Myth.* London: The Bodly Head.

Kirsch, I., T. Moore, A. Scoboria, and S. Nicholls. 2002. The Emperor's New Drugs: An Analysis of Antidepressant Medication Data Submitted to the U.S. Food and Drug Administration. *Prevention & Treatment* 5 (1). doi:.10.1037/1522-37 36.5.1.523a

Kjaersgaard, T. 2015. Enhancing Motivation by Use of Prescription Stimulants: The Ethics of Motivation Enhancement. *American Journal of Bioethics Neuroscience* 6 (1): 4–10.

Klein, N. 2008. *The Shock Doctrine: The Rise of Disaster Capitalism.* London: Penguin.

Koebler, J. 2012. Criminal Minds. *US News*. November 9. http://www.usnews.com/news/articles/2012/11/09/criminal-minds-use-of-neuroscience-as-a-defense-skyrockets

Krainova, N. 2011. In Soviet Relapse, Critics Sent to Psychiatric Hospitals. *Moscow Times*. June 29. http://www.themoscowtimes.com/news/article/in-soviet-relapse-critics-sent-to-psychiatric-hospitals/439672.html

Lachs, S. 1994. Coming Down from Zen Clouds: A Critique of the Current State of American Zen. http://www.thezensite.com/ZenEssays/CriticalZen/ComingDownfromtheZenClouds.htm.

Lachs, S. 2002. Richard Baker and the Myth of the Zen Roshi. http://www.thezensite.com/

Laing, R. 1990. *The Divided Self: An Existential Study in Sanity and Madness*. London: Penguin.

Lamparello, A. 2011. Why Wait until the Crime Happens? Providing for the Involuntary Commitment of Dangerous Individuals without the Showing of Mental Illness. *Seton Hall Law Review* 41 (3): 876–911.

Larson, E. 2010. Biology and the Emergence of the Anglo-American Eugenics Movement. In D. Alexander, and R. Numbers, eds., *Biology and Ideology: From Descartes to Dawkins*, 165–191. Chicago: University of Chicago.

Lawler, P. 2005. *Stuck with Virtue: The American Individual and Our Biotechnological Future*. Wilmington: ISI.

Lechner, S. 2014. Why Moral Bioenhancement Is a Bad Idea and Why Egalitarianism Would Make It Worse. *American Journal of Bioethics* 14 (4): 31-32.

Leitt, S., and S. Dubner. 2007. *Freakonomics: A Rogue Economist Explores the Hidden Side of Everything*. London: Penguin.

Lewis, B. 2003. Review of *Better Than Well: American Medicine Meets the American Dream*. *Literature and Medicine* 22 (2): 269–272.

Liao, S. 2012. Why Children Need to be Loved. *Critical Review of International Social and Political Philosophy* 15 (3): 347–358.

Lucke, J., S. Bell, B. Partridge, and W. Hall. 2011. Debating the Neuroenhancement Bubble. *American Journal of Bioethics Neuroscience* 2 (4): 38–43.

Lustig, A. 2008. Enhancement Technologies and the Person: Christian Perspectives. *Journal of Law, Medicine & Ethics* 36 (1): 41–50.

Lustig, A. 2009. Are Enhancement Technologies "Unnatural"? Musings on Recent Christian Conversations. *American Journal of Medical Genetics* 151C (1): 81–88.

Macdonald, K., and T. Macdonald. 2010. The Peptide That Binds: A Systematic Review of Oxytocin and its Prosocial Effects in Humans. *Harvard Review of Psychiatry* 18 (1): 1–21.

MacIntyre, A. 1981. *After Virtue: A Study in Moral Theory*. Indiana: Notre Dame.

MacLeod, C. 2014. Smog Chokes China as Public, Experts Demand Change. *USA Today*. February 24. http://www.usatoday.com/story/news/world/2014/02/24/china-pollution-smog/5781919/

Madelon, M., M. Riem, M. IJzendoorn, M. Tops, M. Boksem, S. Rombouts, and M. Bakermans-Kranenburg. 2013. Oxytocin Effects on Complex Brain Networks are Moderated by Experiences of Maternal Love Withdrawal. *European Neuropsychopharmacology* 23 (10): 1288–1295.

Mann, K., A. Bladstrom, L. Torup, A. Gual, and W. van den Brink. 2013. Extending the Treatment Options in Alcohol Dependence: A Randomized Controlled Study of as-Needed Nalmefene. *Biological Psychiatry* 73 (8): 706–713.

Mar, V. 2008. Pope Says Pollution Is a Sin. http://greenourish.blogspot.co.uk/2008/03/pope-says-pollution-is-sin.html.

Marshall, F. 2014. Would Moral Bioenhancement Lead to an Inegalitarian Society? *American Journal of Bioethics* 14 (4): 29–30.

Martinez, V., Z. Gerdtzen, B. Andrews, and J. Asenjo. 2010. Viral Vectors for the Treatment of Alcoholism: Use of Metabolic Flux Analysis for Cell Cultivation and Vector Production. *Metabolic Engineering* 12 (2): 129–137.

McCabe, D., and A. Castel. 2008. Seeing Is Believing: The Effect of Brain Images on Judgments of Scientific Reasoning. *Cognition* 107 (1): 343–352.

McDonagh, E. 1982. *The Making of Disciples: Tasks of Moral Theology*. Wilmington: Michael Glazier Press.

McGilchrist, I. 2009. *The Master and His Emissary: The Divided Brain and the Making of Western Culture*. London: Yale University Press.

Merleau-Ponty, M. 2000. *Phenomenology of Perception*. Trans. C. Smith. London: Routledge.

Miah, A. 2008. Engineering Greater Resilience or Radical Transhuman Enhancement? *Studies in Ethics, Law, and Technology* 2 (1): 1–18.

Midgley, M. 1984. *Wickedness: A Philosophical Essay*. London: Routledge.

Midgley, M. 2002. *Beast and Man: The Roots of Human Nature*. London: Routledge.

Midgley, M. 2004. *Myths We Live By*. London: Routledge.

Miller, G. 2012. The Pediatric Physician's Role in Modifying Childhood Behavior: Vendor or Gatekeeper? Facilitator or Judge? http://bioethics.as.nyu.edu/object/bioethics.events.20120330.conference

Miller, W. 1998. Researching the Spiritual Dimensions of Alcohol and Other Drug Problems. *Addiction* 93 (7): 979–990.

Morgen, Keith. 2009. Finding a Voice, Crafting a Purpose: Introduction to the Special Issue on Spirituality. *Journal of Addictions & Offender Counseling* 30 (1): 1–3.

Mundy, A. 2001. *Dispensing with the Truth*. New York: St. Martin's Press.

Munro, R. 2002. Dangerous Minds: Political Psychiatry in China Today and its Origins in the Mao Era. *Human Rights Watch*. http://www.hrw.org/reports/2002/china02/china0802.pdf

Muraven, M., and R. Baumeister. 2000. Self-Regulation and Depletion of Limited Resources: Does Self-Control Resemble a Muscle? *Psychological Bulletin* 126 (2): 247–259.

Nietzsche, F. 1969. *Thus Spoke Zarathustra*. Trans. R. Hollingdale. London: Penguin.

Nietzsche, F. 1990. *Twilight of the Idols*. Trans. R. Hollingdale. London: Penguin.

O'Connell, T. 1998. *Making Disciples: A Handbook of Christian Moral Formation*. New York: Crossroad Herder.

O'Connor, C., and H. Joffe. 2012. Media Representations of Early Human Development: Protecting, Feeding and Loving the Developing Brain. *Social Science & Medicine* 97 (c): 297–306. doi:.10.1016/j.socscimed.2012.09.048

O'Connor, C., and H. Joffe. 2013a. The Brain in the Public Sphere. http://faraday.st-edmunds.cam.ac.uk/uab/index.php

O'Connor, C., and H. Joffe. 2013b. How Has Neuroscience Affected Lay Understandings of Personhood? A Review of the Evidence. *Public Understanding of Science* 22 (3): 254–268.

Osbeck, L., and D. Robinson. 2005. Philosophical Theories of Wisdom. In R. Sternberg, and J. Jordan, eds., *A Handbook of Wisdom: Psychological Perspectives*, 61–83. New York: Cambridge University Press.

Ozer, D., and V. Benet-Martinez. 2006. Personality and the Prediction of Consequential Outcomes. *Annual Review of Psychology* 57 (1): 401–421.

Parens, E., ed. 1998. *Enhancing Human Traits: Ethical and Social Implications*. Washington, D.C.: Georgetown University Press.

Parens, E. 2012. The Second Wave. http://bioethics.as.nyu.edu/object/bioethics.events.20120330.conference

Pargament, K. 2001. *The Psychology of Religion and Coping: The Theory, Research, Practice.* London: Guilford Press.

Pargament, K., D. Ensing, K. Falgout, H. Olsen, B. Reilly, K. Van Haitsma, and R. Warren. 1990. God help me: (I): Religious Coping Efforts as Predictors of the Outcomes to Significant Negative Life Events. *American Journal of Community Psychology* 18 (6): 793–824.

Pascual-Leone, A. 2001. The Brain That Plays Music and Is Changed by It. *Biological Foundations of Music* 930 (1): 315–329.

Persson, I. 2012. Could It Be Permissible to Prevent the Existence of Morally Enhanced People? *Journal of Medical Ethics* 38 (11): 692–693.

Persson, I., and J. Savulescu. 2008. The Perils of Cognitive Enhancement and the Urgent Imperative to Enhance the Moral Character of Humanity. *Journal of Applied Philosophy* 25 (3): 162–177.

Persson, I., and J. Savulescu. 2011a. The Turn for Ultimate Harm: A Reply to Fenton. *Journal of Medical Ethics* 37 (7): 441–444.

Persson, I., and J. Savulescu. 2011b. Getting Moral Enhancement Right: The Desirability of Moral Enhancement. *Bioethics* 27 (3): 124–131.

Persson, I., and J. Savulescu. 2012. *Unfit for the Future: The Need for Moral Enhancement.* Oxford: Oxford University Press.

Peterson, C., and M. Seligman. 2004. *Character Strengths and Virtues: A Handbook and Classification.* Oxford: Oxford University Press.

Planes, A. 2012. Our Frightening Future, Brought To You By Microsoft. January 18. http://www.fool.com/investing/general/2012/01/18/our-frightening-future-brought -to-you-by-microsoft.aspx.

Powell, R. 2013. The Biomedical Enhancement of Moral Status. *Journal of Medical Ethics* 39 (2): 65–66.

Powell, S. 2014. SSRIs as a Component of, Rather Than a Means to, Moral Enhancement. *American Journal of Bioethics Neuroscience* 5 (3): 1–2.

Prado-Lima, P. 2009. Pharmacological Treatment of Impulsivity and Aggressive Behavior. *Revista Brasileira de Psiquiatria* 31 (s2): 58–65.

Priester, P., J. Scherer, J. Steinfeldt, A. Jana-Masri, T. Jashinsky, J. Jones, and C. Vang. 2009. The Frequency of Prayer, Meditation, and Holistic Interventions in Addictions Treatment: A National Survey. *Pastoral Psychology* 58 (3): 315–322.

Rakić, V. 2014a. Voluntary Moral Bioenhancement Is a Solution to Sparrow's Concerns. *American Journal of Bioethics* 14 (4): 37–38.

Rakić, V. 2014b. Voluntary Moral Enhancement and the Survival-at-Any-Cost Bias. *Journal of Medical Ethics* 40 (4): 246–250.

Rakić, V. 2014c. We Can Make Room for SSRIs. *American Journal of Bioethics Neuroscience* 5 (3): 1–2.

Rakić, V. 2015. We Must Create Beings with Moral Standing Superior to our Own. *Cambridge Quarterly of Healthcare Ethics* 24 (1): 58–65.

Rampton, S., and J. Stauber. 2001. *Trust Us, We're Experts!* New York: Putnam.

Ranganatha, S., A. Caria, R. Veit, T. Gaber, G. Rota, A. Kuebler, and N. Birbaumer. 2007. FMRI Brain-Computer Interface: A Tool for Neuroscientific Research and Treatment. *Computational Intelligence and Neuroscience*. doi:.10.1155/2007/25487

Reactions Weekly. 2014. Nalmefene: Risk of Psychiatric Symptoms. 1511 (1): 2.

Rhodes, E. 2014. Evaluating Nudge Techniques. *Psychologist* 27 (8): 572–573.

Rosenthal, R. 1998. Covert Communication in Classrooms, Clinics, and Courtrooms. *Eye on Psi Chi* 3 (1): 18–22.

Rosenthal, R., and L. Jacobson. 1968. *Pygmalion in the Classroom*. New York: Holt, Rinehart & Winston.

Rosenthal, R., and L. Jacobson. 1992. *Pygmalion in the Classroom* (expanded ed.). New York: Irvington.

Rutecki, G. 2010. Forced Sterilization of Native Americans: Late Twentieth Century Physician Cooperation with National Eugenic Policies. https://cbhd.org/content/forced-sterilization-native-americans-late-twentieth-century-physician-cooperation-national-

Sartre, J. 1998. *Being and Nothingness*. Trans. H. Barnes. London: Routledge.

Savage, S. 2013. Head and Heart in Preventing Religious Radicalization. In F. Watts, and G. Dumbreck, eds., *Head and Heart: Perspectives from Religion and Psychology*, 157–194. Philadelphia: Templeton Foundation Press.

Savulescu, J. 2001. Procreative Beneficence: Why We Should Select the Best Children. *Bioethics* 15 (5): 413–426.

Savulescu, J. 2005. New Breeds of Humans: The Moral Obligation to Enhance. *Ethics. Law and Moral Philosophy of Reproductive Biomedicine* 1 (1): 36–39.

Savulescu, J. 2006. Justice, Fairness, and Enhancement. *Annals of the New York Academy of Sciences* 1093:321–338. doi:.10.1196/annals.1382.021

Savulescu, J. 2009. Unfit for Life: Genetically Enhance Humanity or Face Extinction. http://humanityplus.org/2009/11/genetically-enhance-humanity-or-face-extinction/.

Savulescu, J. 2010. Genetically Unfit: Enhance Humanity or Face Extinction. http://podcasts.ox.ac.uk/unfit-life-genetically-enhance-humanity-face-extinction.

Savulescu, J. 2012. Label with Care. *Sydney Morning Herald*. June 17. http://www.smh.com.au/it-pro/label-with-care-20120616-20gvv.html

Savulescu, J. 2013. Pills That Improve Morality. http://tedxtalks.ted.com/video/The-need-for-moral-enhancement.

Savulescu, J., and I. Persson. 2012. Moral Enhancement, Freedom and the God Machine. *Monist* 95 (3): 399–421.

Schermer, M. 2008. Enhancements, Easy Shortcuts, and the Richness of Human Activities. *Bioethics* 22 (7): 355–363.

Schermer, M. 2013. On the Argument That Enhancement Is "Cheating." *Journal of Medical Ethics* 34 (2): 85–88.

Sharon, J. 2013. Haredim Plan Protest Against Women of the Wall. *Jerusalem Post*. June 6. http://www.jpost.com/National-News/Haredim-plan-mass-protest-against-Women-of-the-Wall-315691.

Sherman, N. 2000. Wise Emotions. In *Understanding Wisdom: Sources, Science and Society*, 320–333, ed. W. Brown. Philadelphia: Templeton Foundation Press.

Shook, J. 2012. Is Ethical Theory Relevant to Neuroethical Evaluations of Enhancing Moral Brains? http://bioethics.as.nyu.edu/object/bioethics.events.20120330.conference

Sinnott-Armstrong, W. 2012. Is There One Moral Brain? http://bioethics.as.nyu.edu/object/bioethics.events.20120330.conference

Snoek, A. 2014. Do We Have a Right to Drink? On Australian Thugs and French Hedonists. http://blog.practicalethics.ox.ac.uk/2014/01/do-we-have-a-right-to-drink-on-australian-thugs-and-french-hedonists/.

Solomon, L. 2013. A New Approach to Jews with Addictions. *Sun Sentinel*. February 3. http://articles.sun-sentinel.com/2013-02-03/news/fl-jewish-recovery-20130203_1_jews-and-alcohol-benzion-twerski-frequency-of-jewish-alcoholism.

Soyka, M. 2014. Nalmefen Eine Neue Pharmakotherapeutische Option bei Alkoholabhängigkeit. *Der Nervenarzt* 85:578–582.

Sparrow, R. 2010. Better than Men? Sex and the Therapy/Enhancement Distinction. *Kennedy Institute of Ethics Journal* 20 (2): 115–144.

Sparrow, R. 2011. A Not-So-New Eugenics: Harris and Savulescu on Human Enhancement. *Hastings Center Report* 41 (1): 32–42.

Sparrow, R. 2014a. Egalitarianism and Moral Bioenhancement. *American Journal of Bioethics* 14 (4): 20–28.

Sparrow, R. 2014b. Better living through chemistry? A reply to Savulescu and Persson on "Moral Enhancement." *Journal of Applied Philosophy* 31 (1): 23–32.

Spence, D. 2014. Bad Medicine: Nalmefene in Alcohol Misuse. *British Medical Journal* 348: g1531. doi: .10.1136/bmj.g1531

Stein, D. 2012. Psychopharmacological Enhancement: A Conceptual Framework. *Philosophy, Ethics, and Humanities in Medicine* 7 (5). doi:.10.1186/1747-5341-7-5

Struthers, W., and R. Schuchdardt. 2013. The Persuasive Power of Brain Scans. http://faraday.st-edmunds.cam.ac.uk/uab/index.php.

Stryker, S., and R. Serpe. 1982. Commitment, Identity Salience, and Role Behavior: Theory and Research Example. In W. Ickes, and E. Knowles, eds. *Roles, Personality and Social Behavior*, 199–218. New York: Springer-Verlag.

Szasz, T. 2010. *The Myth of Mental Illness: Foundations of a Theory of Personal Conduct.* New York: Harper.

Takahashi, A., I. Quadros, R. de Almeida, and K. Miczek. 2011. Brain Serotonin Receptors and Transporters: Initiation vs. Termination of Escalated Aggression. *Psychopharmacology* 213 (2–3): 183–212.

Tallis, R. 2008. *The Kingdom of Infinite Space: A Fantastical Journey Around Your Head.* London: Atlantic Books.

Tallis, R. 2012. *Aping Mankind: Neuromania, Darwinitis, and the Misrepresentation of Humanity.* Durham: Acumen.

Temkin-Greener, H., S. Kunitz, D. Broudy, and M. Haffner. 1981. Surgical Fertility Regulation Among Women on the Navaho Indian Reservation, 1972–1978. *American Journal of Public Health* 71 (4): 403–407.

Trout, J. 2008. Seduction without Cause: Uncovering Explanatory Neurophilia. *Trends in Cognitive Sciences* 12 (8): 281–282.

Trungpa, C. 2003. *Training the Mind and Cultivating Loving Kindness.* Boston: Shambhala.

Turgut, P. 1998. "Honour" Killings Still Plague Turkish Province. *Toronto Star.* May 14. http://www.gendercide.org/case_honour.html

Vannoy, S. 2005. *Evaluating the impact of a meditation curriculum on anger, hostility, and egoism with incarcerated adults* (Doctoral dissertation, University of Wisconsin, Madison). UMI No. 3186120.

Voltaire, F. 2006. *Candide.* Trans. T. Cuffe. London: Penguin.

Wagner, B. 1977. Lo the Poor and Sterilized Indian. *America* 136: 75.

Wagner, G. 2001. What Is the Promise of Developmental Evolution? Part II: A Causal Explanation of Evolutionary Innovations May Be Impossible. *Journal of Experimental Zoology* 291 (4): 305–309.

Wallach, W. 2012. *The Illusion of the Technological Moral Fix.* http:// bioethics.as.nyu.edu/object/bioethics.events.20120330.conference.

Wasserman, D. 2013. When Bad People Do Good Things: Will Moral Enhancement Make the World a Better Place? *Journal of Medical Ethics* 40 (6): 374–375.

Watts, F. 2010. Psychology and Theology. In P. Harrison, ed., *The Cambridge Companion to Science and Religion*, 190–206. Cambridge: Cambridge University Press.

Wee, S., M. Hicks, B. De, J. Rosenberg, A. Moreno, S. Kaminsky, K. Janda, R. Crystal, and G. Koob. 2012. Novel Cocaine Vaccine Linked to a Disrupted Adenovirus Gene Transfer Vector Blocks Cocaine Psychostimulant and Reinforcing Effects. *Neuropsychopharmacology* 37:1083–1091. doi:.10.1038/npp.2011.200

Wilkinson, T. 2008. Thou Shalt Honor thy Mother Earth. March 14. http:// articles.latimes.com/2008/mar/14/world/fg-pollute14.

Wilson, A. 2014. Egalitarianism and Successful Moral Bioenhancement. *American Journal of Bioethics* 14 (4): 35–36.

Wiseman, H. 2013. Head, Heart, and Wisdom. In F. Watts, and G. Dumbreck, eds., *Head and Heart: Perspectives from Religion and Psychology*, 247–272. Philadelphia: Templeton Foundation Press.

Wiseman, H. 2014a. Moral Enhancement—"Hard" and "Soft" Forms. *American Journal of Bioethics* 14 (4): 48–49.

Wiseman, H. 2014b. SSRIs as Moral Enhancement Interventions: A Practical Dead End. *American Journal of Bioethics Neuroscience* 5 (3): 1–10.

Wiseman, H. 2014c. SSRIs and Moral Enhancement: Looking Deeper. *American Journal of Bioethics Neuroscience* 5 (4): W1–7.

Wiseman, H. 2015. Enhancing Motivation with a Tablet ... Wouldn't You? *American Journal of Bioethics Neuroscience* 6 (1): 31–33.

Wolpe, P. 2002. Treatment, Enhancement, and the Ethics of Neurotherapeutics. *Brain and Cognition* 50 (3): 387–395.

Wright, H. 1987. *Self-Talk, Imagery and Prayer in Counseling*, vol. 2. Milton Keynes: Word Publishing.

Wright, I. 2002. Conflict of Interest and the *British Journal of Psychiatry*. *British Journal of Psychiatry* 180 (1): 82–83.

Wright, I. 2002. Just How Tainted Has Medicine Become? *Lancet* 359 (9313): 1167.

Wright, T. 2010. *Virtue Reborn*. London: SPCK.

Young, L., J. Camprodon, M. Hauser, A. Pascual-Leone, and R. Saxe. 2010. Disruption of the Right Temporoparietal Junction with Transcranial Magnetic Stimulation Reduces the Role of Beliefs in Moral Judgments. *Proceedings of the National Academy of Sciences of the United States of America* 107: 6753–6758. doi:.10.1073/pnas.0914826107

Young, M., D. Armas, and L. Cunningham. 2009. Using Meditation in Addiction Counseling. *Journal of Addictions & Offender Counseling* 32 (1): 58–71.

Zak, P. 2004. Neuroeconomics. *Philosophical Transactions of the Royal Society of London* 359:1737–1748. doi:.10.1098/rstb.2004.1544

Zak, P. 2010. The Science of Trust. http://tedxtalks.ted.com/video/ TEDxConstitutionDrive-Paul-Zak.

Zak, P. 2011a. Can a molecule make us moral? http://www.cnn.com/2011/12/27/ opinion/zak-moral-molecule/.

Zak, P. 2011b. Trust, Morality—and Oxytocin? http://www.ted.com/talks/ paul_zak_trust_morality_and_oxytocin.

Zak, P. 2012. *The Moral Molecule: The New Science of What Makes Us Good or Evil*. London: Bantam Press.

Zak, P. 2013. *The Moral Molecule: The Source of Love and Prosperity* (audio CD). Brilliance Corporation.

Zak, P., and F. Ahlam. 2006. Neuroactive Hormones and Interpersonal Trust: International Evidence. *Economics and Human Biology* 4 (3): 412–429.

Zak, P., K. Borja, W. Matzner, and R. Kurzban. 2005. The Neuroeconomics of Distrust: Physiologic and Behavioral Differences between Men and Women. *American Economic Review* 95 (2): 360–363.

Zak, P., and A. Fakhar. 2006. Neuroactive Hormones and Interpersonal Trust: International Evidence. *Economics and Human Biology* 4 (3): 412–429.

Zak, P., R. Kurzban, S. Ahmadi, R. Swerdloff, and J. Park. 2009. Testosterone Administration Decreases Generosity in the Ultimatum Game. *Public Library of Science One* 4 (12): 1–7.

Zak, P., R. Kurzban, and W. Matzner. 2005a. Oxytocin Is Associated with Human Trustworthiness. *Hormones and Behavior* 48 (5): 522–527.

Zak, P., A. Stanton, and S. Ahmadi. 2007. Oxytocin Increases Generosity in Humans. *Public Library of Science One* 2 (11): 1–5.

Zald, D., R. Cowan, P. Riccardi, R. Baldwin, M. Ansari, R. Li, E. Shelby, C. Smith, M. McHugo, and R. Kessler. 2008. Midbrain Dopamine Autoreceptor Availability Is

Inversely Associated with Novelty Seeking Traits in Humans. *Journal of Neuroscience* 28 (53): 14372–14378.

Zimbardo, P. 1971. The power and pathology of imprisonment. *Congressional Record.* (Serial No. 15, 1971–10–25). Hearings before Subcommittee No. 3, of the Committee on the Judiciary, House of Representatives, Ninety-Second Congress, *First Session on Corrections, Part II, Prisons, Prison Reform and Prisoner's Rights: California.* Washington: U.S. Government Printing Office.

Zimbardo, P. 2007. *The Lucifer Effect: Understanding How Good People Turn Evil.* New York: Random House.

Zyl, L. 2013. Review of Character as Moral Fiction. *Journal of Applied Philosophy* 31 (1): 104–106.

Index

Basic Bioethics

Arthur Caplan, editor

Books Acquired under the Editorship of Glenn McGee and Arthur Caplan

Peter A. Ubel, *Pricing Life: Why It's Time for Health Care Rationing*

Mark G. Kuczewski and Ronald Polansky, eds., *Bioethics: Ancient Themes in Contemporary Issues*

Suzanne Holland, Karen Lebacqz, and Laurie Zoloth, eds., *The Human Embryonic Stem Cell Debate: Science, Ethics, and Public Policy*

Gita Sen, Asha George, and Piroska Östlin, eds., *Engendering International Health: The Challenge of Equity*

Carolyn McLeod, *Self-Trust and Reproductive Autonomy*

Lenny Moss, *What Genes Can't Do*

Jonathan D. Moreno, ed., *In the Wake of Terror: Medicine and Morality in a Time of Crisis*

Glenn McGee, ed., *Pragmatic Bioethics, 2d edition*

Timothy F. Murphy, *Case Studies in Biomedical Research Ethics*

Mark A. Rothstein, ed., *Genetics and Life Insurance: Medical Underwriting and Social Policy*

Kenneth A. Richman, *Ethics and the Metaphysics of Medicine: Reflections on Health and Beneficence*

David Lazer, ed., *DNA and the Criminal Justice System: The Technology of Justice*

Harold W. Baillie and Timothy K. Casey, eds., *Is Human Nature Obsolete? Genetics, Bioengineering, and the Future of the Human Condition*

Robert H. Blank and Janna C. Merrick, eds., *End-of-Life Decision Making: A Cross-National Study*

Norman L. Cantor, *Making Medical Decisions for the Profoundly Mentally Disabled*

Margrit Shildrick and Roxanne Mykitiuk, eds., *Ethics of the Body: Post-Conventional Challenges*

Alfred I. Tauber, *Patient Autonomy and the Ethics of Responsibility*

David H. Brendel, *Healing Psychiatry: Bridging the Science/Humanism Divide*

Jonathan Baron, *Against Bioethics*

Michael L. Gross, *Bioethics and Armed Conflict: Moral Dilemmas of Medicine and War*

Karen F. Greif and Jon F. Merz, *Current Controversies in the Biological Sciences: Case Studies of Policy Challenges from New Technologies*

Deborah Blizzard, *Looking Within: A Sociocultural Examination of Fetoscopy*

Ronald Cole-Turner, ed., *Design and Destiny: Jewish and Christian Perspectives on Human Germline Modification*

Holly Fernandez Lynch, *Conflicts of Conscience in Health Care: An Institutional Compromise*

Mark A. Bedau and Emily C. Parke, eds., *The Ethics of Protocells: Moral and Social Implications of Creating Life in the Laboratory*

Jonathan D. Moreno and Sam Berger, eds., *Progress in Bioethics: Science, Policy, and Politics*

Eric Racine, *Pragmatic Neuroethics: Improving Understanding and Treatment of the Mind-Brain*

Martha J. Farah, ed., *Neuroethics: An Introduction with Readings*

Jeremy R. Garrett, ed., *The Ethics of Animal Research: Exploring the Controversy*

Books Acquired under the Editorship of Arthur Caplan

Sheila Jasanoff, ed., *Reframing Rights: Bioconstitutionalism in the Genetic Age*

Christine Overall, *Why Have Children? The Ethical Debate*

Yechiel Michael Barilan, *Human Dignity, Human Rights, and Responsibility: The New Language of Global Bioethics and Bio-Law*

Tom Koch, *Thieves of Virtue: When Bioethics Stole Medicine*

Timothy F. Murphy, *Ethics, Sexual Orientation, and Choices about Children*

Daniel Callahan, *In Search of the Good: A Life in Bioethics*

Robert Blank, *Intervention in the Brain: Politics, Policy, and Ethics*

Gregory E. Kaebnick and Thomas H. Murray, eds., *Synthetic Biology and Morality: Artificial Life and the Bounds of Nature*

Dominic A. Sisti, Arthur L. Caplan, and Hila Rimon-Greenspan, eds., *Applied Ethics in Mental Healthcare: An Interdisciplinary Reader*

Barbara K. Redman, *Research Misconduct Policy in Biomedicine: Beyond the Bad-Apple Approach*

Russell Blackford, *Humanity Enhanced: Genetic Choice and the Challenge for Liberal Democracies*

Nicholas Agar, *Truly Human Enhancement: A Philosophical Defense of Limits*

Bruno Perreau, *The Politics of Adoption: Gender and the Making of French Citizenship*

Carl Schneider, *The Censor's Hand: The Misregulation of Human-Subject Research*

Lydia S. Dugdale, *Dying in the Twenty-First Century: Toward a New Ethical Framework for the Art of Dying Well*

Harris Wiseman, *The Myth of the Moral Brain: The Limits of Moral Enhancement*